Probability and Mathematical Statistics (Continued)

MUIRHEAD • Aspects of Multivariate Statistical Theory
PARZEN • Modern Probability Theory and Its Applications
PURI and SEN • Nonparametric Methods in General Linear Models
PURI and SEN • Nonparametric Methods in Multivariate Analysis
RANDLES and WOLFE • Introduction to the Theory of Nonparametric Statistics
RAO • Linear Statistical Inference and Its Applications, *Second Edition*
RAO and SEDRANSK • W.G. Cochran's Impact on Statistics
ROHATGI • An Introduction to Probability Theory and Mathematical Statistics
ROHATGI • Statistical Inference
ROSS • Stochastic Processes
RUBINSTEIN • Simulation and The Monte Carlo Method
SCHEFFE • The Analysis of Variance
SEBER • Linear Regression Analysis
SEBER • Multivariate Observations
SEN • Sequential Nonparametrics: Invariance Principles and Statistical Inference
SERFLING • Approximation Theorems of Mathematical Statistics
TJUR • Probability Based on Radon Measures
WILLIAMS • Diffusions, Markov Processes, and Martingales, Volume I: Foundations
ZACKS • Theory of Statistical Inference

Applied Probability and Statistics

ABRAHAM and LEDOLTER • Statistical Methods for Forecasting
AGRESTI • Analysis of Ordinal Categorical Data
AICKIN • Linear Statistical Analysis of Discrete Data
ANDERSON, AUQUIER, HAUCK, OAKES, VANDAELE, and WEISBERG • Statistical Methods for Comparative Studies
ARTHANARI and DODGE • Mathematical Programming in Statistics
BAILEY • The Elements of Stochastic Processes with Applications to the Natural Sciences
BAILEY • Mathematics, Statistics and Systems for Health
BARNETT • Interpreting Multivariate Data
BARNETT and LEWIS • Outliers in Statistical Data, *Second Edition*
BARTHOLOMEW • Stochastic Models for Social Processes, *Third Edition*
BARTHOLOMEW and FORBES • Statistical Techniques for Manpower Planning
BECK and ARNOLD • Parameter Estimation in Engineering and Science
BELSLEY, KUH, and WELSCH • Regression Diagnostics: Identifying Influential Data and Sources of Collinearity
BHAT • Elements of Applied Stochastic Processes, *Second Edition*
BLOOMFIELD • Fourier Analysis of Time Series: An Introduction
BOX • R. A. Fisher, The Life of a Scientist
BOX and DRAPER • Evolutionary Operation: A Statistical Method for Process Improvement
BOX, HUNTER, and HUNTER • Statistics for Experimenters: An Introduction to Design, Data Analysis, and Model Building
BROWN and HOLLANDER • Statistics: A Biomedical Introduction
BUNKE and BUNKE • Statistical Inference in Linear Models, Volume I
CHAMBERS • Computational Methods for Data Analysis
CHATTERJEE and PRICE • Regression Analysis by Example
CHOW • Econometric Analysis by Control Methods
CLARKE and DISNEY • Probability and Random Processes: A First Course with Applications, *Second Edition*
COCHRAN • Sampling Techniques, *Third Edition*
COCHRAN and COX • Experimental Designs, *Secon*
CONOVER • Practical Nonparametric Statistics, *Sec*

Applied Probability and Statistics (Continued)

CONOVER and IMAN • Introduction to Modern Business Statistics
CORNELL • Experiments with Mixtures: Designs, Models and The Analysis of Mixture Data
COX • Planning of Experiments
DANIEL • Biostatistics: A Foundation for Analysis in the Health Sciences, *Third Edition*
DANIEL • Applications of Statistics to Industrial Experimentation
DANIEL and WOOD • Fitting Equations to Data: Computer Analysis of Multifactor Data, *Second Edition*
DAVID • Order Statistics, *Second Edition*
DAVISON • Multidimensional Scaling
DEMING • Sample Design in Business Research
DILLON and GOLDSTEIN • Multivariate Analysis: Methods and Applications
DODGE • Analysis of Experiments with Missing Data
DODGE and ROMIG • Sampling Inspection Tables, *Second Edition*
DOWDY and WEARDEN • Statistics for Research
DRAPER and SMITH • Applied Regression Analysis, *Second Edition*
DUNN • Basic Statistics: A Primer for the Biomedical Sciences, *Second Edition*
DUNN and CLARK • Applied Statistics: Analysis of Variance and Regression
ELANDT-JOHNSON and JOHNSON • Survival Models and Data Analysis
FLEISS • Statistical Methods for Rates and Proportions, *Second Edition*
FLEISS • The Design and Analysis of Clinical Experiments
FOX • Linear Statistical Models and Related Methods
FRANKEN, KÖNIG, ARNDT, and SCHMIDT • Queues and Point Processes
GALAMBOS • The Asymptotic Theory of Extreme Order Statistics
GIBBONS, OLKIN, and SOBEL • Selecting and Ordering Populations: A New Statistical Methodology
GNANADESIKAN • Methods for Statistical Data Analysis of Multivariate Observations
GOLDSTEIN and DILLON • Discrete Discriminant Analysis
GREENBERG and WEBSTER • Advanced Econometrics: A Bridge to the Literature
GROSS and CLARK • Survival Distributions: Reliability Applications in the Biomedical Sciences
GROSS and HARRIS • Fundamentals of Queueing Theory, *Second Edition*
GUPTA and PANCHAPAKESAN • Multiple Decision Procedures: Theory and Methodology of Selecting and Ranking Populations
GUTTMAN, WILKS, and HUNTER • Introductory Engineering Statistics, *Third Edition*
HAHN and SHAPIRO • Statistical Models in Engineering
HALD • Statistical Tables and Formulas
HALD • Statistical Theory with Engineering Applications
HAND • Discrimination and Classification
HILDEBRAND, LAING, and ROSENTHAL • Prediction Analysis of Cross Classifications
HOAGLIN, MOSTELLER and TUKEY • Exploring Data Tables, Trends and Shapes
HOAGLIN, MOSTELLER, and TUKEY • Understanding Robust and Exploratory Data Analysis
HOEL • Elementary Statistics, *Fourth Edition*
HOEL and JESSEN • Basic Statistics for Business and Economics, *Third Edition*
HOGG and KLUGMAN • Loss Distributions
HOLLANDER and WOLFE • Nonparametric Statistical Methods

(continued on back)

Statistical Analysis of Finite Mixture Distributions

D. M. TITTERINGTON

University of Glasgow

A. F. M. SMITH

University of Nottingham

U. E. MAKOV

Chelsea College, London

John Wiley & Sons

Chichester · New York · Brisbane · Toronto · Singapore

Reprinted March 1987

Library of Congress Cataloging in Publication Data:

Titterington, D. M.
Statistical analysis of finite mixture distribution.

Bibliography: p.
Includes index.
1. Mixture distributions (Probability theory)
I. Smith, A. F. M. II. Makov, U. E. III. Title.
QA276.7.T57 1985 519.2 85-6434

ISBN 0 471 90763 4

British Library Cataloguing in Publication Data:

Titterington, D. M.
Statistical analysis of finite mixture distributions.—(Wiley series in probability and mathematical statistics (Applied section))
1. Distribution (Probability theory)
I. Title II. Smith, A. F. M. III. Makov, U. E.
519.5′32 QA 273.6

ISBN 0 471 90763 4

Printed and bound in Great Britain

Contents

PREFACE ix

CHAPTER 1 INTRODUCTION 1

1.1 *Basic definitions and concepts* 1
1.2 *Statistical problems* 3
1.2.1 *Forms of sampling* 3
1.2.2 *The number of components* 4
1.2.3 *Unknown parameters* 5
1.2.4 *Discriminant analysis and classification* 5
1.2.5 *Sequential aspects* 6
1.3 *Other forms of mixture* 6
1.4 *The literature* 7

CHAPTER 2 APPLICATIONS OF FINITE MIXTURE MODELS 8

2.1 *Direct applications* 8
2.2 *Indirect applications* 22

CHAPTER 3 MATHEMATICAL ASPECTS OF MIXTURES 35

3.1 *Identifiability* 35
3.1.1 *Introduction and definition* 35
3.1.2 *Theorems and applications* 36
3.1.3 *Further results and literature* 41
3.2 *Information* 42
3.3 *Miscellany* 48
3.3.1 *Multimodality* 48
3.3.2 *Mixtures with negative weights* 50
3.3.3 *Properties of general mixtures* 50

CHAPTER 4 LEARNING ABOUT THE PARAMETERS OF A MIXTURE 52

4.1 *Graphical methods* 52
4.1.1 *Methods based on density functions* 52
4.1.2 *Methods based on the cumulative distribution function* 58
4.1.3 *Methods for mixtures of discrete and multivariate distributions* 67
4.2 *The method of moments* 71
4.2.1 *Introduction* 71
4.2.2 *Mixtures of two densities* 72
4.2.3 *Mixtures of k densities* 79
4.3 *Maximum likelihood* 82
4.3.1 *Introduction* 82
4.3.2 *EM and other numerical algorithms* 84
4.3.3 *Theoretical considerations* 91
4.3.4 *Further examples* 97
4.4 *Bayesian methods* 106
4.4.1 *Introduction* 106
4.4.2 *Bayesian approaches to outlier models* 108
4.4.3 *Bayesian cluster analysis* 113
4.5 *Minimum distance estimation based on distribution functions* 114
4.5.1 *Introduction to distance measures* 114
4.5.2 *Estimation of mixing weights based on quadratic distances* 117
4.5.3 *Problems with non-explicit estimators* 121
4.5.4 *What to do with extra categorized data?* 125
4.6 *Minimum distance estimators based on transforms* 126
4.6.1 *Introduction* 126
4.6.2 *Theoretical aspects of the MGF and CF methods* 128
4.6.3 *Illustrations based on the estimation of mixing weights* 130
4.7 *Numerical decomposition of mixtures* 133
4.7.1 *Some introductory methods* 133
4.7.2 *Formal methods for mixtures of exponentials* 137
4.7.3 *Medgyessy's method* 138
4.7.4 *Further examples* 142

CHAPTER 5 LEARNING ABOUT THE COMPONENTS OF A MIXTURE 148

5.1 *Introduction* 148
5.2 *Informal techniques* 149
5.3 *Formal techniques for special cases* 149
5.4 *General formal techniques* 152

5.5 *The structure of modality* 159
5.6 *Assessment of modality* 165
5.7 *Discriminant analysis* 168

CHAPTER 6 SEQUENTIAL PROBLEMS AND PROCEDURES 176

6.1 *Introduction to unsupervised learning problems* 176
6.1.1 *The problem and its Bayesian solution* 176
6.1.2 *Computational constraints and the need for approximations* 179
6.2 *Approximate solutions: unknown mixing parameters* 179
6.2.1 *The two-class problem: Bayesian and related procedures* 179
6.2.2 *The two-class problem: a maximum likelihood related procedure* 183
6.2.3 *Asymptotic and finite-sample comparisons of the quasi-Bayes and Kazakos procedures* 184
6.2.4 *The k-class problem: a quasi-Bayes procedure* 189
6.3 *Approximate solutions: unknown component distribution parameters* 193
6.3.1 *A general recursive procedure for a one-parameter mixture* 193
6.3.2 *Unsupervised learning for signal versus noise* 196
6.3.3 *A quasi-Bayes sequential procedure for the contaminated normal distribution* 199
6.3.4 *A quasi-Bayes sequential procedure for bipolar signal detection and related problems* 201
6.3.5 *Problems with several unknown parameters* 203
6.4 *Approximate solutions: unknown mixing and component parameters* 203
6.4.1 *A review of some pragmatic approaches* 203
6.4.2 *A general recursion for parameter estimation using incomplete data* 205
6.4.3 *Illustrations of the general recursion* 208
6.4.4 *Connections with the EM algorithm* 210
6.5 *Approximate solutions: dynamic linear models* 212
6.5.1 *Dynamic linear models and finite mixture Kalman filters* 212
6.5.2 *An outline of suggested approximation procedures* 214

REFERENCES 216

INDEX 238

Preface

Finite mixture distributions have been used as models throughout the history of modern statistics. We are currently on the point of celebrating the centenary of Newcomb's (1886) application of normal mixtures as models for outliers and Pearson's (1894) classic paper on the decomposition of normal mixtures by the method of moments will soon similarly come of age. The ensuing century has revealed a multitude of fields of application which exemplify features that demand the use of mixture models: measurements are available from experimental units which are known to belong to one of a set of classes, but whose individual class-memberships are unavailable. Typical examples come from fisheries research (fish lengths are provided but their sexual identities are not), sedimentology (the grain size distribution of a sample of sand is known but its constitution in terms of different minerals is not) and medical diagnosis (clinical measurements are available for a set of patients whose disease classifications, however, are not).

Statistical analysis of such data has proved not to be straightforward, for two main reasons. Firstly, explicit formulae generally do not exist for estimators of the various parameters, so that numerical methods are required. This in itself discouraged the treatment of any but the simplest problems before the age of computers. Secondly, theoretical difficulties which arise in certain aspects of the statistical analysis reveal some common mixture problems to be 'non-standard'. As a result, detailed investigation of the analysis of finite mixture problems offers more than just a catalogue of straightforward applications of standard methods to a particular class of statistical models: our statistical approach to sand-sifting will indeed reveal a few special nuggets.

The monograph offers a systematic treatment of the structure of finite mixture distributions, an account of the wide range of applications, a detailed description of the attempts to apply various statistical methodologies to the analysis of data from mixture distributions, and a large, up-to-date bibliography. A special feature is the final chapter, where methods are described for accommodating data sequentially and where the connection is made with the engineering literature on problems involving 'unsupervized learning'.

The book should be of interest to research workers in statistics and to engineers involved in pattern recognition, as well as to investigators in the many fields of application. Although the monograph has not been designed as a textbook, the material could form the basis of a specialized postgraduate course in finite mixtures.

Thanks are due to Professor P. D. M. Macdonald and Professor A. S. Paulson, who contributed valuable information and advice at an early stage in the writing of the book. Grateful acknowledgement is made in the text for permission to use material published elsewhere, and deep appreciation is expressed to Miss E. M. Nisbet, Mrs M. F. Smith and Miss K. Glowczewska for their excellent work in preparing the typescript.

CHAPTER 1

Introduction

1.1 BASIC DEFINITIONS AND CONCEPTS

Suppose that a random variable or vector, X, takes values in a sample space, $\mathscr{X}$, and that its distribution can be represented by a probability density function (or mass function* in the case of discrete $\mathscr{X}$) of the form

$$p(x) = \pi_1 f_1(x) + \cdots + \pi_k f_k(x) \qquad (x \in \mathscr{X}), \tag{1.1.1}$$

where

$$\pi_j > 0, \qquad j = 1, \ldots, k; \qquad \pi_1 + \cdots + \pi_k = 1$$

and

$$f_j(\cdot) \geqslant 0, \qquad \int_{\mathscr{X}} f_j(x)\,\mathrm{d}x = 1, \qquad j = 1, \ldots, k.$$

In such a case, we shall say that X has a *finite mixture distribution* and that $p(\cdot)$, defined by (1.1.1), is a *finite mixture density function.*

The parameters $\pi_1, \ldots, \pi_k$ will be called the *mixing weights* and $f_1(\cdot), \ldots, f_k(\cdot)$ the *component densities* of the mixture. It is straightforward to verify that (1.1.1) does, indeed, define a p.d.f.

In many situations, $f_1(\cdot), \ldots, f_k(\cdot)$ will have specified parametric forms and the right-hand side of (1.1.1) will have the more explicit representation

$$\pi_1 f_1(x|\boldsymbol{\theta}_1) + \cdots + \pi_k f_k(x|\boldsymbol{\theta}_k), \tag{1.1.2}$$

where $\boldsymbol{\theta}_j$ denotes the parameters occurring in $f_j(\cdot)$. We shall denote the collection of all distinct parameters occurring in the component densities by $\boldsymbol{\theta}$, and the complete collection of all distinct parameters occurring in the mixture model by $\boldsymbol{\psi}$.

For example, a frequently used two-component mixture model has the form

$$p(x|\boldsymbol{\psi}) = \pi\phi(x|\mu_1, \sigma) + (1 - \pi)\phi(x|\mu_2, \sigma),$$

*We shall use the term 'probability density function' in both the continuous and discrete cases, abbreviating it in the usual way to p.d.f.

where $\phi(x|\mu_j, \sigma)$, $j = 1, 2$, denotes a univariate normal density with mean μ_j and variance σ^2. In this case, $\pi_1 = \pi$, $\pi_2 = 1 - \pi$, $\boldsymbol{\theta}_1 = (\mu_1, \sigma)$, $\boldsymbol{\theta}_2 = (\mu_2, \sigma)$, $\boldsymbol{\theta} = (\mu_1, \mu_2, \sigma)$, and $\boldsymbol{\psi} = (\pi, \mu_1, \mu_2, \sigma)$.

There is no requirement that the component densities appearing in (1.1.2) should all belong to the same parametric family, but in most applications this will be the case. The finite mixture density function will then have the form

$$p(x|\boldsymbol{\psi}) = \sum_{j=1}^{k} \pi_j f(x|\boldsymbol{\theta}_j), \tag{1.1.3}$$

where $f(\cdot|\boldsymbol{\theta})$ denotes a generic member of the parametric family. Almost all of our detailed discussion in what follows is directed towards this model.

In the case of a finite mixture model defined by (1.1.3), each of $\boldsymbol{\theta}_1, \ldots, \boldsymbol{\theta}_k$ is an element of the same parameter space, Θ, say. It follows that $\boldsymbol{\pi} = (\pi_1, \ldots, \pi_k)$ may be thought of as defining a probability distribution over Θ, with $\pi_j = \Pr(\boldsymbol{\theta} = \boldsymbol{\theta}_j)$, $j = 1, \ldots, k$. If $G_\pi(\cdot)$ denotes the probability measure over Θ defined by $\boldsymbol{\pi}$, then (1.1.3) may be formally rewritten as

$$p(x|\boldsymbol{\psi}) = \int_{\Theta} f(x|\boldsymbol{\theta})\, \mathrm{d}G_\pi(\boldsymbol{\theta}). \tag{1.1.4}$$

Although we shall largely confine our attention to the situation where $G_\pi(\cdot)$ defines a finite, discrete measure over Θ, the form of (1.1.4) suggests an obvious generalization to *general mixture densities* simply by allowing $G_\pi(\cdot)$ to be a more general form of measure over Θ.

As a slight digression, we also note that the form (1.1.4) explains an alternative nomenclature which is sometimes encountered—that of *compound distributions.* The distribution on $\mathscr{X}$ summarized by $f(\cdot|\boldsymbol{\theta})$ is *compounded* with the distribution over Θ given by $G(\cdot)$.

In the next chapter, we shall give a detailed overview of a number of applications and uses of finite mixture models. It will then be seen that finite mixture models are used in two rather different ways (although the dividing line is not always clear and, in any case, is related more to expository convenience than to philosophical niceties). We shall refer to these two broad classes of usage as *direct* and *indirect applications* of finite mixture models.

By a *direct application*, we have in mind a situation where we believe, more or less, in the existence of k underlying *categories*, or *sources*, such that the experimental unit on which the observation X is made belongs to one of these categories. We do not, however, observe directly the source of X. In this form of application, $f_j(\cdot)$ summarizes the probability distribution of X given that the observation actually derives from category j, and π_j denotes the probability that the observation comes from this source.

By an *indirect application*, we have in mind a situation where the finite mixture form is simply being used as a *mathematical device* in order to provide an indirect means of obtaining a flexible, tractable form of analysis. For example, a two-component normal mixture where one component has an inflated variance may be used to approximate an intractable heavy-tailed distribution. Similarly, kernel methods of density estimation employ a finite mixture as a smooth curve-fitting device.

We shall consider (at length!) both types of application. The main reason for introducing the dichotomy is that the flavour and emphasis of the inference problems posed and statistical techniques employed will tend to be rather different in the two cases. It may help to keep this in mind as we attempt to survey the astonishing diversity of approaches and procedures that have, in one context or another, been used in the analysis of finite mixture models.

1.2 STATISTICAL PROBLEMS

1.2.1 Forms of sampling

The main topic of this book is the statistical analysis of data from mixture models. The starting point for any particular investigation will therefore be a consideration of the form in which data are obtained.

In most applications, the data take the form of a random sample of observations $X_1 = x_1, \ldots, X_n = x_n$, where the distribution of each X is described by a parametric finite mixture density of the form (1.1.3). Most statistical methods will then take as their starting point the *likelihood function*

$$L_0(\boldsymbol{\psi}) = \prod_{i=1}^{n} p(x_i \mid \boldsymbol{\psi}) = \prod_{i=1}^{n} \left[\sum_{j=1}^{k} \pi_j f(x_i \mid \boldsymbol{\theta}_j) \right]. \tag{1.2.1}$$

In some direct applications of finite mixture models, in addition to the random sample from the mixture distribution there may also be random samples available of observations known to derive from individual underlying categories. For example, in studies of fish populations (Example 2.1.2) we may have samples of fish lengths from fish of *known* age, in addition to a sample of lengths from the mixed population.

If we denote such additional data by

$$\{x_{jh}: j = 1, \ldots, k, h = 1, \ldots, n_j\},$$

where at least one of the n_j is non-zero, then the overall likelihood provided by both the uncategorized and categorized observations is given by

$$L_1(\boldsymbol{\psi}) = L_0(\boldsymbol{\psi}) \prod_{j=1}^{k} \prod_{h=1}^{n_j} f_j(x_{jh} \mid \boldsymbol{\theta}_j). \tag{1.2.2}$$

Moreover, if the categorized observations can be assumed to arise independently, with incidence rates $\pi_1, \ldots, \pi_k$ for the individual categories, then this provides further information about the mixing weights and the appropriate likelihood is

$$L_2(\boldsymbol{\psi}) = L_1(\boldsymbol{\psi}) \prod_{j=1}^{k} \pi_j^{n_j}. \tag{1.2.3}$$

As we shall see later, it is important to clarify which of (1.2.1), (1.2.2), or (1.2.3) is appropriate in any particular application. In particular, if information is available about categorized observations it is important to use (1.2.2) rather than (1.2.1), since the additional information in the former may be substantial. On the other hand, information about categorized observations is often obtained by *selecting* $n_1, \dots, n_k$, in which case (1.2.3) is not applicable.

We shall adopt Hosmer's (1973a) notation M0, M1 and M2 for the three data structures giving rise to L_0, L_1 and L_2, respectively.

In the remaining sections, we shall briefly indicate some of the general statistical problems that arise in the context of finite mixture models. These, of course, are determined in any particular context by the form of application and our degrees of ignorance about the various features of the model.

1.2.2 The number of components

In many contexts, uncertainty about the number of components leads to statistical problems closely related to *cluster analysis*, possibly with strong assumptions about parametric structure. For example, it is often assumed that the densities are normal (univariate or multivariate, as appropriate). In other cases, a considerable amount of data is available from the mixture, so that, in effect, we 'know' the form of $p(x)$ in (1.1.1). Given parametric assumptions about the underlying components, the problem then becomes one of *curve fitting*.

Sometimes, we are concerned to find the mixture with the fewest components that still provides a satisfactory fit to the data. In particular, we commonly wish to know whether to assume that there are two underlying components or just a single component. Given a particular, proposed mixture model with assumed parametric forms for the component densities, a hypothesized mixture having fewer components may be regarded as the imposition of a 'null hypothesis' on the original model framework. The problem of comparing the two mixture models may therefore seem to be that of *testing between nested hypotheses*.

However, it soon becomes clear that traditional testing recipes are not so easily applied in this context. Consider, for example, the apparently straightforward problem of testing between the following hypotheses:

$$\begin{aligned} \mathrm{H}_0: &\quad p(x) = \phi(x \mid \mu, \sigma), \\ \mathrm{H}_1: &\quad p(x) = \pi\phi(x \mid \mu_1, \sigma_1) + (1-\pi)\phi(x \mid \mu_2, \sigma_2), \end{aligned}$$

where $x \in \mathbb{R}$, and the parameters under both H_0 and H_1 are assumed unknown. In other words, we are simply asking whether there is a single normal component, or two.

If we were thinking in traditional terms, it would be natural—particularly if a large sample were available—to consider applying the generalized likelihood ratio test, referring, for significance, to a table of the χ_r^2 distribution, where r is equal to the number of constraints imposed on H_1 in order to produce H_0.

This leads to immediate difficulties, since there is not a unique way of obtaining H_0 from H_1! We could impose

$$\pi = 0 \qquad \text{(1 constraint)}$$

or

$$\mu_1 = \mu_2, \sigma_1 = \sigma_2 \qquad \text{(2 constraints).}$$

What are the appropriate degrees of freedom for the chi-squared distribution? Or perhaps—as we shall see in Section 5.4—we should not be trying to apply the traditional procedure at all.

1.2.3 Unknown parameters

Assuming a given values of k—even if only as a provisional step in an analysis pertaining to questions raised in Section 1.2.2—we shall need, in a parametric formulation, to make inferences about the unknown parameters of the mixture model. Several different cases arise.

In some direct applications, it is possible to carry out detailed studies of the individual component distributions separately from the mixture problem. Thus, for example, the forms of grain size distributions for various individual minerals can be established in the laboratory as a preliminary to the analysis of an ash deposit composed of a mixture of minerals (Example 2.1.1). In the context of (1.1.2), we would then have *known components*, $f_1(\cdot|\boldsymbol{\theta}_1), \ldots, f_k(\cdot|\boldsymbol{\theta}_k)$, and inference problems would relate solely to the *unknown mixing weights*, $\pi_1, \ldots, \pi_k$.

In other applications, there may be considerable uncertainty about the $\boldsymbol{\theta}_j$'s, whereas the π_j's may be considered known. Thus, in fish population studies it might well be the case that the sex distribution in the shoal is known, whereas ascertaining the sex of any particular fish involves a time-consuming dissection, so that detailed knowledge of the $f_j(\cdot|\boldsymbol{\theta}_j)$'s is not available. Assuming a parametric form for the latter, we have a case of *known mixing weights* and *unknown component parameters*, $\boldsymbol{\theta}_1, \ldots, \boldsymbol{\theta}_k$. Of course, both the π_j's *and* the $\boldsymbol{\theta}_j$'s may be unknown and this is probably the most common situation in applications.

Once we have decided upon a parametric approach—and, of course, there are *non-parametric* alternatives in the case of unknown component densities—there are a remarkable variety of estimation methods that have been applied to finite mixture problems. In particular, the *method of moments, maximum likelihood, minimum chi-square, least squares* and *Bayesian approaches* have all been considered in one application or another, as have various other less well-known procedures. We shall consider all these approaches in Chapter 4. In passing, our detailed investigations may help to explain why the theoretical and computational complications of analysing finite mixture data tend to call forth such a variety of responses.

1.2.4 Discriminant analysis and classification

Traditionally, the objective of *statistical discriminant analysis* is to use available data, typically in the form of training samples of categorized observations from the individual components, in order to construct a rule for assigning future individual observations from the mixture to one of the underlying categories.

Sometimes, the problem is explicitly posed as a decision problem in which the 'costs' of *misclassification* are taken into account in formulating the *classification rule*.

In any case, the classification rule, once derived, is taken as a fixed procedure to be applied to all future cases. In many applications, however, the target population of uncategorized cases will be much larger than the initial (typically rather small) training samples of categorized cases. It is of interest in such cases to investigate whether the information acquired from the uncategorized cases can be used to update or amend the classification procedure and, if so, whether the incorporation of this information can lead to substantial improvements in the performance of the procedure. We shall discuss this topic in Section 5.7.

1.2.5 Sequential aspects

In many applications, *sequential problems* arise directly because observations actually become available in sequence and it is required to both *classify* the current observation and to use it to *update* estimates of unknown aspects of the mixture. Sometimes—as, for example, in the case of tracking a signal of uncertain origin, as described in Example 2.1.6—there is an additional need for *fast computation*, since observations may arrive at a very rapid rate and require 'instantaneous' processing, in conditions where little or no memory is available for storing past data and decisions. This will lead naturally to a requirement for efficient *recursive* procedures.

However, it is easy to see that 'natural' approaches to sequential learning may well lead to computational and storage explosions and thus pose interesting new problems of finding efficient approximate recursive procedures. Essentially, the difficulty arises because n observations from a k component mixture generate k^n possible 'histories' of how the observations may have arisen if category membership is taken into account.

Even in situations where the data arise in a non-sequential manner, sequential methods may also be useful as an analytic device for overcoming the severe computational problems which preclude direct analysis by non-sequential procedures.

Chapter 6 will be devoted to a detailed discussion of these sequential aspects of finite mixture problems.

1.3 OTHER FORMS OF MIXTURE

In the introduction to Section 1.1, we remarked that we would mainly be considering mixtures of component densities all belonging to the same parametric family. Examples of mixtures of components from different families may be found in Davis (1952), Aitchison (1955), Cohen (1965), and Ashton (1971).

1.4 THE LITERATURE

As the list of references indicates, there is a very large literature on methodology for and applications of finite mixture models, originating in the nineteenth century and growing rapidly since the advent of computers. We have tried to refer to as many of the technical papers as possible at appropriate points in the text and here we merely list some of the expository literature. Some of this has appeared in the context of special applications, such as pattern recognition (Fu, 1968; Duda and Hart, 1973; Young and Calvert, 1974), mathematical geology (Clark, 1976), fisheries (Macdonald and Pitcher, 1979) and remote sensing (Odell and Basu, 1976). In the statistical literature there are general methodological contributions such as those of Behboodian (1975) and Redner and Walker (1984), together with detailed investigations of special cases such as discrete mixtures (Blischke, 1965) and normal mixtures (Holgersson and Jorner, 1979). There is also information about many special cases interspersed in the volumes of Johnson and Kotz (1969, 1970a, 1970b, 1972). In addition, Ord (1972) includes a chapter about finite mixtures and Blischke (1978) provides a helpful general discussion. Finally, there is the specialized monograph on finite mixtures by Everitt and Hand (1981).

CHAPTER 2

Applications of finite mixture models

2.1 DIRECT APPLICATIONS

In this section we shall describe in detail several areas of application of finite mixture models where the model is intended as a direct representation of the underlying physical phenomenon. At the end of the section we present, in tabular form (Table 2.1.3), an extensive summary of references to these and other direct applications.

Example 2.1.1 Grain sizes in samples of particles

Figure 2.1.1 shows grain size distributions in ash particles deposited at various distances downwind of the volcanic eruption of Mount St. Helens on 18 May 1980 (Brazier *et al.*, 1983). The samples are analysed by sieving, and this results in a distribution measured by percentages of total weight in various ranges of size, as indicated in the diagrams. The abscissa is given in units of $\phi = -\log_2 r$, where r is the grain diameter in millimetres. At each of the three distances there is evidence of a mixture of two components, although the mixing weights change from place to place. The physical explanation for the mixture offered by Brazier *et al.* (1983) is that some of the fine ash aggregates to form the larger particles that contribute to the left-hand component.

The general field of sedimentology gives rise to a great deal of data of this type, as Table 2.1.3 reveals. Care must be taken in formal statistical analysis, however. The data in Figure 2.1.1 are percentages of weight and not particle frequencies, so that the diagrams cannot be treated as histograms. Some attempts have been made to propose equivalent frequencies (Jones, 1969; Clark, 1976) in order that, say, chi-squared tests of goodness-of-fit can be applied. None of these attempts seems to have been successful.

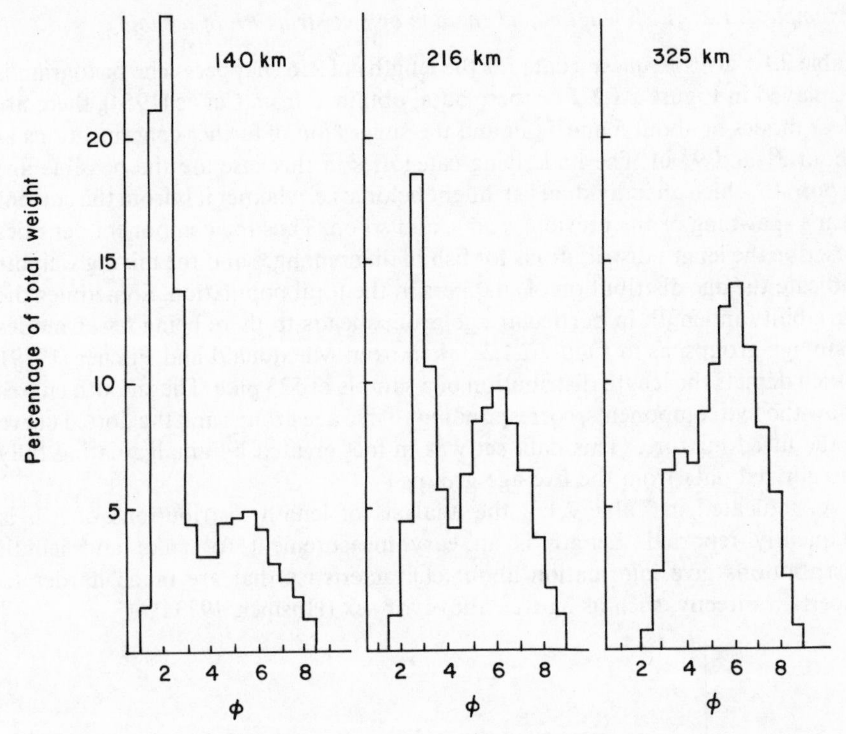

Figure 2.1.1 Grain size distributions at various distances downwind from Mount St Helens, deposited from the eruption of 18 May 1980. Adapted with permission from Brazier *et al.*, 1983 and *Nature*

Table 2.1.1 Snapper lengths: frequency data. Reproduced by permission of CSIRO Editorial and Publication Service from Cassie (1954)

Length (in)	Frequency	Length	Frequency	Length	Frequency	Length	Frequency
$2\frac{7}{8}$	6	$5\frac{3}{8}$	17	$7\frac{7}{8}$	9	$10\frac{3}{8}$	2
$3\frac{1}{8}$	7	$5\frac{5}{8}$	17	$8\frac{1}{8}$	6	$10\frac{5}{8}$	1
$3\frac{3}{8}$	9	$5\frac{7}{8}$	14	$8\frac{3}{8}$	4	$10\frac{7}{8}$	1
$3\frac{5}{8}$	3	$6\frac{1}{8}$	11	$8\frac{5}{8}$	3	$11\frac{1}{8}$	0
$3\frac{7}{8}$	3	$6\frac{3}{8}$	8	$8\frac{7}{8}$	3	$11\frac{3}{8}$	1
$4\frac{1}{8}$	4	$6\frac{5}{8}$	4	$9\frac{1}{8}$	2	$11\frac{5}{8}$	0
$4\frac{3}{8}$	6	$6\frac{7}{8}$	7	$9\frac{3}{8}$	2	$11\frac{7}{8}$	1
$4\frac{5}{8}$	11	$7\frac{1}{8}$	11	$9\frac{5}{8}$	4	$12\frac{1}{8}$	0
$4\frac{7}{8}$	26	$7\frac{3}{8}$	11	$9\frac{7}{8}$	3	$12\frac{3}{8}$	1
$5\frac{1}{8}$	24	$7\frac{5}{8}$	11	$10\frac{1}{8}$	2	$12\frac{5}{8}$	1

Example 2.1.2 Fish lengths and the age or sex structure of a shoal

Table 2.1.1 gives *frequency* data for the lengths of 256 snappers. The histogram is displayed in Figure 2.1.2. For these data, obtained from Cassie (1954), there are clear modes at about 5 and $7\frac{3}{4}$ in and the suggestion of further concentrations at about $3\frac{1}{4}$ and $9\frac{3}{4}$ in. The underlying categories in this case are the possible age groups to which an individual fish might belong: i.e. whether it is from the current year's spawning or the previous year's, and so on. Thus the component densities describe the length distributions for fish of different ages and the mixing weights indicate the age distribution of snappers in the total population. Sometimes the variability in length in particular age groups leads to there being fewer modes than age groups, as in Figure 2.1.3, taken from Macdonald and Pitcher (1979), which depicts the length distribution of a sample of 523 pike. The smooth curves show the five components, corresponding to five age groups and the dotted curve is the fitted mixture. (This data set was in fact created by amalgamating fully categorized data from the five age groups.)

As indicated in Table 2.1.3, the analysis of length distributions of fish is frequently reported. Length is an easy measurement to make and length distributions give information about characteristics that are much harder to ascertain directly, such as age, as above, or sex (Hosmer, 1973a).

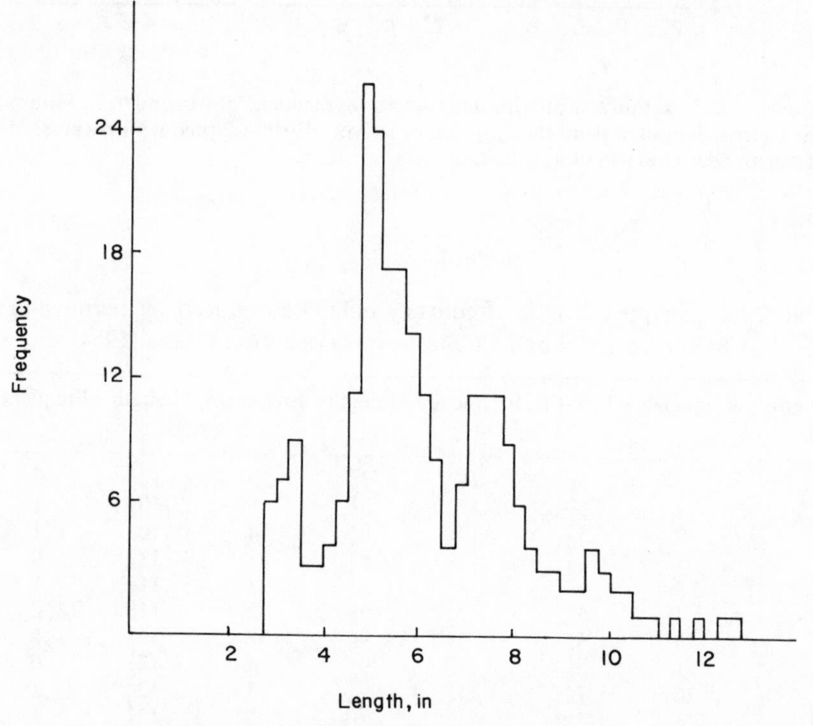

Figure 2.1.2 Histogram of snapper lengths

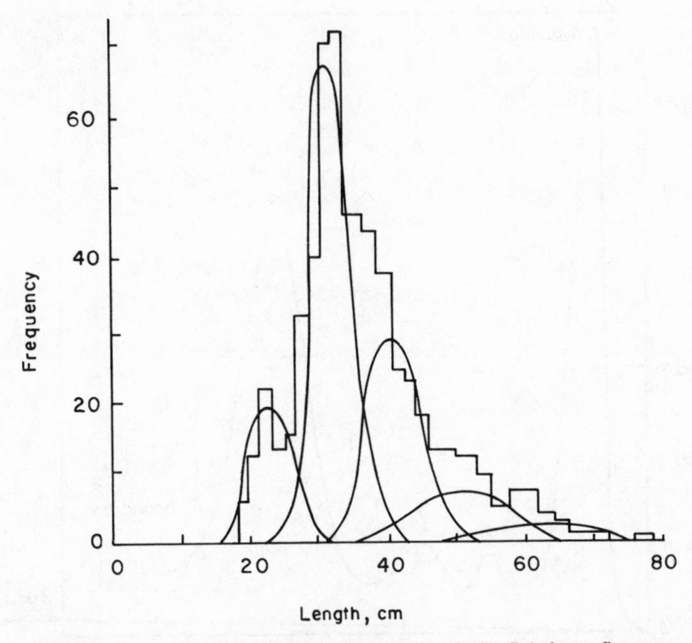

Figure 2.1.3 Pike lengths from a sample of 523 fish from five age groups. From Macdonald and Pitcher, 1979, reproduced by permission of the Journal of the Fisheries Research Board of Canada

Example 2.1.3 Electrophoresis

Electrophoresis is the study of the 'mobility' of proteins from an initial boundary with a buffer solution when subjected to an electrical potential difference. In one apparatus (Longsworth, 1942) there are two boundaries, one in each arm of a U-tube. When the charge is applied, migration occurs downwards in one tube and upwards in the other. Different proteins migrate at different speeds so that a protein mixture can be separated out into its components by this method. Ideally, all molecules of a particular protein would move at the same speed. A plot of protein concentration against distance from the initial boundary would then show a series of spikes which identify the different proteins. The heights of the spikes show the relative concentrations. In practice, convection and other distortions lead to a concentration curve that is smooth and typically looks like a mixture density with symmetric components. Figure 2.1.4 shows a typical example, taken from Tiselius and Kabat (1939). The relative mobilities of different proteins tend to be well known so that, when the modes are clear, as in Figure 2.1.4, they can be attributed to the correct proteins. The main objectives of the analysis of these curves are to assess the relative concentrations of the proteins (estimate the mixing weights) and, sometimes, to look for the tell-tale signs of unforeseen constituents.

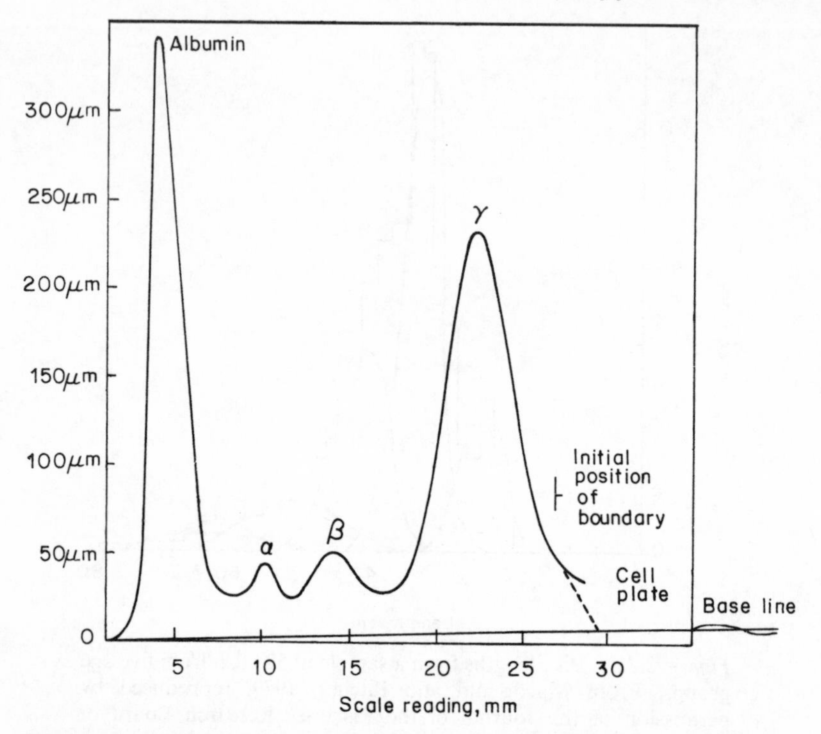

Figure 2.1.4 Electrophoresis diagram of anti-egg albumin rabbit serum, showing peaks corresponding to albumin, α-, β-, and γ-globulin. Reproduced from Tiselius and Kabat, *The Journal of Experimental Medicine, 1939*, **69**, 127, by copyright permission of the Rockefeller University Press

In this example, the problem is that of decomposing a given *curve* into its components and mixing weights. More detailed discussion of methods for doing this is given in Section 4.7. Similar problems occur with diffusion patterns, results from ultracentrifuges, chromatographic scannings, and absorption spectroscopy (see Noble, Hayes, and Eden 1959, and Fraser and Suzuki, 1966).

Example 2.1.4 Switching regressions

Goldfeld and Quandt (1973) discuss the following model for a housing market in disequilibrium. In any month, the relationship between explanatory variables, $\mathbf{x}$, and the number of houses whose construction is started, y, takes the form

$$y = \mathbf{x}^{\mathrm{T}}\boldsymbol{\theta} + \varepsilon,$$

where the first component of $\mathbf{x}$ is a dummy variable taking the value 1 and ε is a random variable with mean zero and variance σ^2.

It is thought that $\boldsymbol{\theta}$ takes one of two unknown values $\boldsymbol{\theta}_1$ and $\boldsymbol{\theta}_2$, depending on whether the market is in a supply phase or a demand phase. If we regard the phase in operation in any given month to be the underlying categorical variable, we have a two-component mixture model. In practice it may not be clear which phase is in operation at any given month. We might then consider the phases in different months to be independent, which gives the basic mixture model or, perhaps more realistically, we might consider a model incorporating serial dependence from month to month. One such is the Markov model discussed in Example 4.3.10. A model incorporating even more complicated dependence or a model involving change points might also be considered.

The switching regression model has been proposed for a variety of problems in economics, as indicated in Table 2.1.3.

Example 2.1.5 Medical diagnosis and prognosis

As we shall see, much of the literature on mixtures is restricted, at least so far as the practical details are concerned, to the univariate case. However, the data available on each observational unit are often multivariate. In particular this is usually the case with medical data, where the mixture data come from patients whose disease category is unconfirmed but on whom several, and often very many, clinical tests have been made.

This is the case, for example, with a set of patients suffering from the hypertensive condition known as Conn's syndrome. Further subclassification of the patients can be made into those for whom the major cause is an adenoma, which is a benign tumour, and those for whom this is not the case. However, this part of the diagnosis, which is critical as far as choice of treatment is concerned, is non-trivial medically, so that there are often cases of Conn's syndrome in the databank which are uncategorized. The type of data available on Conn's patients is indicated in Table 1.6 of Aitchison and Dunsmore (1975). An important feature is the high dimensionality. Multivariate normality is assumed for the logarithms of the data and an approximate procedure for incorporation of unconfirmed cases is discussed by Titterington (1976); see also Section 5.7 and Example 6.1.3. For similar examples, see Skene (1978), Makov (1980a), and Table 2.1.3.

Example 2.1.6 Tracking in a multitarget environment

The process of tracking a target in a multitarget environment involves the reception of signals over noisy channels. The nature of each particular observation is therefore uncertain and could be any one of the following: (a) noise alone (cut-off communication link); (b) false alarm (thermal or process noise, or clutter from a target not being tracked); or (c) the target actually being tracked (see Chang and Srinath, 1976, and a review paper by Bar-Shalom, 1978).

Sources (a) and (b) are often both referred to as 'noise' and source (c) as 'signal'—hence the terminology 'single versus noise' commonly discussed in engineering literature. Typically, such problems require the sequential processing of observations. Bayesian and non-Bayesian approaches to such problems are discussed in Section 6.3.2.

Example 2.1.7 Remote sensing

Artificial satellites are used to generate data from which estimates can be made of the relative acreages allocated to various crops. Energy is recorded in four spectral wave bands for each square pixel of ground observed from the satellite. Usually, areas of categorized pixels, covered by known crops, are used to provide estimates of the underlying component densities, which can then be used as a basis for estimating the mixing weights associated with the set of unknown pixels. These weights give the required relative acreages (Tubbs and Coberly, 1976). Sometimes the objective is that of 'image segmentation', which seeks to identify areas of contiguous pixels that are devoted to the same crop. This pattern-recognition activity is strongly related to cluster analysis, and mixture models are often used in this context (see Example 2.2.4, Section 4.3.4, and Sclove, 1983, who discusses these problems in the context of Landsat data). Issue 12 of Volume A5 of *Communications in Statistics* is largely devoted to remote sensing; see also Nagy (1972) and Example 6.1.1 later.

Example 2.1.8 The hypsometric curve of the earth

This is the title given to the density curve of height above mean sea level over the surface of the earth. The 'data' available, shown in Table 2.1.2 and taken from Tanner (1962), consist of estimates of the *percentages* of the earth's surface area on which the elevation lies within the range indicated. The corresponding histogram of percentages is given in Figure 2.1.5. The figures are fairly crude estimates and the percentages cannot be translated into sample sizes. This suggests that parametric statistical analysis may not be possible but that, instead, numerical or graphical fitting of a parametric mixture density may be more appropriate.

There are two clear modes. Whether or not it is reasonable to hope that more detailed structure can be discerned is open to doubt, although Tanner (1962) mentions another two, more complicated, decompositions.

The first consists of three lognormal components with means at -4.5, 0.2, and 0.5 km and mixing weights (0.61, 0.23, 0.16). The other, involving four equal-

Table 2.1.2

Elevation range (km)	Percentage of area	Elevation range (km)	Percentage of area
$(-\infty, -6)$	1.0	(0, 1)	20.8
$(-6, -5)$	16.4	(1, 2)	4.5
$(-5, -4)$	23.3	(2, 3)	2.2
$(-4, -3)$	13.9	(3, 4)	1.1
$(-3, -2)$	4.8	(4, 5)	0.4
$(-2, -1)$	3.0	$(5, \infty)$	0.1
$(-1, 0)$	8.5		

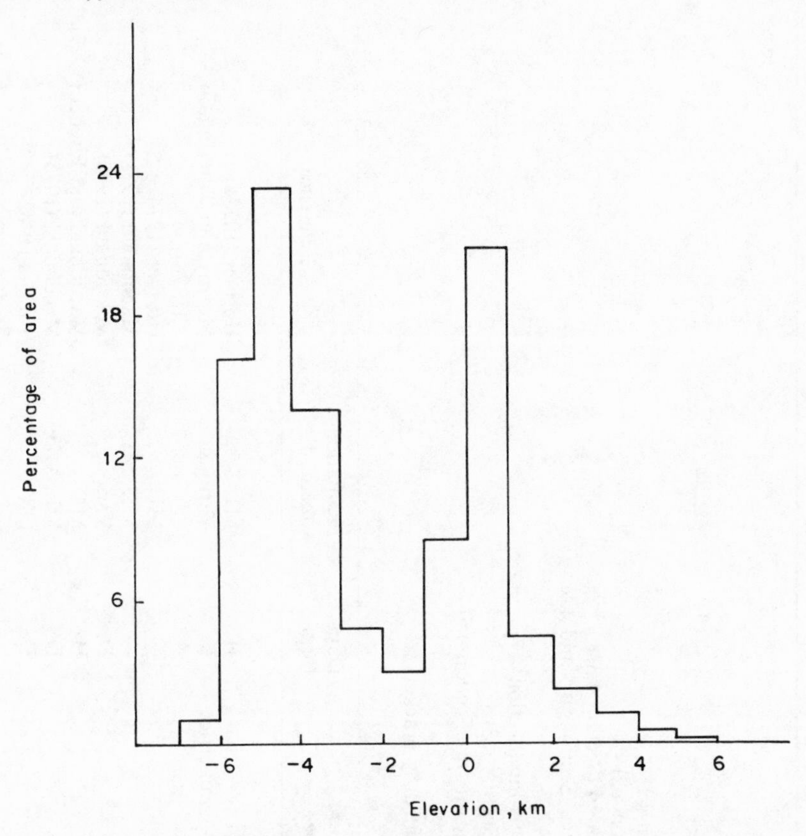

Figre 2.1.5 Distribution of the earth's surface area in terms of elevation (the hypsometric curve of the earth)

variance normal components, has means at – 5.2, – 4.8, 0.3, and 0.35 km, with mixing weights (0.50, 0.10, 0.20, 0.20). The latter is obtained by a semigraphical procedure discussed later in Section 4.1. Tanner admits the empirical and somewhat subjective nature of his method but he proposed interpretations of the components as, respectively, the ocean basins, the sedimentary fans at the base of continental shelves, the continental-shelf/coastal regions, and the continental interiors.

Example 2.1.9 Structure of a stellar cluster

In a photograph of a cluster of stars, the stars of interest are typically 'mixed' with a background, or foreground, of other stars that do not belong to the cluster. It is of interest to estimate the size and shape of the cluster, including the proportion of all the stars in the picture which do belong to the cluster. A natural model is that of a mixture of a Poisson point process and a bivariate normal

Table 2.1.3 References on direct applications

Form of data (if available)	D—numerical data H—histogram C—data curve R—referenced
Fitting procedure (where applicable)	GR—graphical (Section 4.1) MM—moments (Section 4.2) ML—maximum likelihood (Section 4.3) MD—minimum distance (Sections 4.5 and 4.6) B—Bayesian (Section 4.4) FD—function decomposition (Section 4.7) CA—cluster analysis (Section 4.3) CF—curve fitting (least squares; Section 4.5)
Mixture type	General mixtures indicated by brackets

Application	Mixture type	Data form	Estimation method	Reference
FISHERIES RESEARCH				
Halibut lengths (by sex)	Normals	H	ML	Hosmer (1973a)
	Normals/non-parametrics	R	Various	Murray and Titterington (1978)
Halibut lengths, two ages (by sex)	Regressions	—	ML	Hosmer (1974)
Pike lengths	Normals	—	MD, ML	Macdonald (1971)
(by age group)		D	Various	Macdonald (1975)
		H	MD	Macdonald and Pitcher (1979)
Minnow lengths	Normals	D	ML, MD	Macdonald (1975)
Porgy fork lengths (by age)	Normals	D	GR	Tanaka (1962)
			GR	Bhattacharya (1967)
		R	Various	Clark (1976)

Herring lengths (by spawning)	Normals	—	GR	Buchanan-Wollaston and Hodgson (1928)
		H	GR	Harding (1949)
Trout lengths	Normals	—	MM, ML	Dick and Bowden (1973)
Snapper lengths (by age)	Normals	D	GR	Cassie (1954)
Rainbow trout	Normals	H	GR	Everett (1973)
ECONOMICS				
Wage regressions	Regressions	—	MD	Quandt and Ramsey (1978)
Housing regressions	Regressions	—	ML	Quandt (1972)
	Markov regressions	—	ML	Goldfeld and Quandt (1973)
Numbers of purchases	Compound Poisson compound negative binomial	D	ML	Paull (1978)
MEDICINE				
Clinical test scores (head injury outcome)	Latent class	—	ML	Skene (1978)
Clinical test scores	MV normal	R	Sequential Bayes	Titterington (1976)
Mast cell counts (cot deaths)	Non-parametrics	R	—	Silverman (1978)
Age of onset of arthritis	Gammas	R	MM	Masuyama (1977)
Clinical test scores (haemophilia carriership)	BV normals	R	Various	Makov (1980a)
		D	ML, CA	Ganesalingam and McLachlan (1979b)
Renal xenon levels (sources)	Exponential decay curves	—	FD	Sandor, Sridhar, and Hollenberg (1978)
DNA	Normals	D	FD	Gregor (1969)
Conception times	Exponentials (censoring)	D	ML	Suchindran and Lachenbruch (1974)
Teeth pressures (outliers)	Normals	D	MM	Marks and Rao (1979)
Cholesterol triglyceride data	BV normals	—	FD(GR)	Tarter and Silvers (1975)
Chromosome association	(Beta binomial)	R	MM, ML	Skellam (1948)
Allele proportions in evolution	Discrete (Markov)	—	—	Blakley (1967)

Table 2.1.3 (*Contd.*)

Application	Mixture type	Data form	Estimation method	Reference
Clinical test scores (diabetes)	MV normals	R	ML, CA	Symons (1981)
Plasma glucose levels (two populations)	Normals	H	ML	Raper *et al.* (1982)
Blood glucose levels (diabetes)	Normals	H	GR, ML	Rushforth *et al.* (1971)
Blood pressure	Normals	H	ML	Clark *et al.* (1968)
Clinical measurements	Various	C	—	Grannis and Lott (1978)
PSYCHOLOGY				
Response times	Lognormals	D	—	Broadbent (1966)
			GR	Cox (1966)
Reaction times	Discrete	H	—	Thomas (1969)
PALAEONTOLOGY				
Foraminifer diameter (types)	Normals	H	MM	Ghose (1970)
		R	Various	Clark (1976)
Egyptian mandible sizes (sex)	Normals	D	MM	Martin (1936)
ELECTROPHORESIS, etc.				
Absorption spectrum	Normals	C	FD	Fraser and Suzuki (1966)
Protein concentrations	Normals	C	FD	Tiselius and Kabat (1939)
Protein concentrations	Normals	C	FD	Berry and Chanutin (1955)
Protein concentrations	Normals	C	FD	Wild (1965)
Gas chromatograms	Normals	—	FG/GR	Smith and Bartlet (1961)
SEDIMENTOLOGY/GEOLOGY				
Reflectivity (coal types)	Normals	H	MD	Mundry (1972)
Limestone cross-bedding	Von Mises	H	ML	Jones and James (1972)
Tin assays	Normals	D	MM	M. W. Clark (1977)
Peat-ash data	Normals	D	GR	Hald (1952)
			ML	Hasselblad (1966)
			MM	Everitt and Hand (1981)
		R	Various	Clark (1976)

Sand grain sizes	Normals	D	GR	Tanner (1959)
Sand grain sizes	Normals	C	FD	Van Andel (1973)
Sand grain sizes	Symmetric + skew	C	FD	Bagnold (1941)
Sand grain structure	Normals	—	MD	Clark (1976)
Metal concentrations	Lognormals	C	GR	Lepeltier (1969)
Particle sizes	(Quartic exponentials)	D	CF	Dallaville, Orr and Blocker (1951)
Sand dune composition	Normals	—	FD	Folk (1971)
Volcanic ash deposits	—	H*	—	Brazier *et al.* (1983)
	*Strictly, percentage of total weight.			
BOTANY				
Pollen grains	Normals	D	GR(FD)	Usinger (1975)
Pollen grains	Multinomials	D	ML, MD	Gordon and Prentice (1977)
Iris data	BV normals	R	CA	Scott and Symons (1971)
	or MV normals		ML	Wolfe (1970)
			ML	O'Neill (1978)
			Bayes CA	Binder (1978a)
			ML	Everitt and Hand (1981)
Height differences in plant pairs (Darwin's data)	Normals	D	Bayes	Box and Tiao (1968)
			ML	Aitkin and Tunnicliffe Wilson (1980)
Plant heights	Normals	D	MM, ML	Rao (1948)
			GR	Tanaka (1962)
		R	Various	Clark (1976)
Pollen counts	(Compound multinomials)	D	MM	Mosimann (1962)
Flowering times (plant strains)	Normals	D	MM, ML	Tan and Chang (1972b)
AGRICULTURE				
Crop concentrations	Known components	—	ML	Peters and Coberly (1976)
Barley yields	Factorial model	R	Bayes CA	Binder (1978a)
Panicle lengths (rice strains)	Normals	D	MM, ML	Tan and Chang (1972b)
ZOOLOGY				
Bird speeds (groups)	—	H	CA	Larkin (1979)
Trypanosome lengths (strains)	Normals	D	MM	Pearson (1914)
		R	Various	Clark (1976)
		D	GR	Fowlkes (1979)
		D	ML	Everitt and Hand (1981)

Table 2.1.3 (*Contd.*)

Application	Mixture type	Data form	Estimation method	Reference
Crab forehead sizes	Normals	D	MM	Pearson (1894)
		R	Various	Clark (1976)
Dental distances (prawn sexes)	Normals	D	MM	Pearson (1894)
		R	Various	Clark (1976)
Copepod lengths (sex)	Normals	H	GR	Harding (1949)
Rat species skull sizes	MV normals	—	ML	Do and McLachlan (1984)
Egg counts	Poissons	D	MM	Muench (1938)
Mice death times	Binomials	D	MM	Muench (1936)
Swimming directions of daphniae	Von Mises	R	ML	Mardia and Sutton (1975)
Lamphrey lengths	Normals	H	FD	Taylor (1965)
Plankton frequencies	Normals	D	GR	Cassie (1962)
Granule sizes in hen oviduct	Normals	H	ML	Covey-Crump (1970)
Corn borer larvae	(Neyman A)	D	MM	Neyman (1939)
	(Neyman)	D	MM	Beall (1940)
	(Compound Poissons)	D	MM	McGuire, Brindley, and Bancroft (1957)
Larvae	(Poisson, Pascal)	D	*Ad hoc*	Katti and Gurland (1961)
FAILURE TIMES, etc.				
Valve lifetimes (types)	Exponentials (censored)	D	—	Davis (1952)
			ML	Everitt and Hand (1981)
Transmitter receiver lifetimes	Exponentials (censored)	D	ML	Mendenhall and Hader (1958)
Bus failure times	Normal + exponential	H	—	Davis (1952)
Electron tubes	Weibulls	D	GR	Kao (1959)
Telephone call lengths	Lognormals	D	GR	Fowlkes (1979)
Laser lifetimes	Normals	R	GR	Fowlkes (1979)
Reliability	Weibull/exponentials	—	ML	Mandelbaum and Harris (1982)

MISCELLANEOUS				
Surface defects	Poissons	D	MM	Rider (1961)
				Cohen (1965)
Evening temperatures (two seasons)	Normals	D	MM	Charlier and Wicksell (1924)
Stack loss data (outliers)	Regressions	R	ML	Aitkin and Tunnicliffe Wilson (1980)
Death notice frequencies (seasons)	Poissons	D	MM	Schilling (1947)
		D	ML	Hasselblad (1969)
		D	MM	Everitt and Hand (1981)
Male deaths, by cause and occupation	Empirical Bayes	D	ML	Laird (1978b)
Geodetic measurements	Normals	H	MM	Gridgeman (1970)
Comet frequencies	Poissons	D	MM	Schilling (1947)
Jury decisions	Binomials	D	MM, MD ML	Gelfand and Solomon (1975)
Word frequencies	(Compound Poisson)	D	ML	Sichel (1975)
Yeast cell frequencies	(Neyman A)	D	MM	Neyman (1939)
Genetic reproduction	Discrete	—	ML, CA	Bryant and Williamson (1978)
Number of servers in queue, busy period	(Compound geometric)	—	—	Daniels (1961)
Turbulence	(Compound normal)	—	—	Barndorff-Nielsen (1979)
Hypsometric curve	Normals	D	GR	Tanner (1962)
Smoke particle sizes	Theoretical curves	D	—	Lipscomb, Rubin, and Sturdivant (1947)
Radioactive tracer	Exponential decays	—	FD	Brownell and Callahan (1963)
			FD	Gardner (1963)
Spray intensities	Truncated normals or Cauchys	—	MD	Wilkins (1961)
Sea-ice thickness	Unspecified	C	—	Wadhams (1981)
Arsenical response	Probit models	D	GR, ML	Ashford and Walker (1972)
Gaps in traffic	Exponentials	D	MM, ML	Ashton (1971)
Crime frequencies	Geometrics	D	ML	Harris (1983)
Precipitation	Normals	—	ML	Leytham (1984)

scatter. The parameters of the normal distribution and the mixing weight are all of interest. The component representing the Poisson process could either be approximated by a circular normal distribution, with mean at the centre of the photograph and with large variance parameters, or be described by the uniform distribution over the photograph. Whether or not subsequent analysis is easy depends partly on whether the cluster is well contained by the photograph. If not, then difficulties arise from the need to consider truncated distributions.

The scope and frequency of applications is indicated in Table 2.1.3, compiled from a wide variety of fields. Most references concern finite mixtures, but a few applications of general mixtures are also included. A striking feature is the overwhelming preponderance of normal and lognormal mixtures. Many of these concern univariate mixtures and very often only two components are involved.

2.2 INDIRECT APPLICATIONS

In this section we describe a number of applications of finite mixture models where the underlying 'categories' do not necessarily have a direct physical interpretation.

Example 2.2.1 The contaminated normal model, outliers and robustness

According to Beckman and Cook (1983), outliers can be classified into *discordant observations* (those which appear 'surprising or discrepant' to the investigator) and *contaminants* (those which are not 'realizations from the target distributions'). One approach to the latter has been to use a two-component mixture model of the form

$$p(x) = \pi f_1(x) + (1 - \pi) f_2(x) \qquad (x \in \mathbb{R}),$$

with π close to 1, $f_1(\cdot)$ the p.d.f. of interest, and $f_2(\cdot)$ the contaminating density function. The idea is that most observations belong to the first component subpopulation but, occasionally, the observation is a rogue value that comes from the second component. Usually the contaminants cannot be identified as such but they can be 'accommodated' (another term from Beckman and Cook, 1983) by suitably estimating the parameters associated with f_1 within the above model. The *contaminated normal* model is the version in which f_1 and f_2 are both normal densities. Usually these densities have the same mean but different standard deviations, with $\sigma_2 > \sigma_1$, so that the contaminants are likely to be the extreme observations and are therefore likely to have substantial 'influence' on statistics such as the sample mean. Versions in which the two densities differ only in their means have also been considered (Dixon, 1950; Abraham and Box, 1978; see also Section 4.4). Excellent accounts of the many proposed approaches to modelling and studying outliers are given by Barnett and Lewis (1978) and Beckman and Cook (1983).

Mixture models for contamination involving more general forms of f_2 have proved an important stimulus for the development of so-called *robust estimation* procedures, which seek to reduce the 'influence' of the outlying observations corresponding to contaminants. Huber (1964) introduced the idea of an M-estimator of location, defined as that $\theta = \hat{\theta}$ which minimizes, for a given sample, $x_1, \ldots, x_n$:

$$\sum_{i=1}^{n} \rho(x_i - \theta),$$

for a given even function $\rho(\cdot)$. The $\rho(\cdot)$ which minimizes the supremum (over mixtures with f_1 normal and f_2 symmetric) of the asymptotic variance of the corresponding $\hat{\theta}$ is

$$\begin{aligned} \rho(t) &= \tfrac{1}{2}t^2 && (t \leqslant k) \\ &= k|t| - \tfrac{1}{2}k^2 && (k < t). \end{aligned}$$

The proposed M-estimator is thus the maximum likelihood estimator corresponding to a density with normal centre and double-exponential tails. The cut-off point k depends on $1 - \pi$, the mixing weight corresponding to the contaminating component f_2.

The mixture formulation of the contamination model also motivates another important concept of robust estimation, the *breakdown point* of an estimator, which is, in effect, the largest value of $(1 - \pi)$ for which the estimator still behaves reasonably well; see Huber (1981, p. 13) for a technical definition and detailed discussion.

Robust estimation of correlation coefficients with a contaminated multivariate normal distribution is discussed by Bebbington (1978) and Titterington (1978).

Bayesian approaches to outliers using mixture models have been proposed by a number of authors, including Box and Tiao (1968), Guttman, Dutter, and Freeman (1978), Freeman (1981), and Pettit and Smith (1984, 1985). A detailed discussion is given in Section 4.4.

Example 2.2.2 Normal mixtures as checks on robustness

A slightly different usage of the word 'robustness' from that in Example 2.2.1 occurs in the context of 'robustness studies'. Normal mixtures have often been used in the evaluation of the properties, in non-normal conditions, of standard test statistics arising in 'normal' inference. In most of these studies, the distributions of the statistics are evaluated under the assumption that the true underlying distribution is a specified two-component normal mixture. Typically, exact, but complicated, expressions are provided for the distributions, moments, or cumulants of the statistics, along with reports and conclusions from numerical studies. Thus, for example, the celebrated Princeton Robustness Study (Andrews *et al.*, 1972) investigated in detail the 'lack of robustness' of the sample mean and the behaviour of robust estimators such as the M-estimators mentioned in Example 2.2.1. Further detailed references are given in Table 2.2.1.

Table 2.2.1 References on robustness studies using mixtures

Topic	References
Sample mean and/or covariance matrix	*Univariate*: Hyrenius (1950), Johnson, McGuire, and Milliken (1978), Mudholkar and Trivedi (1981) *Multivariate*: Tan and Chang (1972b), Tan (1978, 1980)
One-sample t statistic	Baker (1932), Hyrenius (1950), Subrahmaniam, Subrahmaniam, and Messeri (1975), Lee and Gurland (1977)
Sample linear regression F statistic	Subrahmaniam, Subrahmaniam, and Messeri (1975)
One-way ANOVA F statistic	Subrahmaniam, Subrahmaniam, and Messeri (1975)
Sample correlation coefficient	Hyrenius (1952), Kocherlakota and Kocherlakota (1981)
Discriminant functions	Hosmer (1978b), Lachenbruch and Broffitt (1980)
Quadratic forms	Subrahmaniam (1972)
Variance components	Tiao and Ali (1971)
Two-sample test statistics	Lee and D'Agostino (1976)
Hotelling's T^2	Mardia (1972, 1975)
Double sampling	Blumenthal and Govindarajulu (1977)

Example 2.2.3 Gaussian sums

In Dalal (1978) and Elashoff (1972) a mixture of two equal-variance normal distributions is used as a model for an error distribution. Sorenson and Alspach (1971) use more complicated, but still finite, normal mixtures as approximations to other distributions. As remarked by Ferguson (1983), an arbitrary density on the real line may be closely approximated by a normal mixture, a result that finds application in recursive estimation procedures within models for dynamic systems.

Consider, for example, the first-order autoregression $\{\theta_t\}$, defined by

$$\theta_t = a\theta_{t-1} + w_t, \qquad t = 1, 2, \ldots, \tag{2.2.1}$$

where θ_t is assumed unobservable, but we can observe x_t such that

$$x_t = H\theta_t + v_t, \qquad t = 1, 2, \ldots. \tag{2.2.2}$$

Note that both (2.2.1) and (2.2.2) are linear in $\{\theta_t\}$.

The noise processes (w_t) and (v_t) are taken to be uncorrelated sequences, each of uncorrelated random variables with zero means and known, constant variances:

$$\operatorname{var} w_t = q, \qquad \operatorname{var} v_t = r.$$

The parameters a and H are assumed known and we consider, for simplicity, the case of *scalar* variables. If

(a) $\theta_0 \sim N(m_0, C_0)$ and
(b) the error variables $\{w_t\}$ and $\{v_t\}$ are all normally distributed.

then, for all t, the conditional distribution of θ_t, given $x_1, \ldots, x_t$, is

$$\theta_t \sim N(m_t, C_t),$$

where m_t and C_t are given by the following simple recursions:

$$m_t = am_{t-1} + K_t(x_t - Hm_{t-1})$$
$$C_t = a^2C_{t-1} + r - K_t^2B_t$$

where

$$K_t = (a^2C_{t-1} + r)H/B_t$$

and

$$B_t = H^2(a^2C_{t-1} + r) + q, \qquad t = 0, 1, \ldots.$$

General forms of these recursions are known in the control engineering literature as the Kalman filter equations (see Section 6.5.1). Such recursions have received increasing attention in the statistical literature following the pioneering work of Harrison and Stevens (1976).

As Sorenson and Alspach (1971) point out, the neat recursions are not appplicable if the assumptions of linearity or normality do not hold. They observe, however, that, if the linearity is satisfactory, the up-to-date density function of θ_t can still be calculated exactly if the distributions of the errors and of θ_0 are all independent normal mixtures, with known parameters. There are, however, practical drawbacks. It turns out that, for any t, the density for θ_t, given $x_1, \ldots, x_t$, is a normal mixture of $n_1(t)$ components, say. If then the distributions of v_t and w_{t+1} contain, respectively, n_2 and n_3 components, then the density for θ_{t+1}, given $x_1, \ldots, x_{t+1}$, is that of a normal mixture with $n_1(t) \times n_2 \times n_3$ components.

Unless $n_1(t) = n_2 = n_3 = 1$ it is clear that the number of terms involved will become unmanageable. To make the procedure practicable, the number of terms in the mixture has to be controlled, for instance by neglecting terms with small weights and renormalizing, or by combining terms which overlap strongly. Sorenson and Alspach (1971) give several illustrations of such procedures. A general account of related problems will be given in Chapter 6; see also Alspach and Sorenson (1972), Alspach (1974, 1977), and Namera and Stubberud (1983).

Example 2.2.4 Cluster analysis and latent structure models

The objective of cluster analysis (Gordon, 1981) is to construct a classification of a set of n observations, possibly vectors, into 'interesting' subsets. In particular, we may try to discover whether there are a certain number of well-defined groups or to derive 'optimal' groupings of the observations into a specified number of clusters. One approach for the case of k clusters is to assume a mixture model, with k components, for the overall distribution of the data. We then try to estimate the mixture density and perhaps to assign observations to component densities. We might also wish to assess the suitability of the assumption of k components.

Seminal papers, using multivariate normal components, include Wolfe (1970), Scott and Symons (1971), Marriott (1975), and Symons (1981). A general discussion of the approach appears in Everitt (1980). Binder (1978a) applies Bayesian methods to the problem (see also Sections 4.3 and 4.4).

The problem of deciding upon the number of components (clusters) involved is typically very difficult, and we shall review in detail, in Chapter 5, some of the proposals that have been made.

A closely related application of finite mixture distributions occurs in *latent structure* analysis, for which there is a large literature, particularly in publications devoted to applications of statistics in the social sciences. Useful recent references are Goodman (1974) and Fielding (1977). For some k, the density of the random variable or vector of interest is assumed to have the form

$$p(x\,|\,\boldsymbol{\psi}) = \sum_{j=1}^{k} \pi_j f_j(x\,|\,\boldsymbol{\theta}_j) \qquad (x \in \mathscr{X}),$$

where the mixture used usually has the special feature that the components themselves are independence models. Thus, if x has d components, $x^{(1)}, \ldots, x^{(d)}$,

$$f_j(x\,|\,\boldsymbol{\theta}_j) = \prod_{l=1}^{d} f_{jl}(x^{(l)}, \boldsymbol{\theta}_{jl}), \qquad j = 1, \ldots, k. \tag{2.2.3}$$

The model therefore assumes that there are k hypothetical *latent classes*, which may or may not have real physical meaning, and that all correlations among the components of x are caused by ignorance about the identity of the latent class. The statistical problem is to discover how many latent classes are required to explain the covariance structure observed in the data. There is an obvious resemblance, in spirit, to factor analysis and, indeed, latent structure analysis has most frequently been used as an analogue of factor analysis for categorical data (see, also, Bartholomew, 1980). As in factor analysis, some physical interpretation is often sought for the latent classes.

An interesting example is given by Skene (1978), who extends the method to deal simultaneously with m sets of data, each from a different group of patients. He assumes that the component densities are the same for all groups but that the sets of mixing weights may differ from group to group. As an example, Skene (1978) considers patients who form two groups, those who die or survive, subsequent to suffering severe head injury. Three latent classes were found to be adequate in this case. One of them was strongly associated with death, another with survival, and the third was mildly associated with both. The need for this third class reflects the fact that very clear diagnosis is not always possible between these two groups. Another illustration is given in Dawid and Skene (1979). Skene (1980) notes that missing data are fairly easily dealt with, given the independence model (2.2.3).

Example 2.2.5 Modelling prior densities

A feature of many Bayesian statistical analyses is the adoption of a prior density from a so-called *conjugate* family: (see, for example, DeGroot, 1970). Suppose $\theta(\in\Omega)$ is the parameter of interest and that the experiment yields data x. Inference then proceeds on the basis of Bayes' theorem,

$$\pi(\theta|x) \propto p(x|\theta)\pi(\theta) \qquad (\theta\in\Omega),$$

where $\pi(\cdot|x)$ denotes the posterior density and $\pi(\cdot)$ the prior. In many situations—e.g. when $p(x|\theta)$ belongs to the exponential family—if $\pi(\cdot)$ is chosen from an appropriate family of distributions, then typically $\pi(\cdot|x)$ also belongs to that family, with the prior-to-posterior transformation simply described in terms of the sufficient statistics. For instance, if x represents the results of a sequence of independent Bernoulli trials, with success probability θ, it is easily seen that the beta densities form such a (conjugate) family. However, an obvious extension is possible, in that, if $\pi(\cdot)$ is chosen from the class of k-component mixtures of betas, then $\pi(\cdot|x)$ also belongs to that class. A far richer class of priors can therefore be considered without any real increase in the difficulty of the prior-to-posterior calculations. Diaconis and Ylvisaker (1985) contrast coin tossing with coin spinning and argue that, if θ is the probability of 'head', then a beta prior centred on '$\frac{1}{2}$' might be suitable for tossed coins, whereas spun coins tend to give 'head' with relative frequency either near $\frac{1}{3}$ or near $\frac{2}{3}$, so that a much more suitable prior can be chosen from the class of two-component beta mixtures. The posterior density would also be such a mixture, most likely reinforcing one or other of the 'modes'. In similar spirit to remarks made in Example 2.2.3., Diaconis and Ylvisaker (1985) comment that any prior density on $(0, 1)$ can be arbitrarily well approximated by a finite mixture of betas, and they generalize the discussion to exponential family distributions (see also Dalal and Hall, 1983, and other papers referenced by Diaconis and Ylvisaker, 1985).

Extension to the use of general mixtures as priors leads to the notion of multistage priors (see Lindley and Smith, 1972, and Hoadley, 1969).

Example 2.2.6 Empirical Bayes

The so-called *empirical Bayes* method is an approach to the estimation of an unknown prior density, or parameters thereof, using non-Bayesian inference. Suppose that there are n independent sets of data $x_1, \ldots, x_n$ generated by distributions with densities $p(\cdot|\theta_1), \ldots, p(\cdot|\theta_n)$, where $\theta_1, \ldots, \theta_n$ are unknown but are regarded as a random sample from the prior, $\pi(\cdot)$. Then the marginal density for the *observed* quantity, x_i, is

$$\int p(x_i|\theta)\pi(\theta)\mathrm{d}\theta = p(x_i|\pi),$$

say. For each i, $p(x_i|\pi)$ is clearly a (general) mixture, with the prior as the mixing density. Inference about π is then based on properties of the data induced by these marginal densities. For instance, the product

$$\prod_{i=1}^{n} p(x_i|\pi)$$

leads to parametric and non-parametric maximum likelihood estimation (see Robbins, 1964; Laird, 1978a: Deely and Lindley, 1981; Lindsay, 1983a); and Example 4.3.12. Lindsay (1983c) looks at a version of this problem in which $\theta_i = (\eta, \xi_i)$, where η is a common parameter of interest and $\xi_1, \ldots, \xi_n$ are regarded as nuisance parameters.

Example 2.2.7 Kernel-based density estimation

Suppose that, given a sample $(x_1, \ldots, x_n)$ of n independent observations, we require an estimate of the underlying probability density function, but we are not willing to use a parametric model. Instead, we can resort to a non-parametric procedure, of which several are described by Tapia and Thompson (1978). In one such approach, the *kernel method*, the underlying density function $p(\cdot)$ is estimated by

$$p(x|\lambda) = (n\lambda)^{-1} \sum_{i=1}^{n} K((x - x_i)/\lambda) \quad (x \in \mathscr{X}).$$

Here $K(\cdot)$ is itself a probability density function, usually symmetric, with its mode at zero. $K(\cdot)$ is called the *kernel function* and $\lambda(>0)$ is the *smoothing parameter* or *bandwidth*. Note that $p(\cdot|\lambda)$ is indeed a density function and can be interpreted as a mixture of n density functions, with equal weights. We may also write

$$p(x|\lambda) = \int \lambda^{-1} K((x-y)/\lambda)\,\mathrm{d}F_n(y) \qquad (x \in \mathscr{X}),$$

where $F_n(\cdot)$ is the empirical distribution function based on the data.

Following on from the original work of Rosenblatt (1956) and Parzen (1962), there is now a huge literature concerning the theory and practical application of the kernel method. In particular, we note that:

(a) As $\lambda \to 0$, $\int_{-\infty}^{u} p(x|\lambda)\,\mathrm{d}x$ 'tends' to $F_n(u)$, subject to various technical conditions.

(b) As $\lambda \to \infty$, the density estimate becomes more and more 'smooth'.

(c) As $n \to \infty$, $p(\cdot|\lambda)$ can usually be made to 'tend' to the true density $p(\cdot)$, in some sense or another, by ensuring that $\lambda \to 0$ at a suitable rate. This and (a) illustrate the 'non-parametric' nature of the method.

(d) For a given set of data, an important problem is to choose λ so as to give a suitable degree of smoothness relative to the amount of bias that can be tolerated. This problem has generated the majority of the research papers in this area.

In the present context, we simply point out that kernel-based density estimates may be *interpreted* as equal-weights mixture densities. They have been *applied* to the analysis of mixture problems by Murray and Titterington (1978), Silverman (1981), Titterington (1983), and Hall and Titterington (1985).

Example 2.2.8 Random variate generation

Suppose that a distribution function $P(\cdot)$ can be approximated by a mixture of two distributions, F_1, F_2, so that

$$P(x) = \pi F_1(x) + (1 - \pi)F_2(x) \qquad (x \in \mathscr{X})$$

with π close to 1. Suppose also that it is cheap to simulate from both the uniform distribution on $(0, 1)$ and from the distribution corresponding to $F_1(\cdot)$, but expensive to simulate from $P(\cdot)$ and F_2. Use of the above decomposition leads to a comparatively cheap way of generating random variates which are distributed approximately according to $P(\cdot)$. The only expensive operation will be simulation from $F_2(\cdot)$, and this will only rarely be necessary if π is close to 1 (see Marsaglia, 1961). Peterson and Kronmal (1982) give a more detailed discussion of the use of finite mixtures in random variate generation.

Example 2.2.9 Approximation of mixture models by non-mixture models

Although in many practical contexts, such as those discussed in Section 2.1, the finite mixture model is intuitively appealing, inference from the data is often non-trivial (an unpleasant fact of life, which provides the major motivation for this book!). However, in some contexts an adequate analysis can be carried out by estimating the mixture density directly without exploiting its decomposition. It is useful to point out, therefore, that there are examples of non-mixture densities which provide good approximations to certain finite mixtures.

(a) Lognormal density

Figure 2.2.1 displays the curves corresponding to the following two densities:

$$p(x) = [2x\sqrt{(2\pi)}]^{-1} \exp[-(\log x - 10)^2/8] \qquad (2.2.4)$$

$$p(x) = [\sqrt{(5\pi)}]^{-1}\{0.9 \exp[-(x - 9.5)^2/5] + 0.1 \exp[-(x - 13.5)^2/5]\}. \qquad (2.2.5)$$

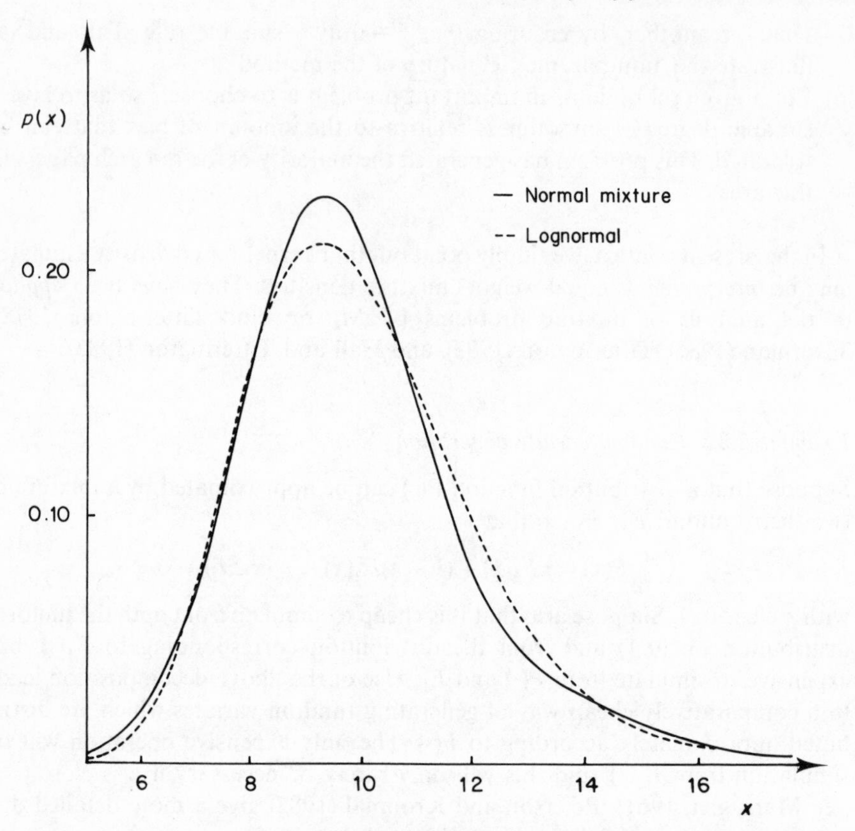

Figure 2.2.1 Plot of densities of the lognormal (2.2.4) and the normal mixture (2.2.5)

The figure stimulates the following two remarks, one rather positive and one rather negative, so far as the treatment of data from (2.2.5) is concerned:

(i) The density can be well estimated using the two-parameter lognormal (2.2.4) instead of the five-parameter normal mixture (2.2.5).
(ii) It will be very difficult to identify the 'correct' model; see Chapter 5 for a further discussion of related problems.

(b) Quartic exponential density

Consider the density

$$p(x) = c(\boldsymbol{\alpha}) \exp[-(\alpha_1 x + \alpha_2 x^2 + \alpha_3 x^3 + \alpha_4 x^4)],$$
$$\alpha_4 > 0, -\infty < x < \infty,$$

where $c(\boldsymbol{\alpha})$ is a normalizing constant. This density clearly has one or three stationary points, depending on the nature of the roots of the cubic equation

$$\alpha_1 + 2\alpha_2 x + 3\alpha_3 x^2 + 4\alpha_4 x^3 = 0,$$

so that the curve can be unimodal or bimodal (see Section 5.5).

Since the distribution is a member of the exponential family, the likelihood, in terms of $\boldsymbol{\alpha}$, is log-concave, so that there is a unique global maximum. Furthermore, if μ_j denotes the maximum likelihood estimator of the jth moment of X, with sample moment m_j, we have

$$\mu_j = m_j, \qquad j = 1, \ldots, 4. \tag{2.2.6}$$

Numerical solution of these equations (Matz, 1978) provides maximum likelihood estimates of the unknown parameters. For the symmetric case, equations (2.2.6) for $j = 1$, 2, 4 are solved.

Early work on the quartic exponential is due to Fisher (1921), O'Toole (1933a, 1933b), Aroian (1948), Dallaville, Orr, and Blocker (1951), and Greenwood and Hartley (1962, p. 465). Dallaville, Orr, and Blocker (1951) used a method of least-

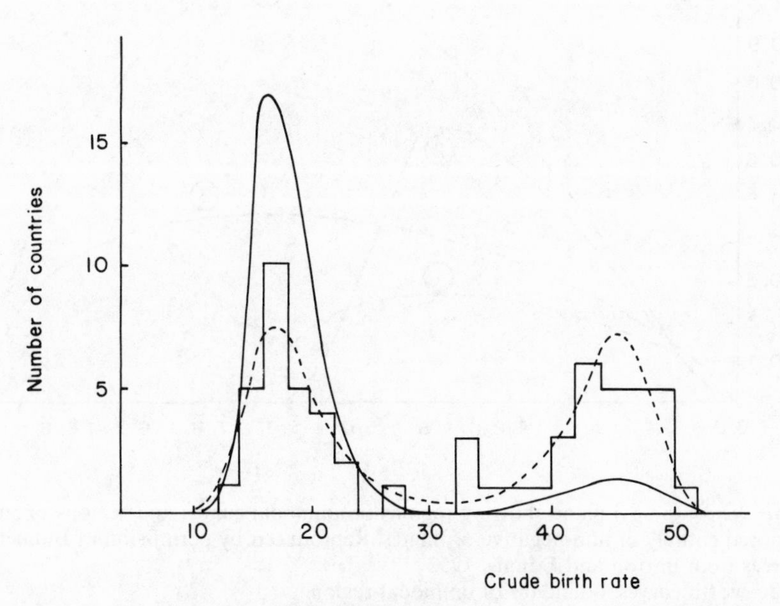

Figure 2.2.2 Quartic exponential fits to data for annual crude birth rates of 59 countries. Reproduced by permission of the American Statistical Association from Cobb, Koppstein, and Chen, 1983

squares fitting (cf. criterion 4.7.1) but forgot the dependence on $\boldsymbol{\alpha}$ of the normalizing constant $c(\boldsymbol{\alpha})$. Aroian (1948) used a moment-based method of generating consistent, asymptotically normal estimators of $\boldsymbol{\alpha}$ which can be calculated explicitly. The estimators are, however, generally thought to be inferior to those from maximum likelihood. Theoretical and empirical evidence for this is given by Cobb, Koppstein, and Chen (1983), from which Figure 2.2.2 is taken. Cobb, Koppstein, and Chen (1983) also describe a generalization of the quartic exponential density which, while retaining exponential-family membership, allows up to q modes, where q is arbitrary. A similar approach, which leads to a log-concave likelihood with grouped data as well, is described by Burridge (1982).

(c) Gram–Charlier expansions

Let

$$p(z) = [1 + \tfrac{1}{6}\sqrt{(\beta_1)}H_3(z) + \tfrac{1}{24}(\beta_2 - 3)H_4(z) + \cdots]\phi(z),$$
$$\beta_1 > 0, \; -\infty < z < \infty, \tag{2.2.7}$$

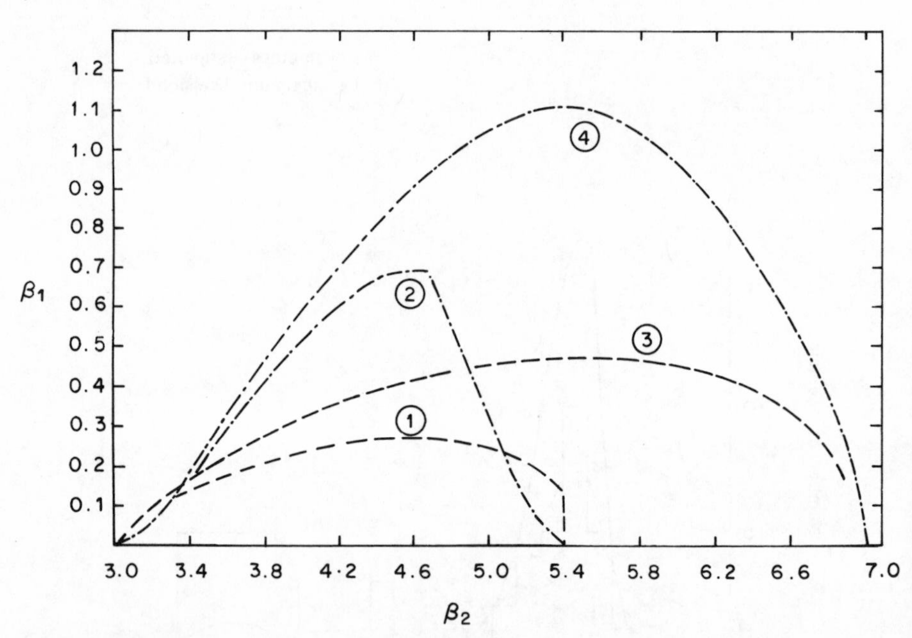

Figure 2.2.3 (β_1, β_2) plane showing regions of unimodal curves and regions of curves composed entirely of non-negative ordinates. Reproduced by permission of Biometrika Trustees from Barton and Dennis, 1952

1. Edgeworth curves: boundary of unimodal region
2. Gram Charlier curves: boundary of unimodal region
3. Edgeworth curves: boundary of positive definite region
4. Gram Charlier curves: boundary of positive definite region

where β_1 and β_2 are the traditional measures of skewness and kurtosis, $H_j(z)$ is the jth-order Hermite polynomial, and $\phi(z)$ denotes the standard normal p.d.f. Equation (2.2.7) gives the leading terms in the *Gram–Charlier expansion* for the density function of a standardized variable; see Johnson and Kotz (1970a, Section 12.4), where the modification known as the *Edgeworth expansion* is also given. For a non-standardized variable one should use

$$\sigma^{-1}p((x-\mu)/\sigma).$$

In practice, truncated versions of the expansions are used, although the resulting function is usually no longer a probability density function. For truncation corresponding to the terms specified by (2.2.7), Figure 2.2.3 (Johnson and Kotz, 1970a, Section 12.4; Barton and Dennis, 1952) indicates the ranges of

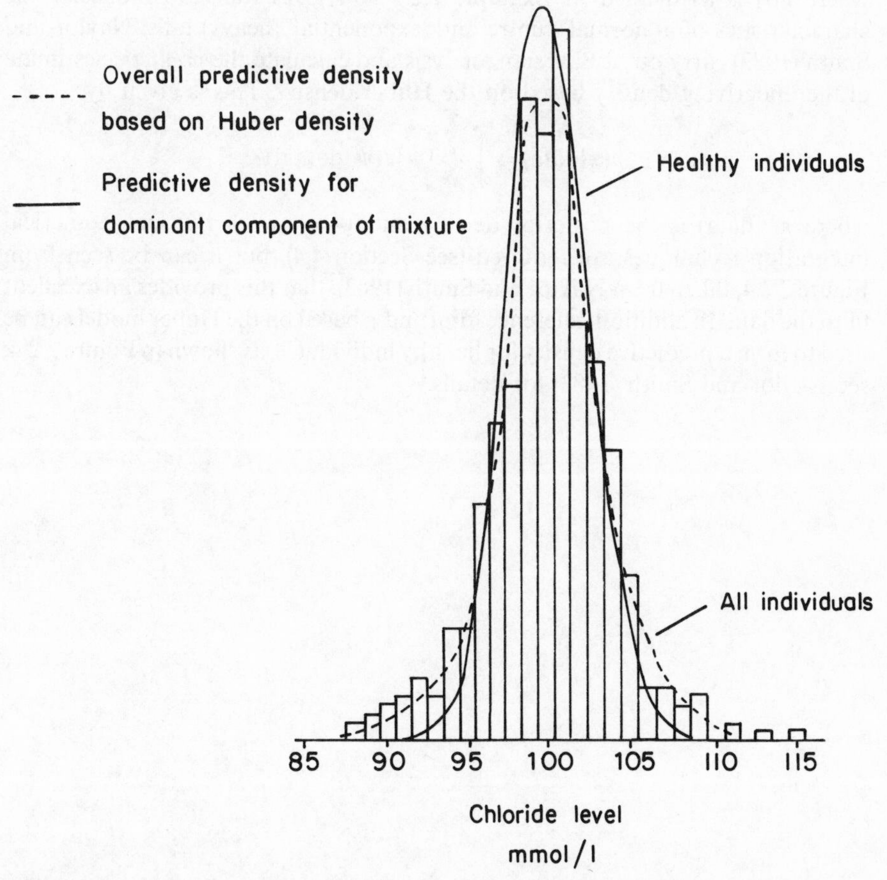

Figure 2.2.4 Predictive densities from two models fitted to histogram data on blood chloride levels. Reproduced by permission of the Institute of Statisticians from Naylor and Smith, 1983

(β_1, β_2) for which non-negativity and/or multimodality is guaranteed. Draper and Tierney (1972) add a small correction.

(d) Approximation of a contaminated normal density by a Huber density

Naylor and Smith (1983) analyse a sample of blood chloride levels from a population which is assumed to contain a proportion, π, of healthy individuals and a small proportion, $1-\pi$, of unhealthy ones. The component distributions are assumed to be $N(\mu, \sigma^2)$ and $N(\mu, \lambda^2\sigma^2)$, respectively, where $\lambda > 1$. An alternative to this four-parameter model is the three-parameter Huber density, defined by

$$p(x|\mu, \sigma, k) \propto \sigma^{-1} \exp[-\rho((x-\mu)/\sigma)],$$

where $\rho(t)$ is as defined in Example 2.2.1 and gives the Huber density the characteristics of a 'normal' centre and 'exponential' (heavy) tails. Naylor and Smith (1983) carry out a Bayesian analysis and calculate the *predictive* estimate of the underlying density based on the Huber density. This is given by

$$p(x|\text{data}) = \int p(x|\boldsymbol{\psi})\pi(\boldsymbol{\psi}|\text{data})\mathrm{d}\boldsymbol{\psi},$$

where $\pi(\cdot|\text{data})$ is the posterior density and $\boldsymbol{\psi} = (\mu, \sigma, k)$. Efficient numerical integration techniques are required (see Section 4.4), but it can be seen from Figure 2.2.4, taken from Naylor and Smith (1983), that this provides an excellent fit to the data. In addition, inference for μ and σ based on the Huber model can be used to form a predictive density for healthy individuals, as shown in Figure 2.2.4; see Naylor and Smith (1983) for details.

CHAPTER 3

Mathematical aspects of mixtures

3.1 IDENTIFIABILITY

3.1.1 Introduction and definition

The assumption of identifiability—i.e. the existence of a unique characterization for any one of the class of models being considered—lies at the heart of most statistical theory and practice. It is important to consider identifiability in practice because, without it, estimation procedures are not likely to be well defined. Furthermore, identifiability is a necessary requirement for the usual asymptotic theory to hold for both the non-sequential and sequential estimation procedures considered in later chapters.

In this section we give a formal definition of identifiability for finite mixtures; we illustrate that non-identifiability can occur and we derive both necessary and sufficient (*and* useful sufficient) conditions for identifiability.

Mixture models can present particular difficulties with identifiability, as the following simple example shows.

Example 3.1.1 Mixtures of two binomials

Let $\mathscr{T}$ denote the class of $\mathrm{Bi}(2,\theta)$ distributions, $0<\theta<1$, and consider an arbitrary mixture of two members of $\mathscr{T}$. Then

$$p(0)=\pi(1-\theta_1)^2+(1-\pi)(1-\theta_2)^2$$

and

$$p(1)=2\pi\theta_1(1-\theta_1)+2(1-\pi)\theta_2(1-\theta_2),$$

where π denotes the mixing weight. There is a third equation, for $p(2)$, but it is of course not independent of the other two. Given the mixture probabilities, $p(0)$, $p(1)$, their representation in terms of (π,θ_1,θ_2) is clearly not unique.

In what follows, we shall consider only mixtures in which all component distributions come from the same parametric class. The mixture densities then take forms such as (1.1.3).

Let

$$\mathscr{T} = \{F(x, \theta), \theta \in \Omega, x \in \mathbb{R}^d\}$$

be the class of d-dimensional *distribution functions* from which mixtures are to be formed. We identify the class of finite mixtures of $\mathscr{T}$ with the appropriate class of distribution functions, $\mathscr{H}$, defined by

$$\mathscr{H} = \left\{ H(x): H(x) = \sum_{j=1}^{k} \pi_j F(x, \theta_j), \pi_j > 0, \sum_{j=1}^{k} \pi_j = 1, \right.$$

$$\left. F(\cdot, \theta_j) \in \mathscr{T}, \text{ all } j, k = 1, 2, \ldots, x \in \mathbb{R}^d \right\},$$

so that $\mathscr{H}$ is the *convex hull* of $\mathscr{T}$.

We shall abbreviate $F(x, \theta_j)$ by $F_j(x)$, and sometimes denote the mixture by

$$H = \sum_{j=1}^{k} \pi_j F_j.$$

In all such expressions, $F_1, \ldots, F_k$ are assumed to be distinct members of $\mathscr{T}$. The following formal definition states that $\mathscr{H}$ is *identifiable* if and only if all members of $\mathscr{H}$ are distinct.

Definition 3.1.1 Identifiability

Suppose H, H' are any two members of $\mathscr{H}$, given by

$$H = \sum_{j=1}^{k} \pi_j F_j, \qquad H' = \sum_{j=1}^{k'} \pi'_j F'_j,$$

and that $H \equiv H'$ if and only if $k = k'$ and we can order the summations such that $\pi_j = \pi'_j$, $F_j = F'_j$, $j = 1, \ldots, k$. Then $\mathscr{H}$ is *identifiable*.

Equivalent definitions of $\mathscr{H}$ and of identifiability may be written down in terms of densities, when they exist, and it is in this context that we present our second example of lack of identifiability.

Example 3.1.2 Mixtures of uniforms

Let $\text{Un}(x; a, b)$ denote the uniform density on (a, b). Then, for instance, $\text{Un}(x; 0, 1) \equiv \pi\text{Un}(x; 0, \pi) + (1 - \pi)\text{Un}(x; \pi, 1)$, for any π between 0 and 1.

3.1.2 Theorems and applications

In line with Example 3.1.1 above, a natural way of expressing the identifiability problem is as follows. Given $F \in \mathscr{H}$, is there a unique solution of the identity

$$\sum_{j=1}^{k} \pi_j F_j \equiv F,$$

for k, $\pi_1,\ldots,\pi_k$ and $F_1,\ldots,F_k \in \mathscr{T}$? (By 'unique', we mean 'unique up to permutation of the subscripts'.) It is perhaps not surprising that linear independence for the members of $\mathscr{T}$ is the key to answering the question. This is given formal expression in Theorem 3.1.1, where the notation $\langle \mathscr{T} \rangle$ is used for the *span* of $\mathscr{T}$ over the real numbers; i.e. the class of all linear combinations of members of $\mathscr{T}$.

Theorem 3.1.1 (Yakowitz and Spragins, 1968)

A necessary and sufficient condition that $\mathscr{H}$ be identifiable is that $\mathscr{T}$ be a linearly independent set over the field of real numbers, $\mathbb{R}$.

Proof (*Necessity*): Suppose that $\mathscr{T}$ is not linearly independent over $\mathbb{R}$, so that, for some $k > 0$, there exists a 'null' linear combination of distinct members of $\mathscr{T}$ such that, for some m, $0 < m < k$, $\sum_{j=1}^{k} \eta_j F_j = 0$, with

$$\begin{aligned} \eta_j &< 0 \quad \text{for} \quad j \leqslant m \\ \eta_j &> 0 \quad \text{for} \quad j > m. \end{aligned}$$

Then

$$\sum_{j=1}^{m} |\eta_j| F_j \equiv \sum_{j=m+1}^{k} |\eta_j| F_j \tag{3.1.1}$$

and, since the $\{F_j\}$ are all distribution functions, it follows that

$$\sum_{j=1}^{m} |\eta_j| = \sum_{j=m+1}^{k} |\eta_j| = b, \text{ say,}$$

on setting all the F_j's equal to 1.

Thus, if we define $\pi_j = |\eta_j|/b$, $j = 1,\ldots,k$, (3.1.1) becomes

$$\sum_{j=1}^{m} \pi_j F_j \equiv \sum_{j=m+1}^{k} \pi_j F_j$$

and we therefore have two distinct representations of the same finite mixture, which implies that $\mathscr{H}$ is not identifiable.

Proof (*Sufficiency*): If $\mathscr{T}$ is linearly independent over $\mathbb{R}$ then it forms a basis for $\langle \mathscr{T} \rangle$. Thus any member of $\langle \mathscr{T} \rangle$ has a unique representation as a linear combination of members of $\mathscr{T}$. Since $\mathscr{H} \subset \langle \mathscr{T} \rangle$, identifiability of $\mathscr{H}$ follows immediately.

Corollary 3.1.1

$\mathscr{H}$ is identifiable if and only if the image of $\mathscr{T}$ under any vector isomorphism on $\langle \mathscr{T} \rangle$ is linearly independent in the image space.

The corollary to Theorem 3.1.1 tends to be easier to apply directly than the theorem itself. In particular, it enables us to work in terms of generating functions, which are often more convenient to handle mathematically than the corresponding distribution functions.

Example 3.1.3 Translation parameter mixtures (Teicher, 1961; Yakowitz and Spragins, 1968; Ord, 1972)

Let F be any univariate distribution function and let $\mathscr{T}$ be the *translation parameter* family induced by F. That is,

$$F(x, \theta) = F(x + \theta) \qquad (x \in \mathbb{R})$$

defines a typical member $F(\cdot, \theta)$ of $\mathscr{T}$, $\theta \in \mathbb{R}$.

The following argument then shows that the set of finite mixtures on $\mathscr{T}$ is identifiable. Denote by $G(z, \theta)$ the characteristic function of $F(x + \theta)$, so that, if $i = \sqrt{(-1)}$,

$$G(z, \theta) = \exp(i\theta z) G(z, 0).$$

To fulfil the conditions imposed in the corollary we must show that, if

$$0 = \sum_{j=1}^{k} \pi_j G(z, \theta_j) = \left[\sum_{j=1}^{k} \pi_j \exp(iz\theta_j) \right] G(z, 0), \qquad \text{for all } z,$$

then $\pi_1 = \cdots = \pi_k = 0$.

Since $G(0, 0) = 1$ and $G(z, 0)$ is continuous, $|G(z, 0)| > 0$ for $z \in (-\delta, \delta)$ for some $\delta > 0$. Thus,

$$0 = \sum_{j=1}^{k} \pi_j \exp(i\theta_j z), \qquad z \in (-\delta, \delta).$$

From this the result follows as in Yakowitz and Spragins (1968).

The class of finite scale mixtures defined by $\{F(x, \theta) = F(x/\theta), \theta \in \Theta\}$ is identifiable if $F(t)$ has some positive power moment (Behboodian, 1975). Yakowitz (1970) extends the result of Example 3.1.3 to the multivariate case and Yakowitz and Spragins (1968) deal with univariate Cauchy mixtures and multivariate normal mixtures in a similar way. For illustration, we consider, in the next example, the case of univariate normal mixtures (see also Robbins and Pitman, 1949).

Example 3.1.4 Mixtures of univariate normals

Suppose $\mathscr{T}$ is the class of univariate normal distribution functions and that we try to use Corollary 3.1.1 in terms of the Laplace transforms (moment-generating functions). As usual, (μ, σ^2) will denote the mean and variance of a typical member of $\mathscr{T}$. If $\mathscr{H}$ is identifiable and if $(\mu_1, \sigma_1^2), \ldots, (\mu_k, \sigma_k^2)$ are distinct pairs, then

$$\sum_{j=1}^{k} \pi_j \exp(\mu_j z + \tfrac{1}{2}\sigma_j^2 z^2) \equiv 0$$

must give $\pi_1 = \cdots = \pi_k = 0$.

For all $z \neq 0$, the pairs $\{(\mu_j z, \sigma_j^2 z^2), j = 1, \ldots, k\}$ are all distinct. It is possible, therefore, to choose k distinct values of z, $z_1, \ldots, z_k$, such that

$$\sum_{j=1}^{k} \pi_j a_{ji} = 0, \qquad i = 1, \ldots, k,$$

where $a_{ji} = \exp(z_i \mu_j + \tfrac{1}{2}\sigma_j^2 z_i^2)$, for all i, j, and such that the $k \times k$ matrix with the (j, i)th entry a_{ji} is non-singular. The result follows.

A corollary of this is that a 'non-degenerate' finite mixture of normal distributions cannot itself be normal, a remark which carries over to other identifiable classes of mixtures, such as the compound Poisson distributions (Teicher, 1960; Ord, 1972).

In many cases, the conditions of Theorem 3.1.1 or its corollary are hard to verify. It is, however, possible to establish *sufficient* conditions for identifiability, exemplified in the following result.

Theorem 3.1.2 (Teicher, 1963)

Let $\mathscr{T}$ be a family of univariate distribution functions with transforms $G(z)$ defined for z belonging to some domain of definition, S_G. It is assumed that the mapping $M: F \to G$ is linear and one-to-one. Suppose that there is a total ordering, denoted by '$\prec$', of $\mathscr{T}$ such that $F_1 \prec F_2$ implies:

(a) $S_{G_1} \subseteq S_{G_2}$.
(b) There is some $z_1 \in \bar{S}_{G_1}$ (the complement of S_{G_1}), with z_1 independent of G_2, such that

$$\lim_{z \to z_1} \frac{G_2(z)}{G_1(z)} = 0.$$

Then the class $\mathscr{H}$ of all finite mixtures of $\mathscr{T}$ is identifiable.

The proof (Teicher, 1963) is by contradiction. Two distinct finite mixture representations are assumed for a member of $\mathscr{H}$, but this is ultimately shown to be untenable.

Theorem 3.1.2 finds direct application in several simple cases such as mixtures of univariate normals and mixtures of gammas (Teicher, 1963), in both cases using the moment-generating function as the transform $G(\cdot)$.

Example 3.1.5 Mixtures of gammas

Suppose $\mathscr{T}$ is the class of distribution functions whose corresponding densities are

$$[\Gamma(\alpha)]^{-1}\theta^{\alpha}x^{\alpha-1}e^{-\theta x}, \qquad x>0, \alpha>0, \theta>0,$$

so that

$$G(z)=\mathbb{E}(e^{xz})=(1-z/\theta)^{-\alpha}, \qquad \text{valid for } z<\theta.$$

If F_1 and F_2 correspond to parameter pairs (α_1, θ_1) and (α_2, θ_2) then a suitable ordering is defined by

$$F_1 \prec F_2 \quad \text{if} \quad \theta_1<\theta_2 \quad \text{or if} \quad \theta_1=\theta_1 \quad \text{and} \quad \alpha_1>\alpha_2.$$

Such an ordering is called a 'lexicographical' ordering. Then

$$S_{G_1}=(-\infty, \theta_1) \subset S_{G_2}=(-\infty, \theta_2)$$

and

$$\lim_{z \to z_1} \frac{G_2(z)}{G_1(z)}=0 \qquad \text{if } z_1 \in S_{G_2} \backslash S_{G_1}.$$

It follows from Theorem 3.1.2 that the class of gamma mixtures is identifiable.

We now return to, and generalize, Example 3.1.1, which represents a case where non-identifiability certainly can occur.

Example 3.1.6 Mixtures of binomials (Teicher, 1961, 1963; Blischke, 1962, Ord, 1972).

Consider the family $\text{Bi}(N, \theta)$, $0<\theta<1$, for fixed N but varying θ. Example 3.1.1 has demonstrated non-identifiability for $N=2$. However, classes of finite mixtures of some subfamilies of binomials are identifiable. In this family, the class of mixtures of at most k members is identifiable if and only if $N \geqslant 2k-1$. To show this, consider two representations of the same mixture:

$$\sum_{j=1}^{c_1} \pi_j F_j(x)=\sum_{j=1}^{c_2} \tilde{\pi}_j \tilde{F}_j(x), \qquad x=0, \ldots, N,$$

with
$$0<\pi_j<1, j=1, \ldots, c_1,$$
$$0<\tilde{\pi}_j<1, j=1, \ldots, c_2,$$
$$\sum_j \pi_j=\sum_j \tilde{\pi}_j=1 \text{ and } c_1, c_2 \leqslant k.$$

Then

$$\sum_{j=1}^{k'} \lambda_j F_j^*(x)=0, \qquad x=0, \ldots, N,$$

where $\{F_j^*(\cdot)\}$ are the k' distinct elements of $\{F_1(\cdot), \ldots, F_{c_1}(\cdot), \tilde{F}_1(\cdot), \ldots, \tilde{F}_{c_2}(\cdot)\}$.

Again we utilize transforms, this time the probability generating function. We have, equivalently,

$$\sum_{j=1}^{k'} \lambda_j(1+\theta_j^* z)^N \equiv 0,$$

from which it follows that

$$\sum_{j=1}^{k'} \lambda_j \theta_j^{*i} = 0, \qquad \text{for } i = 0, \ldots, N.$$

If $N < k' - 1$ there is a non-zero solution for $(\lambda_1, \ldots, \lambda_{k'})$, which rules out identifiability. If, however, $N \geqslant k' - 1$, $\lambda_1 = \cdots = \lambda_{k'} = 0$ is the only solution. To obtain identifiability for the class of mixtures in question, the above must hold for all possible pairs (c_1, c_2). This leads to the necessary and sufficient condition that $N \geqslant 2k - 1$.

Example 3.1.6 considers a special class of binomial distributions, with N known and fixed. Using an argument similar to the above, Teicher (1963) investigates a wider class of binomials.

3.1.3 Further results and literature

(a) Core papers

Seminal papers on identifiability of finite mixtures are those of Teicher (1961, 1963) and Yakowitz and Spragins (1968). In the engineering literature, important general papers are those of Patrick and Hancock (1966) and Yakowitz (1970). Note, however, that the former paper wrongly claims the identifiability of the class of general normal mixtures. Chandra (1977) gives a useful recent survey.

(b) Identifiability of non-finite mixtures

Papers devoted specifically to this topic are those of Patil and Bildikar (1966) and Tallis (1969), for countable mixtures, and Barndorff-Nielsen (1965), Tallis (1969), and Tallis and Chesson (1982), for general mixtures. Teicher (1961) also considers general mixtures, pointing out that the class of general normal mixtures is not identifiable, but that the class of mean mixtures of univariate normals is identifiable.

(c) Multivariate models assuming independence

Given a class $\mathscr{T}$ of univariate distributions, classes of multivariate distributions can be obtained, given independence assumptions, using the product rule. Teicher (1967) shows that 'univariate' identifiability holds if and only if 'multivariate' identifiability does. Without the assumption of independence, neither of these implications need hold, as Rennie (1972) shows by illustration.

(d) Some special cases

Teicher (1961), Ord (1972), and Rennie (1974) discuss mixtures of members of an additively closed one-parameter family, Fraser, Hsu, and Walker (1981) establish identifiability of finite mixtures of von Mises distributions, and Kent (1983) generalizes their result to a wider class of mixtures which has practical uses in modelling directional data.

These positive findings confirm the overall picture that, apart from special cases with finite sample spaces (Example 3.1.1) or very special simple density functions (Example 3.1.2), identifiability of classes of *finite* mixtures is generally assured. For more recent work, see Li and Sedransk (1985).

3.2 INFORMATION

The Fisher information matrix (see, for example, Silvey, 1975, Chapter 2) is a measure of the value of a statistical experiment designed to investigate a parametric model. In the context of this book, calculation of the information matrix will help us to assess whether, for instance, fifty observations from a two-component mixture are more valuable, from the point of view of estimating the component densities, than five from each component separately. This, in turn, might influence decisions as to whether it is worth the cost of trying to identify the true categories of mixture data. The Fisher information matrix is also crucial, of course, in the study of the asymptotic properties of maximum likelihood estimators.

If the experiment consists of n independent observations, we define $I(\psi)$, the Fisher information matrix for a single observation, by

$$nI(\psi) = \mathbb{E}[D_\psi \mathscr{L}(\psi) D_\psi \mathscr{L}(\psi)^{\mathrm{T}}],$$

where the vector ψ contains the unknown parameters, $\mathscr{L}(\psi)$ denotes the loglikelihood, and D_ψ denotes the first derivative operator, with respect to ψ. We shall only consider cases where $I(\psi)$ exists and where it can also be calculated from

$$nI(\psi) = -\mathbb{E}[D_\psi^2 \mathscr{L}(\psi)],$$

where $D_\psi^2 \mathscr{L}$ denotes the matrix of second derivatives of $\mathscr{L}$. In simple parametric families, $[nI(\psi)]^{-1}$ typically provides a matrix lower bound for the covariance matrix of unbiased estimators of ψ, as well as giving the asymptotic covariance matrix of maximum likelihood estimators of ψ. In Section 4.3 we establish that this latter property also holds in many mixture problems. In this section, we consider the problems of calculating $I(\psi)$ and finding out what it tells us about the information content of mixture data.

In Chapter 1 we described three data structures, M0, M1, and M2, made up of various combinations of three types of observation.

Type 0: Observation x from the mixture, with associated p.d.f.

$$p(x \mid \psi) = \sum_{j=1}^{k} \pi_j f_j(x \mid \theta_j)$$

and information matrix $I_0(\psi)$, say.

Type C: Fully categorized 'complete' observation $(x, \mathbf{z})$ from the mixture, with associated joint probability function

$$p(\mathbf{z}, x \mid \boldsymbol{\psi}) = \prod_{j=1}^{k} \pi_j^{z_j} f_j(x \mid \theta_j)^{z_j}$$

and information matrix $I_c(\boldsymbol{\psi})$, where $\mathbf{z}$ is an indicator random vector whose realization identifies the source component of x.

Type j $(j = 1, \ldots, k)$: An observation chosen in advance from the jth subpopulation, with p.d.f.

$$f_j(x \mid \theta_j)$$

and information matrix $I_j(\theta_j)$. Note that, although the notation stresses the fact that such an observation tells us only about θ_j, it is helpful to think of I_j as a matrix of the same order as I_0 and I_c, but consisting largely of zeros.

We should expect a type C observation to be more informative than a type 0 and a type j to be more informative than a type C so far as θ_j is concerned. More formally,

$$p(\mathbf{z}, x \mid \boldsymbol{\psi}) = p(x \mid \boldsymbol{\psi}) p(\mathbf{z} \mid x, \boldsymbol{\psi}),$$

so that

$$\log p(\mathbf{z}, x \mid \boldsymbol{\psi}) = \log p(x \mid \boldsymbol{\psi}) + \log p(\mathbf{z} \mid x, \boldsymbol{\psi})$$

and

$$I_c(\boldsymbol{\psi}) = I_0(\boldsymbol{\psi}) + I_E(\boldsymbol{\psi}),$$

where $I_E(\cdot)$ itself denotes an information matrix. Thus, $I_c(\boldsymbol{\psi}) - I_0(\boldsymbol{\psi})$ is non-negative definite and, in a sense, measures what is 'lost' by not categorizing a mixture observation.

Suppose then we have an experiment with the following *proportions* of observations: ρ_c of type C, ρ_j of type j $(j = 1, \ldots, k)$, and ρ_0 of type 0. Then

$$I(\boldsymbol{\psi}) = \rho_0 I_0(\boldsymbol{\psi}) + \rho_c I_c(\boldsymbol{\psi}) + \sum_j \rho_j I_j(\theta_j). \qquad (3.2.1)$$

Our original data structures correspond to $\rho_0 = 1$ (M0), $\rho_c = 0$ (M1), and $\sum_j \rho_j = 0$ (M2). Note that we can write

$$I_c(\boldsymbol{\psi}) = I_c(\boldsymbol{\pi}) + \sum_{j=1}^{k} \pi_j I_j(\theta_j),$$

where $I_c(\boldsymbol{\pi})$ is the information matrix for a multinomial observation. It is easy to evaluate $I_c(\boldsymbol{\pi})$ and the same is true of $I_j(\theta_j)$ in most cases of interest. This cannot, however, be said of $I_0(\boldsymbol{\psi})$, even in the simplest examples.

Example 3.2.1 Mixture of two known densities

$$p(x \mid \pi) = \pi f_1(x) + (1 - \pi) f_2(x) \qquad (0 < \pi < 1, x \in \mathbb{R}).$$

We shall assume that $f_1(\cdot)$ and $f_2(\cdot)$ are known univariate density functions. Details for multivariate and categorical data mixtures are similar. Then

$$I_0(\pi) = \int \{[f_1(x) - f_2(x)]^2 / p(x|\pi)\}\, dx$$
$$= [s_{11}(\pi) - 1]/(1-\pi)^2, \tag{3.2.2}$$

where

$$s_{jl}(\pi) = \int [f_j(x) f_l(x)/p(x|\pi)]\, dx.$$

For this example, of course, 'type j' data are of no interest and for a type C observation it is straightforward to verify that

$$I_c(\pi) = \pi^{-1}(1-\pi)^{-1}.$$

Not surprisingly, $I_0(\pi) \leqslant I_c(\pi)$, with equality if and only if f_1 and f_2 are completely separate (with disjoint ranges of positive support, so that $f_1(x) f_2(x) \equiv 0$).

The ratio $I_c(\pi)/I_0(\pi)$ can be interpreted as the number of mixture observations required to give as much information as a single fully categorized observation. In this case,

$$I_c(\pi)/I_0(\pi) = \pi^{-1}(1-\pi)/[s_{11}(\pi) - 1]. \tag{3.2.3}$$

Explicit evaluation of $s_{11}(\pi)$ is not possible but if $f_1(x)$ and $f_2(x)$ are normal densities, tables in Behboodian (1972a) allow approximate calculation of $s_{jl}(\pi)$ and direct numerical quadrature is fairly straightforward.

Hill (1963) showed that an alternative expression for (3.2.2) is given by

$$I_0(\pi) = [1 - s_{12}(\pi)]/\pi(1-\pi) \tag{3.2.4}$$

and derived series-expansion representations for $s_{12}(\pi)$. For a mixture of two normal densities with equal variances (σ^2) he obtained the following:

$$s_{12}(\pi) = \sum_{j=0}^{\infty} (-1)^j \exp[\Delta^2 j(j+1)/2] \left\{ \pi^{-1} \left(\frac{1-\pi}{\pi} \right)^j \left[1 - \Phi\left(\frac{2j+1}{2} \Delta + \frac{r}{\Delta} \right) \right] \right.$$
$$\left. + (1-\pi)^{-1} \left(\frac{\pi}{1-\pi} \right)^j \left[1 - \Phi\left(\frac{2j+1}{2} \Delta - \frac{r}{\Delta} \right) \right] \right\},$$

where $\Delta = (\mu_1 - \mu_2)/\sigma$, $r = \log[(1-\pi)/\pi]$ and it is assumed that $\mu_1 > \mu_2$.

If Δ is small then

$$I_0(\pi) = \Delta^2 + o(\Delta^2). \tag{3.2.5}$$

Hill (1963) provides a similar analysis for exponential mixtures and shows that, if $|\Delta| = |\theta_1 - \theta_2|/\theta_1$, where θ_1 and θ_2 are the two rates ($\theta_1 > \theta_2$) then (3.2.5) again holds. To see how poor estimation on the basis of mixture data can be, note that it can be shown, from (3.2.4), that if $\frac{1}{8} \leqslant \Delta \leqslant \frac{1}{4}$ and an estimator $\hat{\pi}$

Table 3.2.1 Values of $s_{11}(\pi)$ and efficiencies of M0 and M2 data

	$\pi=0.1$						$\pi=0.5$					
		ρ_c						ρ_c				
Δ	$s_{11}(\pi)$	0	$\frac{1}{8}$	$\frac{1}{4}$	$\frac{1}{2}$	$\frac{3}{4}$	$s_{11}(\pi)$	0	$\frac{1}{8}$	$\frac{1}{4}$	$\frac{1}{2}$	$\frac{3}{4}$
0.4	1.1341	1.5	13.8	26.1	50.8	75.4	1.0385	3.8	15.9	27.9	51.9	76.0
0.8	1.5782	6.4	18.1	29.8	53.2	76.6	1.1394	13.9	24.7	35.5	57.0	78.5
1.2	2.3787	15.3	25.9	36.5	57.7	78.8	1.2736	27.4	36.4	45.5	63.7	81.8
2.0	4.7016	41.1	48.5	55.9	70.6	85.3	1.5504	55.0	60.7	66.3	77.5	88.8
4.0	9.0717	89.7	91.0	92.3	94.9	97.4	1.9314	93.1	94.0	94.9	96.6	98.3
ρ_c		0	12.5	25.0	50.0	75.0		0	12.5	25.0	50.0	75.0

is to be used whose standard error is at most 0.1, then at least 1600 observations are required!

Example 3.2.2 Mixture of two known normals

Suppose

$$p(x|\pi)=\pi\phi(x)+(1-\pi)\phi(x+\Delta), \qquad 0<\pi<1,$$

where $\phi(\cdot)$ denotes the standard normal density function.

Table 3.2.1 gives values of $s_{11}(\pi)$ for $\pi=0.1$ and 0.5 and for $\Delta=0.4$, 0.8, 1.2, 2, 4. Note (see also Section 3.3.1) that $\pi=0.5$ and $\Delta=2$ lie on a critical boundary between unimodality and strict bimodality. Also provided are the percentage efficiencies of M2 data relative to fully categorized data, defined by

$$[(1-\rho_c)I_0(\psi)+\rho_c I_c(\psi)]/I_c(\psi)$$

for $\rho_c=0, \frac{1}{8}, \frac{1}{4}, \frac{1}{2}, \frac{3}{4}$. The case $\rho_c=0$ corresponds to M0 and the inverses of the percentages in that column give the M0 equivalent of a single fully categorized observation, as defined in (3.2.3). The final row of the table gives the efficiency of the simple estimator of π obtained from the 'complete' part of the M2 data. This value is, of course, ρ_c, and allow us to judge the extra worth of the uncategorized data.

The unbalanced ($\pi=0.1$) mixture produces the lower efficiencies and, if ρ_c and Δ are small, the uncategorized data do not contribute very much information. In the worst case considered ($\rho_c=0$, $\Delta=0.4$), one fully categorized observation gives as much information as $1/0.015=67$ uncategorized ones.

Example 3.2.3 Mixture of two univariate normals

$$p(x|\psi)=\pi_1\phi(x|\mu_1,\sigma_1)+(1-\pi_1)\phi(x|\mu_2,\sigma_2),$$

which we shall, for notational convenience, abbreviate to

$$\pi_1 f_1(x)+\pi_2 f_2(x).$$

If we define $\boldsymbol{\psi} = (\mu_1, \mu_2, \pi_1, \sigma_1^2, \sigma_2^2)^{\mathrm{T}}$, then

$$I_c(\boldsymbol{\psi}) = \mathrm{diag}\,[\pi_1/\sigma_1^2, \pi_2/\sigma_2^2, (\pi_1\pi_2)^{-1}, 2\pi_1/\sigma_1^4, 2\pi_2/\sigma_2^4],$$

so that, for example,

$$I_1(\boldsymbol{\theta}_1) = \mathrm{diag}\,(1/\sigma_1^2, 0, 0, 2/\sigma_1^4, 0).$$

If we further define

$$\mathbf{v}(x) = \{(x-\mu_1)/\sigma_1^2, (x-\mu_2)/\sigma_2^2, (\pi_1\pi_2)^{-1}, [(x-\mu_1)^2 - \sigma_1^2]/2\sigma_1^4,$$
$$[(x-\mu_2)^2 - \sigma_2^2]/2\sigma_2^4\}^{\mathrm{T}}$$

and define

$$I_{\mathrm{E}}(\boldsymbol{\psi}) = \int_{-\infty}^{\infty} \mathbf{v}(x)\mathbf{v}^{\mathrm{T}}(x)\pi_1\pi_2[f_1(x)f_2(x)/p(x\,|\,\boldsymbol{\psi})]\,\mathrm{d}x, \qquad (3.2.6)$$

then

$$I_0(\boldsymbol{\psi}) = I_c(\boldsymbol{\psi}) - I_{\mathrm{E}}(\boldsymbol{\psi}).$$

Note that (3.2.4) is just the special form which this takes for Example 3.2.1. Note also that, as is to be expected, $I_c(\boldsymbol{\psi}) - I_0(\boldsymbol{\psi})$ is non-negative definite. As usual, numerical integration is required in (3.2.6). Behboodian (1972a) outlines several procedures and provides a set of tables from which $I_0(\boldsymbol{\psi})$ can be calculated, approximately, for a wide range of values of the parameters. The tables, given in terms of standardized parameters π_1, $D[=|\mu_2-\mu_1|/2\sqrt{(\sigma_1\sigma_2)}]$, and $r(=\sigma_1/\sigma_2)$, provide a standardized information matrix, $J_0(\boldsymbol{\psi})$, from which we may calculate $I_0(\boldsymbol{\psi}) = WJ_0(\boldsymbol{\psi})W$, where $W = \mathrm{diag}\{\sigma_1^{-1}, \sigma_2^{-1}, 1, \sigma_1^{-2}, \sigma_2^{-2}\}$.

Information matrices for two special cases can be obtained easily from $I_0(\boldsymbol{\psi})$, or from $I(\boldsymbol{\psi})$ of (3.2.1):

(a) $\mu_1 \neq \mu_2$, $\sigma_1 = \sigma_2 = \sigma$. Add the last two rows of $I_0(\boldsymbol{\psi})$ and then the last two columns.

(b) $\mu_1 = \mu_2 = \mu$, $\sigma_1 \neq \sigma_2$. Replace 'last' by 'first' in (a).

For case (a), Tan and Chang (1972a) tabulate $I_0(\boldsymbol{\psi})$ over a range of values of π_1 (between 0 and 0.5), σ, μ, and d, where $\mu = \frac{1}{2}(\mu_1+\mu_2)$ and $d = \frac{1}{2}(\mu_1-\mu_2)$.

Detailed comparison of data structures M0, M1, and M2 can only be carried out numerically. Unless only one parameter is of interest, it is a matter of comparing the 'sizes' of non-negative definite matrices, usually in terms of a real-valued function of the eigenvalues. Hosmer and Dick (1977) use the criteria of trace and determinant of the inverses of information matrices, having the interpretations, respectively, of total and generalized variances. A range of two-component normal mixtures was considered and, qualitatively, the results obtained from the two criteria were roughly the same. Table 3.2.2 displays some of their results with the total variance criterion. The values quoted are the asymptotic efficiencies of M1 and M2, relative to M0. Of the M1 schemes they consider, the results in Table 3.2.2 correspond to equal sample sizes for the fully categorized data ($\rho_1 = \rho_2 = \frac{1}{2}(1-\rho_0)$). For M1 and M2, ρ_0 was chosen from

Table 3.2.2 Asymptotic relative efficiency of M1 and M2 to M0 using the total variance criterion. Reproduced by permission of Gordon and Breach, Science Publishers, Inc., from Hosmer and Dick (1977)

		$\pi_1 = 0.1$		$\pi_1 = 0.3$		$\pi_1 = 0.5$	
μ_2	ρ_0	M1	M2	M1	M2	M1	M2
	0.9	33.3	13.2	4.7	5.6	2.8	3.9
	0.7	50.5	28.4	8.2	9.1	4.7	5.8
1	0.5	73.1	41.3	9.9	11.6	6.0	7.1
	0.3	74.6	52.4	10.6	13.4	6.7	8.1
	0.1	58.9	62.1	8.8	14.9	6.0	8.9
	0.9	6.8	4.2	2.5	3.0	2.6	3.0
	0.7	11.0	6.9	4.0	4.6	4.3	4.7
3	0.5	13.5	8.7	5.0	5.6	5.5	5.8
	0.3	15.1	10.1	5.0	6.4	6.3	6.6
	0.1	15.5	11.3	5.7	7.0	6.5	7.3
	0.9	1.8	1.2	1.1	1.1	1.1	1.1
	0.7	2.6	1.5	1.3	1.3	1.3	1.3
5	0.5	2.9	1.7	1.4	1.5	1.5	1.5
	0.3	3.0	1.9	1.5	1.6	1.6	1.6
	0.1	3.0	2.0	1.4	1.7	1.6	1.7

0.1 (0.2) 0.9; $(\mu_1, \sigma_1, \sigma_2)$ was fixed at (0, 1, 1.5); the values chosen for π_1 were 0.1, 0.3, and 0.5; and the values chosen for μ_2 were 1, 3, and 5.

Since the trace criterion is used, these figures give a rough but direct indication of the M0 sample sizes equivalent to one M1 or M2 observation. For small π_1, μ_2, and ρ_0, these can clearly be very large. Given (π_1, μ_2, ρ_0), the monotonic trend, in ρ_0, of the M2 values is inevitable. Since the fully categorized part of an M1 sample tells us nothing about π_1, there tends to be a fall-off in efficiency as ρ_0 gets small. This is even more dramatic with the generalized-variance criterion. Note that M1 is sometimes better than M2 and sometimes not. For small π_1 and μ_2 even 10 per cent fully categorized data clearly adds a tremendous amount of information. Hosmer and Dick (1977) also show that, if it is possible to obtain fully categorized data at all, then it is best to use M1 with *all* the fully categorized part of the data taken from the component density with the smaller proportion, π_1, if $\pi_1 \leqslant 0.3$, $\rho_0 \geqslant 0.7$, and

$$|\mu_1 - \mu_2| \leqslant 3 \min(\sigma_1, \sigma_2).$$

Example 3.2.4 Mixture of two multivariate normal densities with equal covariance matrices

This example is treated by Chang (1976, 1979). Chang (1976) shows that, if Δ denotes the Mahalanobis distance between the two component densities and

$v(\Delta)$ denotes the asymptotic variance of the estimator of the mixing weight, then $v(\Delta)$ is a decreasing function of Δ. Furthermore, Δ increases as more variables are added, so that $v(\Delta)$ consequently decreases. In the later paper, Chang (1979) shows that, whatever the dimensionality of the data, the information matix can be computed using a series of transformations in conjunction with the information matrix for the trivariate case. The latter involves only univariate integrations.

We have only examined a few special cases, but already there are some clear messages. If the mixing weights are unbalanced and/or the component densities are not well separated, then precise estimation from M0 data is possible only with very large sample sizes, and even a small proportion of fully categorized data can lead to a dramatic improvement.

Finally, we note the evaluation by Whittaker (1973) of the so-called Bhattacharyya information measures for a mixture of two known densities.

3.3 MISCELLANY

3.3.1 Multimodality

The most striking feature of a mixture density curve is often that of multimodality. Indeed, in many applications of mixtures the question of interest is stated as 'Is there evidence of bimodality?' as opposed to 'Is there evidence of more than one subpopulation?'. The titles of the articles by Murphy (1964) and Brazier *et al.* (1983) refer to bimodality rather than to mixtures.

A probability density function $p(x)$ has a *mode* at x^* if x^* gives a local maximum $p(x^*)$ for $p(x)$. A mode may be a stationary point of the density as in the normal case, it may be at the boundary of the sample space as with the exponential case, or the situation may be even less 'regular' as with the double exponential.

For discrete ordered sample spaces, the formal definition of a mode is somewhat awkward (Medgyessy, 1977, Chapter II), but the pictorial description as a point of local maximum among the probability masses describes the concept quite adequately.

A density is *multimodal* if it has more than one mode, *k-modal* if it has k modes and *strictly k-modal* if all the modes are strict local maxima. Figure 3.3.1 gives the density curves for a selection of mixtures of two normal densities with equal variances. Some are bimodal and some are not. If the mixing weights are equal, then it is easy to show that the mixture density is bimodal if and only if $|\mu_1 - \mu_2|/\sigma > 2$, where (μ_1, μ_2) are the two means and σ^2 is the common variance. Figure 3.3.1 also demonstrates *bitangentiality*, a less dramatic departure from unimodality. This occurs if there are two distinct points, x_1, x_2, at which there is a common tangent to the density curve. Thus, bitangentiality is implied by, but does not imply, bimodality. Informally, bimodality implies an extra *hump*, but bitangentiality merely an extra *bump*!

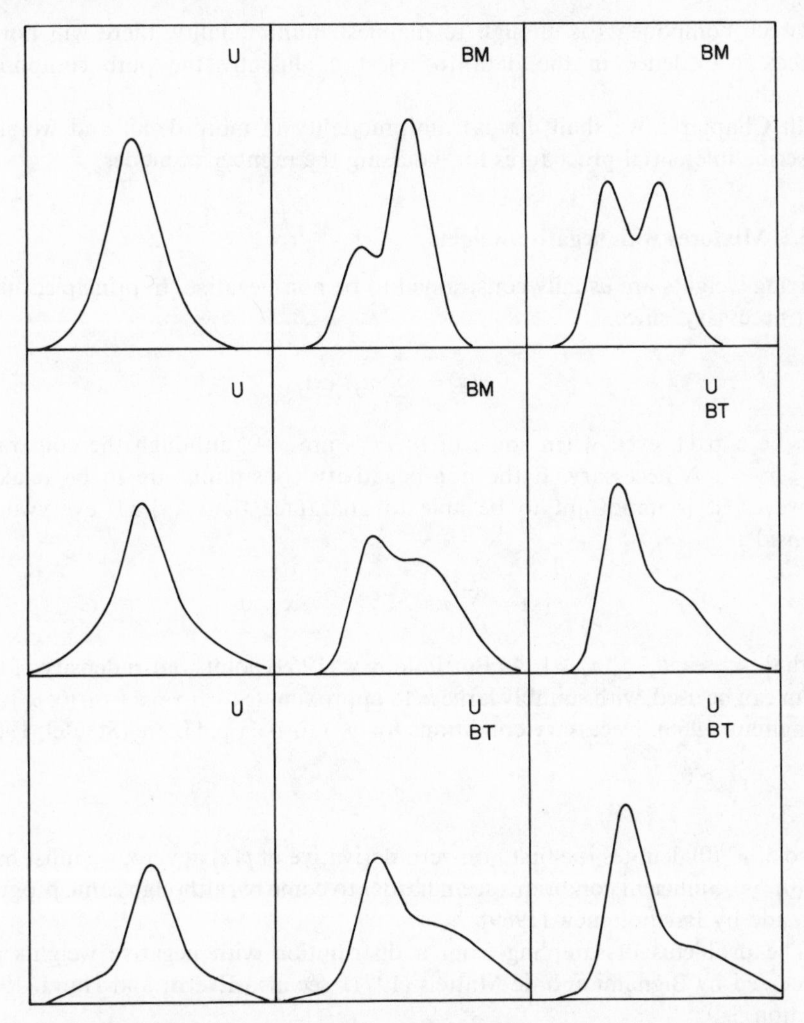

Figure 3.3.1 Densities of some two-component normal mixtures exhibiting unimodality (U), bimodality (BM) and bitangentiality (BT). Diagram by Dr A. W. Bowman

Statisticians have been at some pains to emphasize the distinction between the concepts of multimodality and mixtures. The title of the article by Broadbent (1966) mentions 'bimodality', but 'mixtures' are referred to in the sequel by Cox (1966). Even very recently, Everitt (1981b) reminds his readers that a mixture of two normals, differing in means, can still be unimodal. Unimodality can therefore conceal the existence of two subpopulations. He also remarks that bimodality might sometimes be generated by a single group of individuals, but usually it is assumed that a 'pure' component is unimodal. In defence of those who talk in terms of modes it has to be admitted that, often, unless the separation between components is enough to manifest multimodality, there will not be sufficient evidence in the data to reject confidently the pure component hypothesis!

In Chapter 5 we shall discuss multimodality in more detail and we shall describe inferential procedures for assessing the number of modes.

3.3.2 Mixtures with negative weights

Mixing weights are usually constrained to be non-negative. In principle, this is not necessary, since

$$p(x) = \sum_{j=1}^{k} \pi_j f_j(x)$$

can be a p.d.f. even when some of the π_j's are <0, although the constraint $\sum_{j=1}^{k} \pi_j = 1$ *is* necessary. If the non-negativity constraints are to be relaxed, however, it is important to be able to guarantee that $p(x) \geqslant 0$ everywhere. Consider

$$p(x) = \sum_{j=1}^{k} \pi_j \theta_j e^{-\theta_j x}, \qquad x > 0,$$

with $\theta_1 < \cdots < \theta_k$, $\sum_j^k \pi_j = 1$. As Bartholomew (1969) points out, a density of this form can be used, with suitably large k, to approximate to any p.d.f. on $(0, \infty)$ (see Kingman, 1966). Necessary conditions for $p(\cdot)$ to be a p.d.f. are (Steutel, 1967):

(a) $\sum_j \pi_j \theta_j \geqslant 0$;
(b) $\pi_1 > 0$.

Also, if $p^{(r)}(0)$ denotes the first non-zero derivative of $p(x)$ at zero, we must have $p^{(r)}(0) \geqslant 0$. Sufficient conditions seem harder to come by, although some progress is made by Bartholomew (1969).

The problems of sampling from a distribution with negative weights are discussed by Bignami and de Matteis (1971); see also Everitt and Hand (1981, Section 5.4).

3.3.3 Properties of general mixtures

Provided the number of component densities is not bounded above, certain forms of mixture can be used to provide arbitrarily close approximations to a given probability distribution. We have seen an example of this in Section 2.2 in the kernel-based density estimators. Sometimes it is even possible to provide an exact mixture representation of a given distribution, although this is usually only the case for general mixtures. These include various compound Poisson distributions, such as the contagious disease models (Neyman, 1939; Feller, 1943), the Pascal distribution (Gurland, 1957), some special sums of Poissons (Godambe, 1977), the representation of the negative binomial distribution as a gamma mixture of a Poisson (Ord, 1972, Table 6.1), and the representation of the non-central χ^2 as a Poisson mixture of central χ^2's (Ifram, 1970; see also Everitt and Hand, 1981, Section 3.4). Robbins and Pitman (1949) show how to represent linear combinations, and rational functions, of independent χ^2 random variables as χ^2 mixtures.

Compound multinomial models are studied by Hoadley (1969), mixtures of Dirichlet processes by Antoniak (1974), and compound hypergeometric distributions by Hald (1960). Teichroew (1957) calculates the density function of a scale mixture of normal densities, with a gamma mixing density, and Barndorff-Nielsen, Kent, and Sorensen (1982) consider a general class of normal mixtures.

Further technical properties of general mixtures are discussed by Maceda (1948), Molenaar and van Zwet (1966), Steutel (1967), and Keilson and Steutel (1974).

Sometimes the identification question is inverted. Consider the equation

$$P(x) = \int F(x \mid \theta) \, \mathrm{d}G(\theta),$$

where P, F, and G are distribution functions. Given the forms P and F, is there a distribution G that satisfies the equation? Beale and Mallows (1959) consider this for scale mixtures, showing that the key is whether or not the ratio of the P and F characteristic functions is itself a characteristic function (see also Andrews and Mallows, 1974). Daniels (1961) looks at the question for geometric mixtures. Efron and Olshen (1978) examine, for normal scale mixtures, $P(\cdot)$, the extent to which $P(x)$ can vary, given x, if the values of, say, $P(x_1)$ and $P(x_2)$ are specified.

Suppose $p(x)$ is a mixture of a one-parameter exponential family type with density $f(\cdot \mid \theta)$, in which θ is the natural parameter and $G(\cdot)$ denotes the mixing distribution. Suppose that $\bar{\theta}$ denotes the expected value of θ under G. Then Shaked (1980) shows that very often $p(x)$ has heavier tails than $f(x \mid \bar{\theta})$, in that the sign of $p(x) - f(x \mid \bar{\theta})$ changes twice, in the sequence $+$, $-$, $+$. From a data analytic standpoint, this might provide an informal indicator of the possibility of the observations having come from a mixture. This property of heavier tails has the consequence that, for instance, mixing dilates variance, in that

$$\mathrm{var}_p(x) \geqslant \mathrm{var}_{f \mid \bar{\theta}}(x).$$

See also Molenaar and van Zwet (1966) and Schweder (1981).

CHAPTER 4

Learning about the parameters of a mixture

4.1 GRAPHICAL METHODS

In line with the general increase in popularity of graphical methods, a variety of exploratory procedures based on plots or diagrams have been developed to deal with mixture data. The purposes of such procedures are twofold:

(a) to indicate whether or not data do, in fact, come from a certain type of mixture;

(b) to provide at least crude estimates of the underlying parameters (and we shall preempt Chapter 5, here, in including k, the number of component subpopulations, in our considerations).

Most of the published work is concerned with univariate data and much of it deals only with mixtures of normal or lognormal densities. As we shall see, the majority of these graphical methods are attempts to obtain crude estimates of the unknown parameters of the mixture and, in early applications, were the only form of statistical analysis attempted.

There are two main types of plot for univariate data, depending on whether the density function or distribution function is being depicted. In particular, of course, the former plots include the histogram and the latter the empirical distribution function. In general, in order to be of much use, the former require more data than the latter. We shall consider aspects of both types of plot and, as stated above, most of our discussion will centre on univariate normal mixtures.

4.1.1 Methods based on density functions

Figure 4.1.1 gives two histogram representations of a simulated set of 300 independent observations from a three-component normal mixture. The mixing

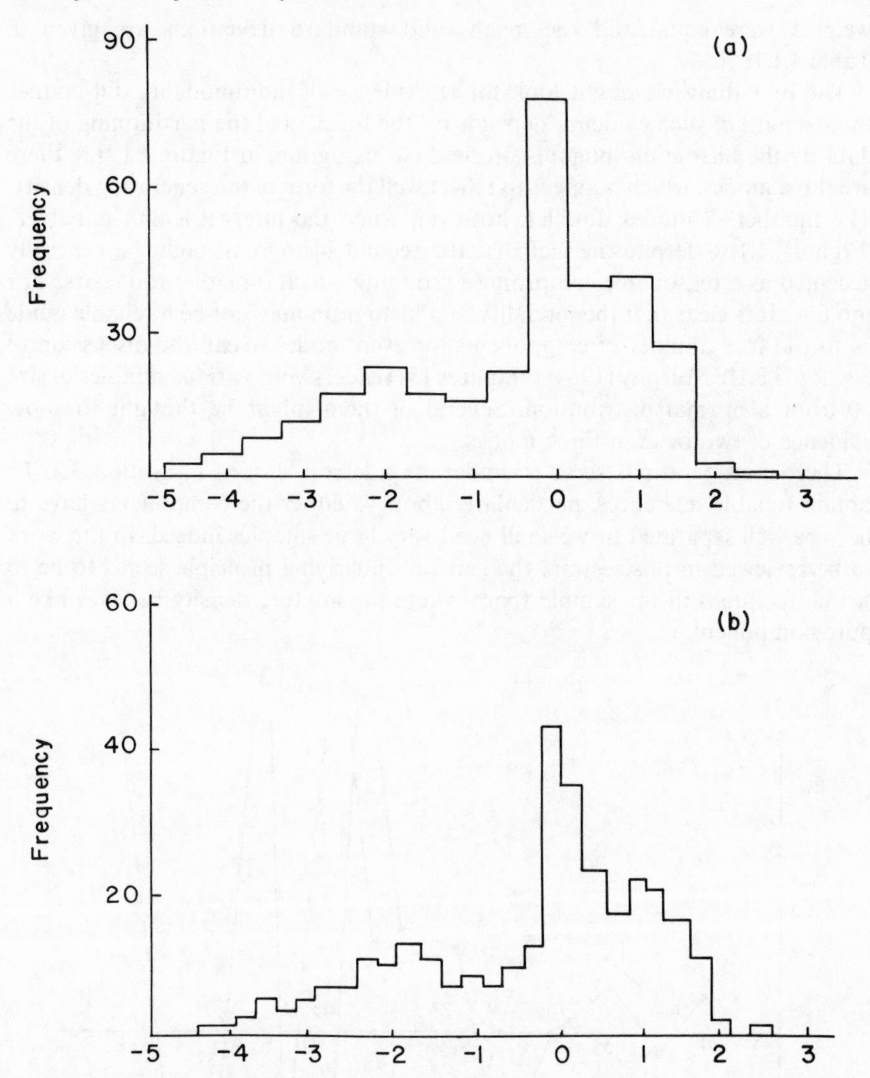

Figure 4.1.1 Two histogram representations of a set of 300 observations from a three-component normal mixture

Table 4.1.1

Component	1	2	3
Mean	−2	0	1
Standard deviation	1	0.25	0.50

weights were equal and the means and standard deviations are given in Table 4.1.1.

The first thing we might look for is evidence of multimodality. Of course, the strength of such evidence depends on the fineness of the partitioning of the data by the histogram intervals. In the first histogram, in Figure 4.1.1(a), there are three modes, which happens to reflect well the form of the generating density. The number of modes doubles, however, when the interval length is halved, (Figure 4.1.1b), despite the fact that the second histogram might be generally accepted as a reasonable compromise grouping which is neither too coarse nor too fine. It is clear that the modality of a histogram may not be a reliable guide as to the true number of components, or even modes (recall the discussion of Section 3.3.1!). Murphy (1964) tantalizes his readers with various samples of size 50 from a normal distribution. Several of them might be thought to show evidence of two or even three modes.

These considerations serve to underline a lesson learned in Section 3.2. To obtain reliable inferences, particularly about k, either the components have to be very well separated or we shall need very large samples indeed. In the work to be reviewed in this section, the general underlying principle seems to be to search for areas in the sample space where the mixture density behaves like a pure component.

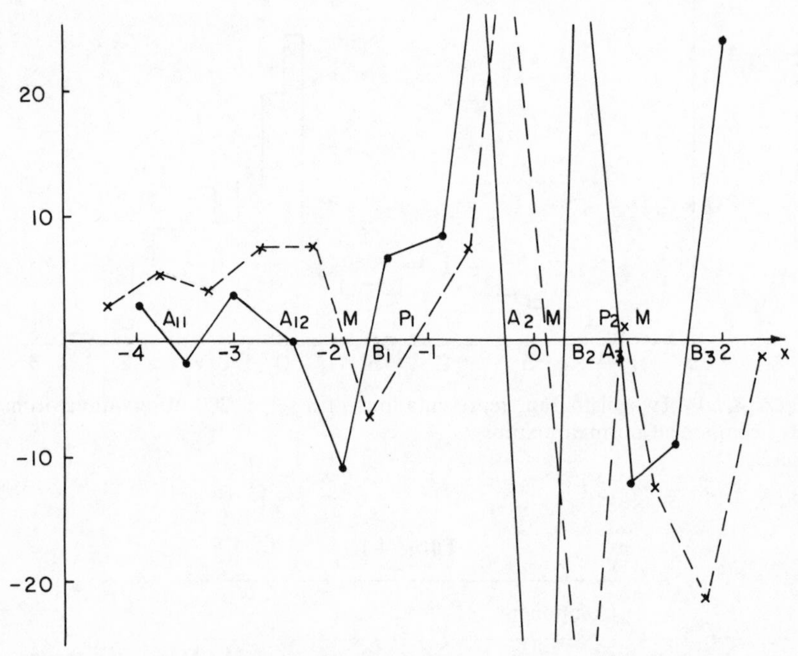

Figure 4.1.2 Plots of first differences (x- - - -x) and second differences (• —— •) based on Figure 4.1.1(a)

In the method of Tanner (1962), the characterization of modes as local maxima (first derivative zero, second derivative negative) and of antimodes as local minima (first derivative zero, second derivative positive) is exploited. First and second differences of histogram counts are used instead of the derivatives, as shown in Figure 4.1.2, constructed from the histogram in Figure 4.1.1(a).

The three zeros of the first-differences plot, which are marked 'M', give crude estimates for the modes, with approximate values

$$(-1.9, 0, 0.9).$$

The distances A_2B_2 and A_3B_3 similarly estimate the distances between points of inflection on the up- and downslopes of the second and third component densities. They estimate, therefore, twice the standard deviations, giving standard deviation estimates of about

$$(0.3, 0.45).$$

For the flatter first component the picture is not so clear. Which of the distances $A_{11}B_1$ and $A_{12}B_1$ should we look at? The two estimates of the first standard deviation are

$$(1.0, 0.4),$$

of which the former happens to be closer to the true value.

The other two zeros of the first-differences plot lead to a partitioning of the sample and thus to estimates of the mixing weights. These are at P_1 and P_2, values of about -1.2 and 0.8. It is then reasonable to 'allocate' all frequencies in the first seven intervals to component 1, along with a proportionate amount from the eight interval. The obvious linearly based proportion is

$$\frac{-1.2-(-1.25)}{-0.75-(-1.25)}, \qquad \text{i.e. } \tfrac{1}{10}.$$

This gives, for π_1, an estimate of

$$\hat{\pi}_1 = (1+3+7+10+16+22+16+\tfrac{15}{10})/300 = 0.255.$$

Correspondingly, we obtain

$$\hat{\pi}_2 = 0.522 \qquad \text{and} \qquad \hat{\pi}_3 = 0.223.$$

Biases are, of course, incurred with these crude estimators as a result of the overlap.

A less crude approach, also based on the histogram data, is that described by, among others, Bhattacharya (1967). The method relies on two facts:

(a) The logarithm of a normal density is a concave quadratic in the variable, so that its derivative is linear, with negative slope.
(b) When there is a lot of data and the grouping imposed by the histogram is quite fine, the histogram heights are roughly proportional to the density.

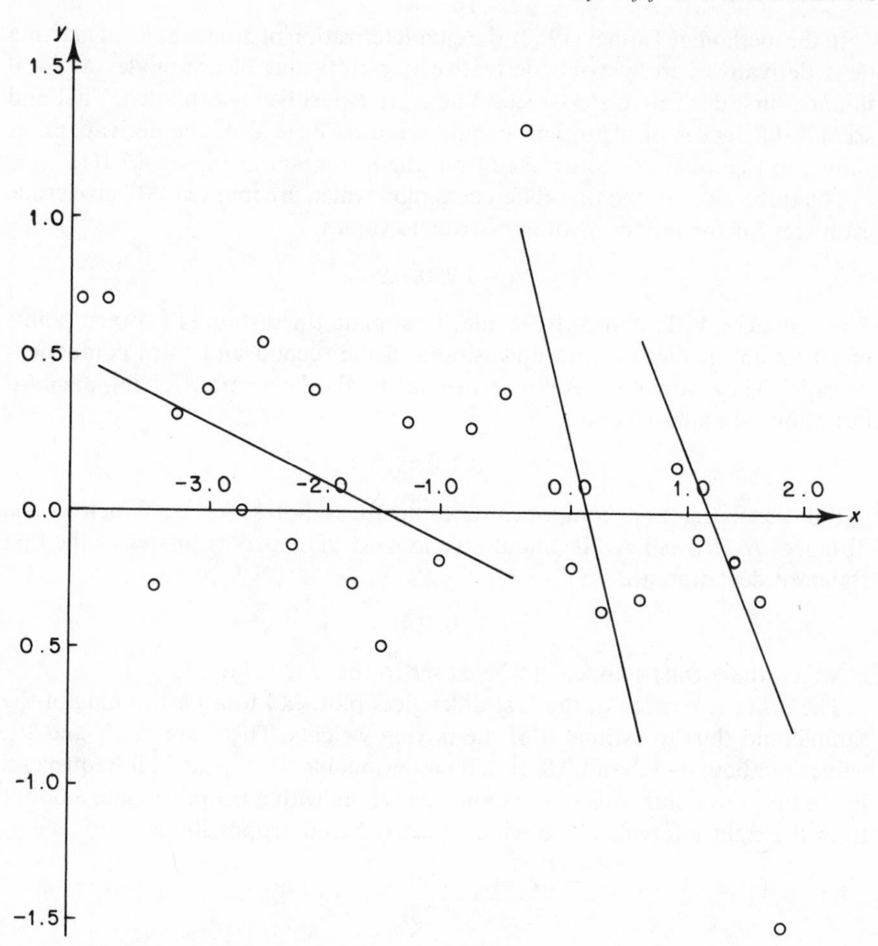

Figure 4.1.3 Bhattacharya plot for data in Figure 4.1.1(b). Lines corresponding to three components were fitted 'by eye'

Thus, a plot of first differences of the logarithms of the histogram frequencies from data from a mixture of well-separated normal components should display a sequence of negatively sloping 'linear' plots, one corresponding to each component. Figure 4.1.3 displays the Bhattacharya plot from the histogram in Figure 4.1.1(b).

Although the picture is not too clear (inevitably), there is some evidence for the presence of three normal components, indicated by the somewhat optimistically drawn lines. Clearly, the positions and orientations of the lines contain information which can be used to provide crude parameter estimates. Under the assumption that a data set arises from an $N(\mu, \sigma^2)$ distribution and that the histogram interval is of width h, Bhattacharya (1967) derives the following estimation procedure for μ and σ from a line fitted as in Figure 4.1.3 (see also Oka, 1954):

$$\hat{\mu} = \lambda + h/2$$
$$\hat{\sigma}^2 = [dh \cot(\theta)/b] - h^2/12,$$

where d and b are the relative scales on the x and y axes, λ is the intercept of the line on the x axis, and θ is the angle between the line and the *negative* direction of the x axis. For Figure 4.1.3 this gives the following crude estimates ($h = 0.25$):

Component	1	2	3
$\hat{\mu}$	−1.45	0.25	1.20
$\hat{\sigma}$	1.00	0.33	0.45

For estimating the mixing weights, Bhattacharya (1967) suggests various methods based on least-squares fitting, to the observed histogram frequencies, of expected frequencies which are calculated under the assumption that the component parameters are correctly estimated by the above method. Explicit estimates are available, as described later in Section 4.5, for general problems when only the mixing weights are unknown. Bhattacharya (1967) includes some 'quick' variations which obviate the matrix inversion necessary for full-blooded least squares, but it seems unlikely that much improvement is to be obtained over the method of Tanner (1962) described earlier, particularly in view of the unreliability of the estimates of the means and variances.

Overlapping of the components clearly biases the estimates. In the Bhattacharya method, as in others, it is sometimes possible to subtract the frequencies likely to have originated from the 'outside' components and to replot the remaining data, from which less-biased estimates can be obtained. In doing this, the degree of overlap is assessed by, say, using $(\hat{\mu}_1, \hat{\sigma}_1)$ and $(\hat{\mu}_3, \hat{\sigma}_3)$ to estimate the frequencies in the overlap region which come from components 1 and 3. They can now be subtracted from the observed frequencies and can also be counted into the estimates of the first and third mixing weights. Admittedly this detracts somewhat from the 'quick, graphical' character of the basic method.

Informal successive subtraction of the components after fitting quadratics to the logarithms of the extreme sets of frequencies is also described by Buchanan–Wollaston and Hodgson (1928). If three frequencies are used, one quadratic fits perfectly. Otherwise a best quadratic may be fitted, by least squares, say. Suppose we obtain the quadratic

$$q(x) = ax^2 + bx + c.$$

The ith contributing hump of the mixture density, scaled to match an n-sample histogram, is of the form

$$n_i f_i(x) = [n_i / \sqrt{(2\pi\sigma_i^2)}] \exp[-(x - \mu_i)^2 / 2\sigma_i^2],$$

where $n_i = \pi_i n$, and its logarithm is

$$-(x-\mu_i)^2/2\sigma_i^2 + \log[n_i/\sqrt{(2\pi\sigma_i^2)}].$$

We may thus identify

$$-1/2\sigma_i^2 \quad \text{with } a$$
$$\mu_i/\sigma_i^2 \quad \text{with } b$$

and

$$-\mu_i^2/2\sigma_i^2 + \log n_i - \log\sqrt{(2\pi\sigma_i^2)} \qquad \text{with } c,$$

which suggests estimates

$$\hat{\sigma}_i^2 = -1/2a \tag{4.1.1}$$
$$\hat{\mu}_i = -b/2a$$

and

$$\log\hat{\pi}_i = c + b^2/4a + \log[n^{-1}\sqrt{(-\pi/a)}].$$

As before, overlaps are bound to cause a problem, and subtraction of components from the extremes is likely to be helpful. Tanaka (1962) outlines the practicalities of carrying this out using a set of quadratic templates of varying curvature (corresponding, as equations (4.1.1) indicate, to different variances). A template is chosen to fit a mode of the log-frequencies as well as possible and the remaining two parameters are fitted as above. This leads to a set of expected frequencies for that component, by which the overall frequencies may be reduced before fitting the next component. As in all these procedures there is some danger of small negative nct frequencies after subtraction but, given the crudeness of the method, this is not worth worrying about.

4.1.2 Methods based on the cumulative distribution function

The alternative to plotting an estimate of the *density* function is to plot the empirical distribution function and see whether it shows evidence of a mixture. When investigating the possibility of a mixture of normals it is natural to use normal probability paper, which leads to a normal quantile–quantile (Q–Q) plot. This plot can be described as a plot of an estimate of $F^{-1}(p)$ against $\Phi^{-1}(p)$ $(0 < p < 1)$, where $F(\cdot)$ is the cumulative distribution function of the mixture and $\Phi(\cdot)$ is that of the standard normal. A sample from a single normal distribution should produce a linear plot, the kinds of plots that are likely from various mixtures of normals being illustrated in Figure 4.1.4. Certain deviations from linearity are characteristic of certain types of mixture, although, as usual, there has to be a fair amount of 'separation' for the pattern to be clear. The four cases depicted in Figure 4.1.4 are as follows:

(a) equally weighted mixture of two normal densities with similar variances but quite different means;
(b) as (a) but with unequal mixing weights;

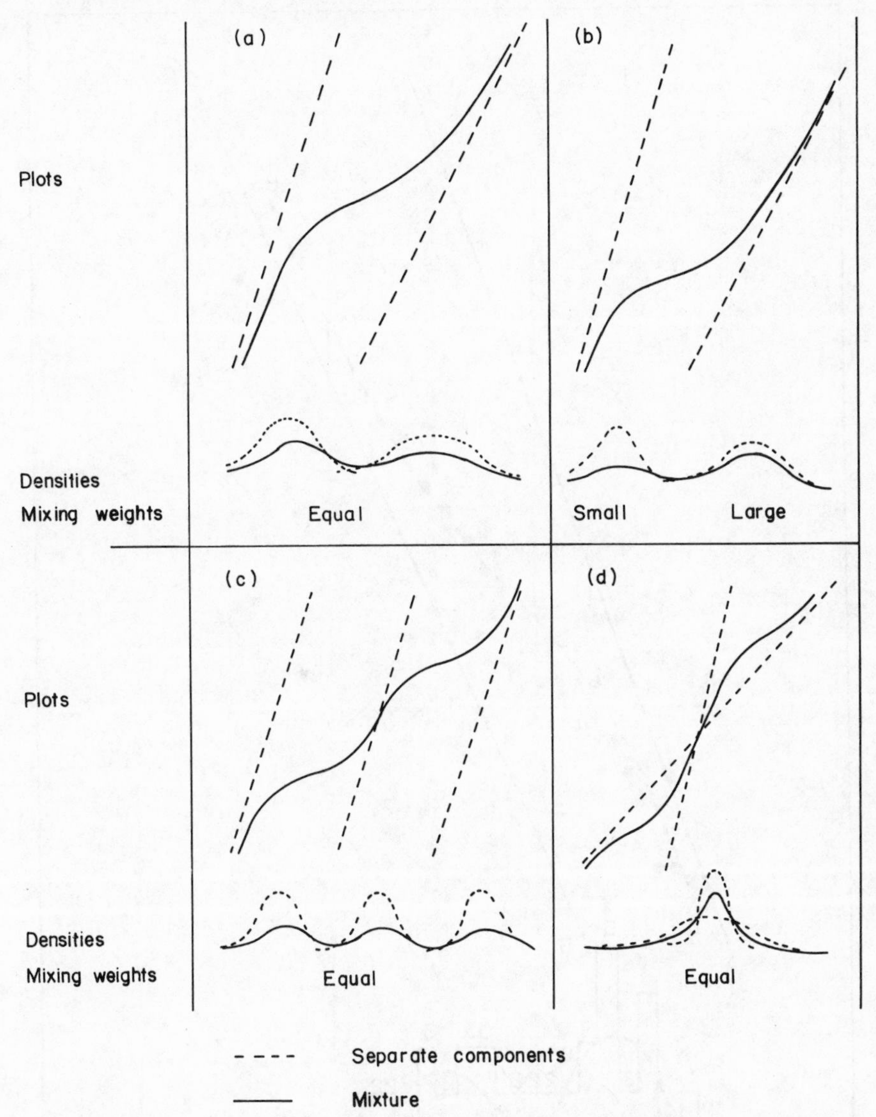

Figure 4.1.4 'Normal' probability plots for some normal mixtures (schematic)

(c) equally weighted mixture of three normal densities which differ mainly in their means;
(d) equal mixture of two normal densities with the same mean but different variances.

Two such plots from data are shown in Figures 4.1.5 and 4.1.6. The former comes from Harding (1949) who provides a very detailed discussion of this

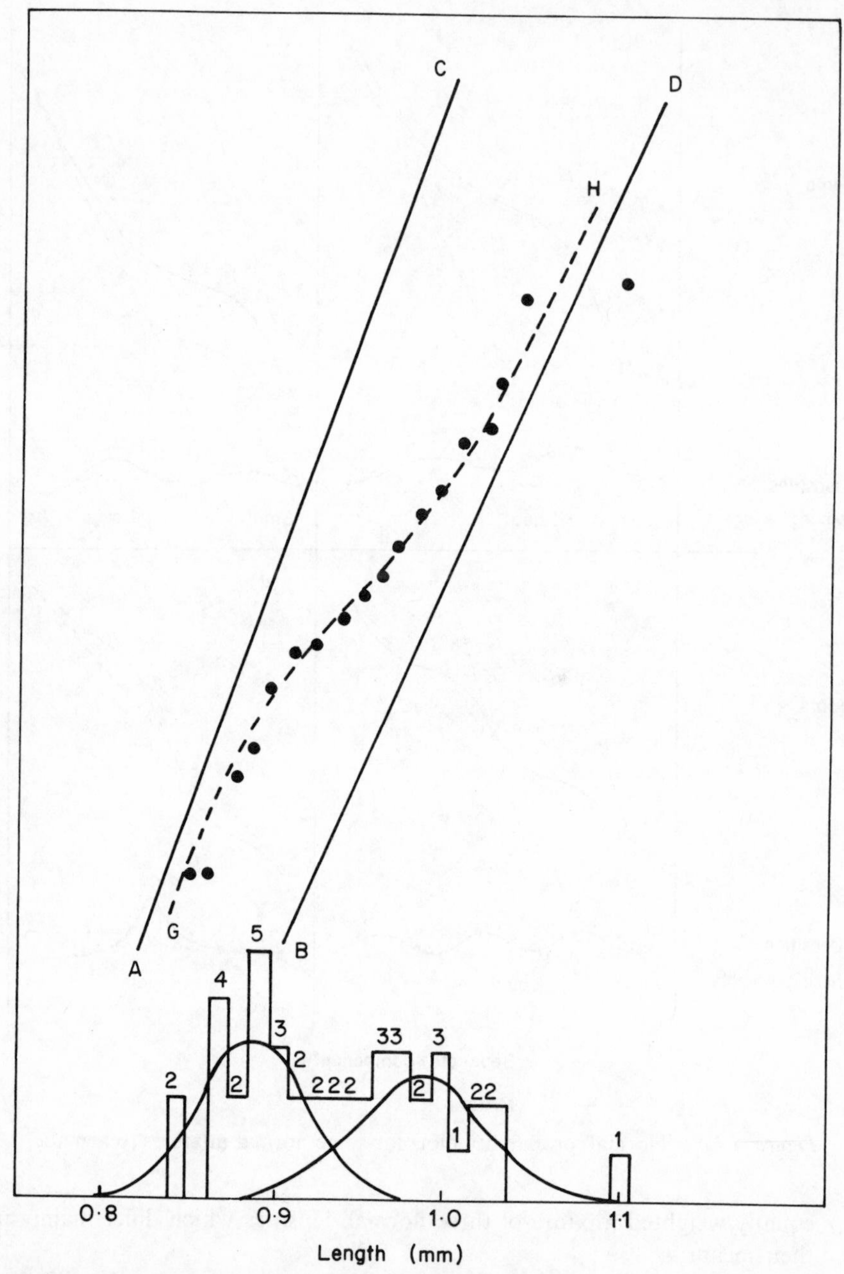

Figure 4.1.5 Histogram, normal Q–Q plot (●), estimated component densities and plots (AC, BD) and plot of estimated mixture (GH), for data on copepod lengths. Reproduced with permission from Harding (1949), 'The use of probability paper for the graphical analysis of polymodal frequency distributions', *J. Marine Biol. Assoc.*, **28**, 141–153, published by Cambridge University Press

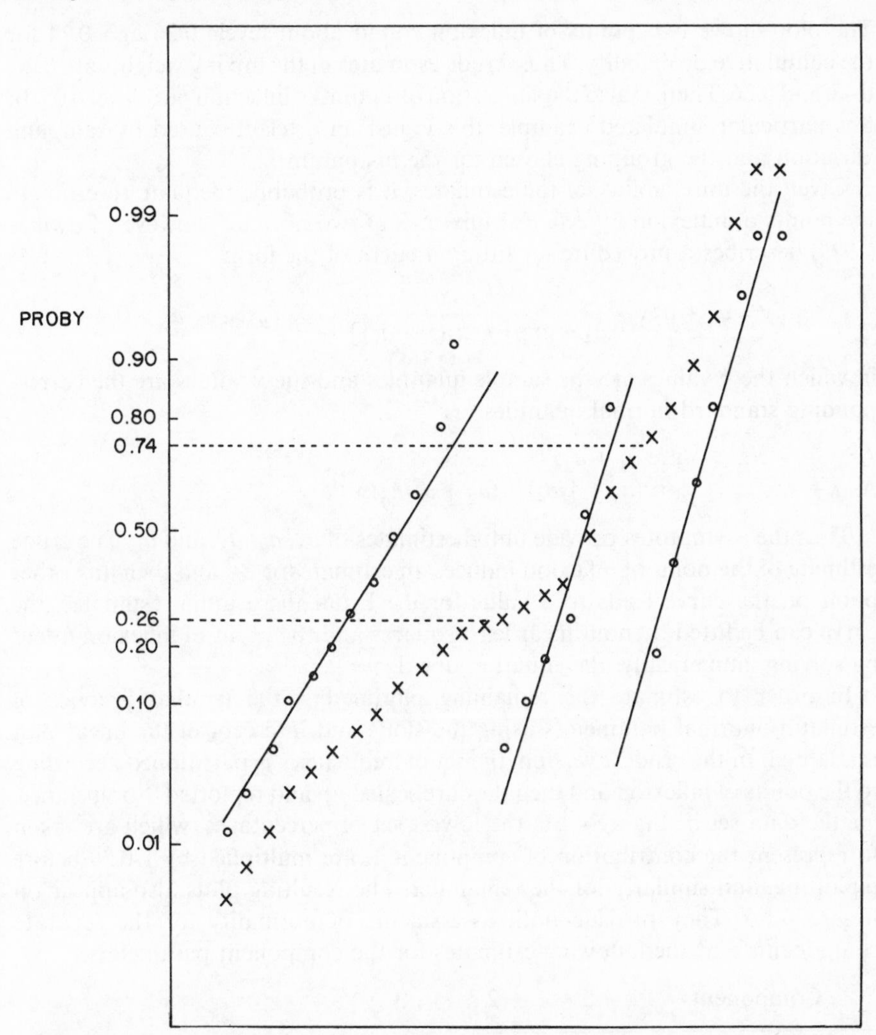

Figure 4.1.6 Normal Q–Q plot and graphical decomposition from data in Figure 4.1.1(b). × :original data; ○: data after partitioning and scaling (see text)

method of plotting in the context of mixtures and who provides several further illustrations. The data comprise the lengths of forty-one miniature copepods (a type of minute crustacean) and the plot, similar to Figure 4.1.4(a), suggests the presence of two groups, having a possible interpretation, for this example, in terms of the two sexes. Figure 4.1.6 shows a plot of the data from the second histogram of Figure 4.1.1. The shape of this plot is not quite as clear as that of the three-component schematic plot in Figure 4.1.4(c). This is partly because of sampling variability (note, in particular, the lack of straightness at the lower end of the plot) and partly due to the inequalities of the underlying variances.

The plot shows two points of inflexion round about levels 0.26 and 0.74 for the cumulative probability. Thus, crude estimates of the mixing weights are 0.26, 0.48, and 0.26. There is also the suggestion of a point of inflexion near level 0.05. In this particular simulated example, this is just an artefact caused by sampling variation and the grouping chosen for the histogram.

Given the unreliability of the estimates, it is probably adequate to estimate the points of inflexion by eye. For mixtures of two normals, however, Fowlkes (1979) describes a procedure for fitting a curve of the form

$$y = a_1 + a_2 x + \frac{1 + a_3 x}{a_4 + \exp[-a_5(x - a_6)]}, \qquad a_2, a_3, a_4, a_5 > 0,$$

in which the y values are the sample quantiles and the x values are the corresponding standard normal quantiles.

As $x \to -\infty$, $\quad y \approx a_1 + a_2 x$.
As $x \to \infty$, $\quad y \approx (a_1 + 1/a_4) + (a_2 + a_3/a_4)x$.

Thus the asymptotes provide initial estimates of a_1, a_2, a_3, and a_4. The crude estimate of the point of inflexion induces an estimate for a_6, and then any other point on the curve leads to a value for a_5. From these initial estimates, the curve can be fitted by non-linear least squares and the point of inflexion found by solving, numerically, the equation $d^2y/dx^2 = 0$.

In order to estimate the remaining parameters, the familiar practice of estimating normal parameters using the slope and intercept of the linear plot is adapted. In the crudest version, the set of frequencies is partitioned according to the points of inflexion and then they are scaled up and replotted. For instance, for the data set in Figure 4.1.6, the lowest set of percentages, which are taken to represent the contribution of component 1, are multiplied by 1/0.26 before replotting, and similarly for the remainder. The resulting plots also appear on Figure 4.1.6. They provide both assessments of normality for the separate components and the following estimates for the component parameters:

Component	1	2	3
μ	−2.32	−0.20	0.95
σ	0.91	0.49	0.46

Overlap causes a considerable problem here, in that the lines from which parameter estimates are obtained are quite hard to choose, and non-linearities, which may not in themselves be due to non-normality, creep into the separate plots. This is particularly clear in the curve at the top end of the plot for the component 1 contribution. However, methods can be devised to counteract this. For instance, the 'excess percentages' at the top end of the component 1 plot can be creamed off and allocated, instead, to component 2. A new plot can then be drawn for the component 2 augmented data and any excess at the top end allocated to component 3, with parameters being reestimated from the

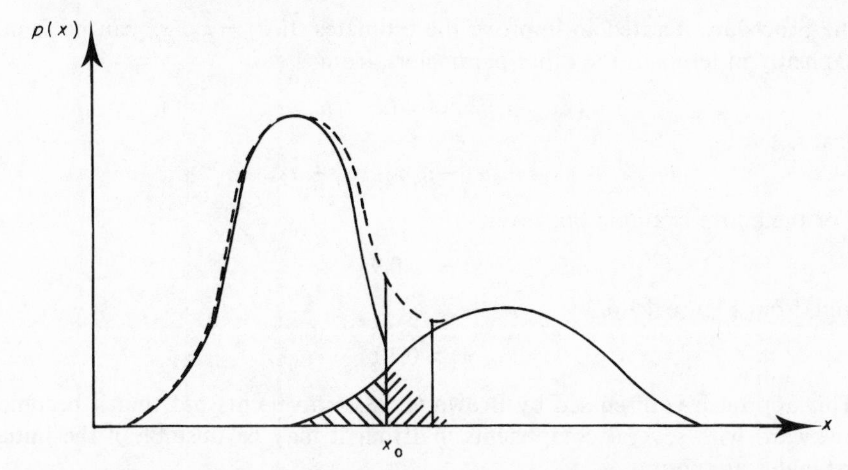

Figure 4.1.7 Use of cut-off x_0 to reduce bias in estimating mixing weight (shaded areas equal)

new plots. This procedure can also be carried out starting at the other end, i.e. with component 3 (see also Cassie, 1954, and Fowlkes, 1979). In general, when drawing lines it is better to fit the points away from the areas of overlap and, to a lesser extent, away from the tails, where there are likely to be few data. (These comments apply also to Bhattacharya's method.)

One gross potential bias in the parameter estimates obtained using the above method is in the estimates of the mixing weights (see Fowlkes, 1979). This estimation is based on estimates of the points of inflexion of the cumulative distribution function; i.e. local minima of the density. Figure 4.1.7 indicates the type of bias that can occur when there is overlap between two density functions with different variances. Clearly, use of the point of inflexion will lead to the estimate of the contribution of the flatter curve being negatively biased. The bias can in principle be removed if the cut-off point is moved to x_0, where, for the two-component case,

$$\pi_1 \int_{-\infty}^{x_0} f_1(x)\,\mathrm{d}x = \pi_2 \int_{x_0}^{+\infty} f_2(x)\,\mathrm{d}x.$$

If $f_1(\cdot)$ and $f_2(\cdot)$ are normal densities, this equation can be written

$$\pi_1 \Phi\left(\frac{x_0-\mu_1}{\sigma_1}\right) = \pi_2\left[1-\Phi\left(\frac{x_0-\mu_2}{\sigma_2}\right)\right]. \tag{4.1.2}$$

In practice we may use initial estimates for the parameters in (4.1.2) to obtain a value for the cut-off point, x_0. When this is done for the first two components of Figure 4.1.6, using the estimates obtained in conjunction with normal tables, we obtain $x_0 = -1.05$, which, from Figure 4.1.6, would give $\hat{\pi}_1 \approx 0.28$, thus reducing the bias somewhat. This can be done at either end and, if necessary, the procedure iterated to improve the estimates. If $\pi_1 = \pi_2$, x_0 can be found explicitly in terms of the other parameters, from

$$(x_0 - \mu_1)/\sigma_1 = -(x_0 - \mu_2)/\sigma_2,$$

that is,

$$x_0 = (\mu_2\sigma_1 + \mu_1\sigma_2)/(\sigma_1 + \sigma_2).$$

For the above example this gives

$$x_0 = -0.95$$

and, from Figure 4.1.6,

$$\hat{\pi}_1 \approx 0.30.$$

This approach is discussed by Brown (1978), who points out that it becomes awkward with several components and that it may be unstable if the initial estimates are poor.

As an alternative to the Q–Q plot, we now consider a variation of a percentile–percentile (P–P) plot, discussed in the context of mixtures by Fowlkes (1979). We plot

$$\Phi((x_{(i)} - \bar{x})/s) - p_i$$

against $(x_{(i)} - \bar{x})/s$, $i = 1, \ldots, n$, where

$$x_{(1)} \leqslant \cdots \leqslant x_{(n)}$$

represent the ordered sample, $(\bar{x}, s)$ are the sample mean and sample standard deviation, and $p_i = (i - \frac{1}{2})/n$, $i = 1, \ldots, n$. This gives a 'sample $\Phi - P$ versus Q plot' (Fowlkes, 1979), the population-based version of which is a plot of

$$\Phi\left(\frac{x - \mu}{\sigma}\right) - F(x) \text{ against } (x - \mu)/\sigma.$$

With normal data, the plot is a horizontal straight line and, as before, deviation from normality is shown up by deviation from this 'null' plot. Figure 4.1.8 shows what happens for the mixtures from Figure 4.1.4. Empirical evidence presented by Fowlkes (1979) suggests that this plot is at least as helpful as the Q–Q plot for detecting mixtures. The method does not seem to provide parameter estimates, however: on the contrary, it requires them!

A caveat that should be issued in this context of graphical identification of the existence of a mixture is that deviation of a plot from the null pattern may have more than one explanation. In Section 2.2 we showed that the density obtained from an unequally weighted mixture of overlapping normal densities could be very similar to a lognormal density. The Q–Q plots and $\Phi - P$ versus Q plots will also be similar.

The use of normal probability paper for mixtures is also illustrated by Lepeltier (1969) with concentration curves of molybdenum in stream-sediment analysis. The same principle of deviations from a null linear plot can be considered for

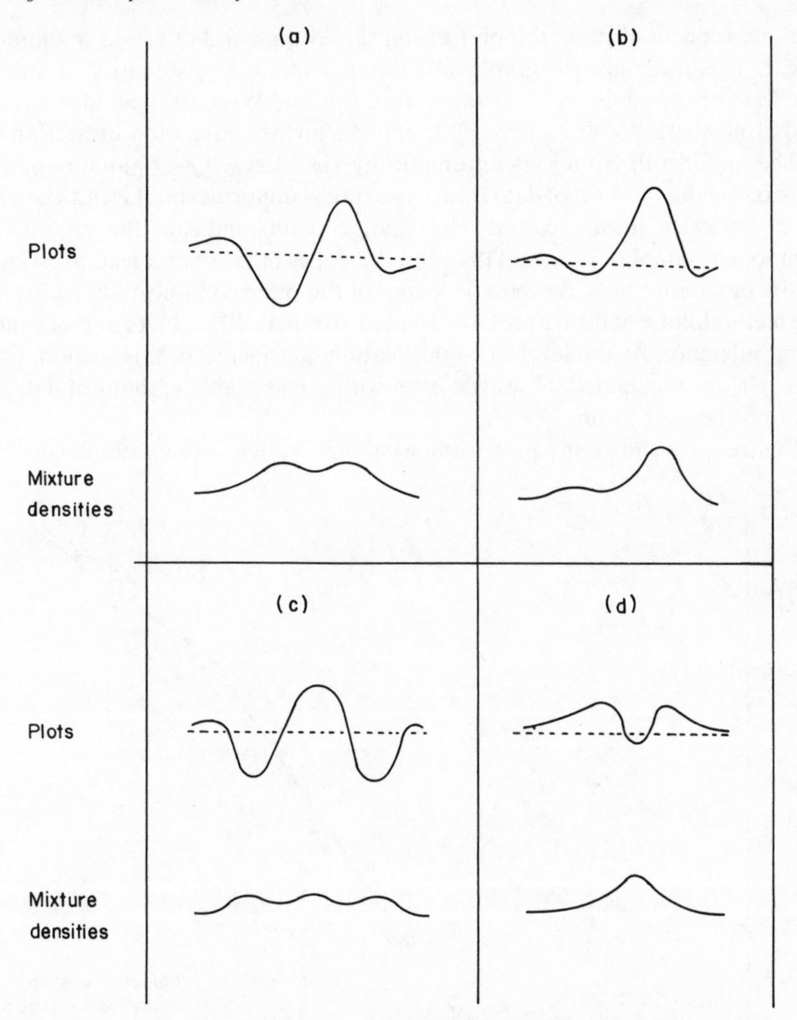

Figure 4.1.8 Fowlkes plots from some normal mixtures (schematic)

some other cases: for example, Harris (1968) looks at mixtures of two normals with roughly equal means but different variances (cf. Figure 4.1.4d). In a similar manner, as described by Ashford and Walker (1972), log-probability paper can be used for graphical decomposition of mixed probit models.

For a Weibull sample, the plot of log (– log (1 – empirical distribution function)) versus log x should be linear. The use of plots based on this for decomposing Weibull mixtures in the same spirit as the normal case above is discussed by Kao (1959) in the context of failures of electron tubes. He considers, in particular, a mixture involving very unequal mixing weights, with an uncommon set of sudden failures intermingled with more usual wear-out failures.

For exponential data, the plot of log (1 – empirical d.f.) versus x should be linear. This leads also to graphical analysis and *ad hoc* estimation techniques (see Mancini and Pilo, 1970; Defares, Sneddon, and Wise, 1973; Sandor, Sridhar, and Hollenberg, 1978; Zierler, 1981; and the further discussion in Section 4.7).

The uniform distributions automatically yield linear Q–Q plots by plotting the c.d.f. against x. A set of data from a mixture of uniforms should lead, therefore, to a piecewise linear scatter. The change points indicate the ends of the component sample spaces and they and the slopes of the scatter lead to estimates of the mixing weights. Systematic fitting of the piecewise linear shape involves the methodology of fitting probability plots (Barnett, 1975, 1976) and of change-point inference. At the level of sophistication considered in this section, fitting by eye is recommended. However, even with a reasonable amount of data, this may not be easy to do.

Figure 4.1.9 shows the plot from sixty independent observations generated

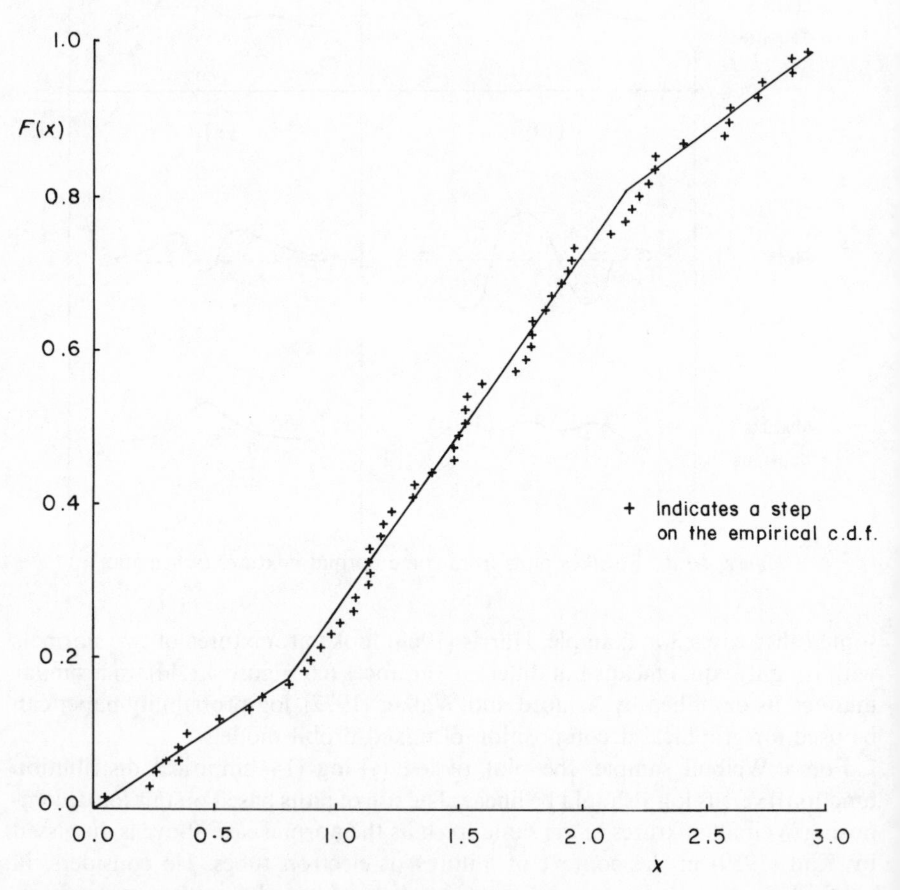

Figure 4.1.9 Plot of empirical and fitted distribution functions for a sample of 60 from a mixture of uniform distributions

from an equally weighted mixture of Un(0, 2) and Un(1, 3), The S-shape suggests that three linear pieces should be fitted, but placing the pieces visually cannot be done very precisely. From the particular piecewise linear fitting depicted on Figure 4.1.9, we have the following estimate of the c.d.f. of x:

$$\begin{aligned} F(x) &= 0.21x, && 0 \leqslant x < 0.79, \\ &= 0.45x - 0.19, && 0.79 \leqslant x < 2.25, \\ &= 0.25x + 0.25, && 2.25 \leqslant x < 3.0 \\ &= 0 && \text{otherwise.} \end{aligned}$$

Of course, as we indicated in Section 3.1, we run into identifiability problems with mixtures of uniforms and the above $F(x)$ could equally well represent the mixture

$$0.17 \times \text{Un}(0, 0.79) + 0.64 \times \text{Un}(0.79, 2.25) + 0.19 \times \text{Un}(2.25, 3).$$

4.1.3 Methods for mixtures of discrete and multivariate distributions

In the discrete case, the following plots could be used to indicate deviation from a pure component distribution.

(a) Binomial

If $p(x) = \binom{N}{x}\theta^x(1-\theta)^{N-x}, x = 0, \ldots, N$, then

$$\frac{p(x+1)}{p(x)} = \frac{x+1}{N-x}\frac{\theta}{1-\theta}, \qquad x = 0, \ldots, N-1.$$

An appropriate 'null' plot, therefore, is that of $p(x+1)/p(x)$ against $(x+1)/(N-x)$.

(b) Poisson

If $p(x) = \mathrm{e}^{-\theta}\theta^x/x!$, $x = 0, 1, \ldots$, then

$$p(x+1)/p(x) = \theta/(x+1), \qquad x = 0, 1, \ldots$$

One could plot $p(x+1)/p(x)$ against $(x+1)^{-1}$ or $p(x)/p(x+1)$ against x.

Such plots could be used to decompose mixtures of binomials and mixtures of Poissons, although there will be the problem of lack of sensitivity unless there are quite a number of large frequencies.

In the multivariate case, the search for graphical methods is more awkward and we shall limit ourselves to the normal case. In principle, we can simply extend the idea of using a plot or plots and detecting suitable deviations from a null pattern that corresponds to multivariate normality. However, some care

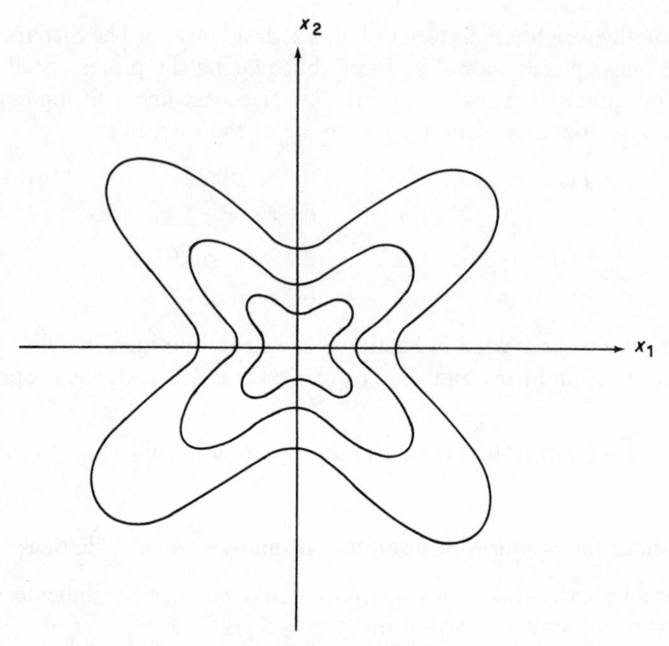

Figure 4.1.10 Sketch contours (not to scale) of a mixture of bivariate normals which has normal marginals:

$$p(\mathbf{x}) = [4\pi\sqrt{(1-\rho^2)}]^{-1}[\exp(-\tfrac{1}{2}\mathbf{x}^{\mathrm{T}}\Sigma_1^{-1}\mathbf{x}) + \exp(-\tfrac{1}{2}\mathbf{x}^{\mathrm{T}}\Sigma_2^{-1}\mathbf{x})]$$

where $\Sigma_j = \begin{pmatrix} 1 & (-1)^{j-1}\rho \\ (-1)^{j-1}\rho & 1 \end{pmatrix}, \quad j = 1, 2$

may be necessary. In particular, it is not sufficient to check univariate normality for each variable in order to be able to reject the possibility of a mixture. For instance, a 'cross-shaped' mixture of two bivariate normals with the same mean vector, the same marginal variances, and correlation coefficients equal in magnitude but opposite in sign, yields normal marginals (Figure 4.1.10). Some useful plots pertaining to multivariate normality are described by Gnanadesikan (1977) and Everitt (1978). Everitt and Hand (1981) suggest using the chi-squared probability plot of the generalized distances of the observations from the sample mean vector. With mixtures, S-shaped deviations from the null linear plot should be apparent.

A potentially more versatile plot is the Andrews curve (Andrews, 1972; Gnanadesikan, 1977). Each observation is depicted as a curve which, for a d-variate observation $\mathbf{x}$, is

$$f_{\mathbf{x}}(t) = \frac{1}{\sqrt{2}}x_1 + x_2 \sin t + x_3 \cos t + x_4 \sin 2t + \cdots \text{ to } d \text{ terms} \qquad (-\pi < t < \pi).$$

It is helpful if the variables $x_1, x_2, \ldots, x_d$ are arranged in decreasing order of informativeness and sometimes a preliminary transformation to principle component scores is made.

One interpretation of a set of Andrews curves from a data set is as an infinity (as t varies) of sets of univariate projections of the data. The intersection of the set of Andrews curves with $t = t_0$ gives a set of realizations of the linear combination

$$\mathbf{h}(t_0)^{\mathrm{T}}\mathbf{x},$$

where

$$\mathbf{h}(t) = \left(\frac{1}{\sqrt{2}}, \sin t, \cos t, \sin 2t, \ldots\right)^{\mathrm{T}}.$$

An informal test of multivariate normality is to assess univariate normality simultaneously for all t. (Since the form of $\mathbf{h}(t)$ obviously restricts the scope of

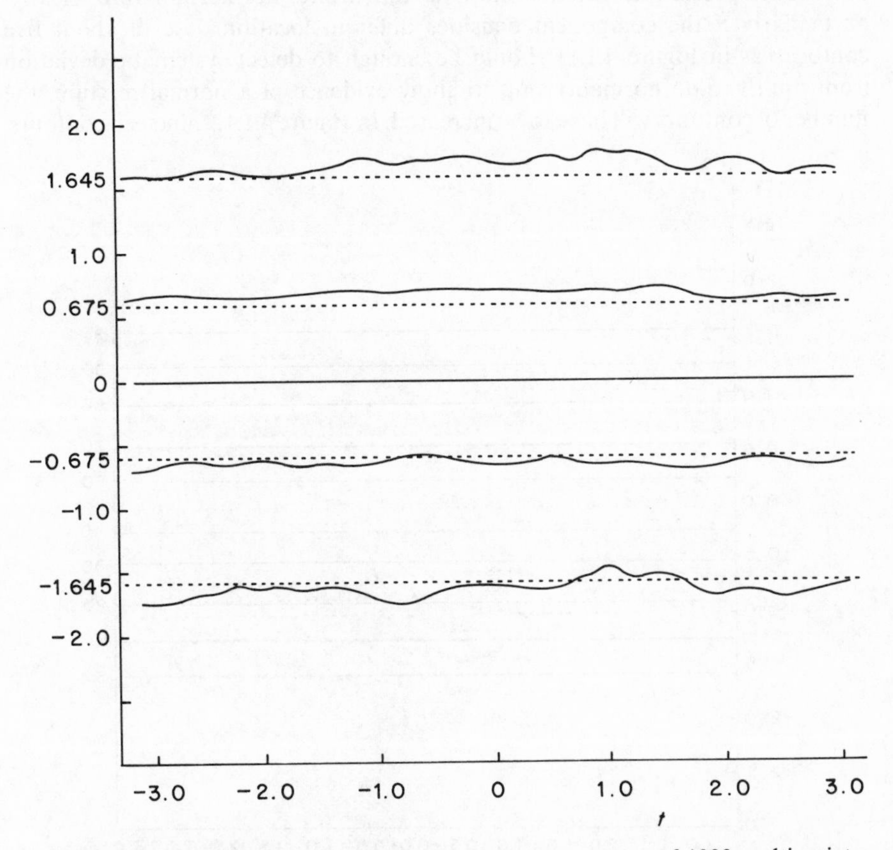

Figure 4.1.11 Quantile contour plots from Andrews curves of 1000 multivariate normal observations, standardized and compared with true percentiles. $\Phi(0.75) = 0.675$; $\Phi(0.95) = 1.645$

the linear combinations considered it is not a secure test of multivariate normality, but it should certainly be better than just looking at the d univariate marginal plots.) With a large set of data the conglomeration of Andrews curves looks impossible to disentangle and Gnanadesikan (1977) outlines a workable procedure in the *quantile contour plot*. Here the values of a few sample percentiles are evaluated for a large number of t values, giving, say, five contour curves. A useful refinement is first to standardize the corresponding univariate data, for each t. If the original data were multivariate normal, the resulting standardized quantile contours should be roughly horizontal straight-line plots at levels indicated by standard normal quantile. Figure 4.1.11 shows the resulting plot from a set of 1000 independent five-dimensional normal random vectors corresponding to percentiles 5, 25, 50, 75, and 95.

Deviation from this null plot indicates deviation from multivariate normality. If there is an underlying normal mixture then, for at least some t values, the univariate projection should show a univariate normal mixture clearly, particularly if the component densities differ in location. Use of about five contours as in Figure 4.1.11 should be enough to detect systematic deviation from multivariate normality but, to show evidence of a normal mixture, the number of contours will have to be increased. In Figure 4.1.12, nineteen contours,

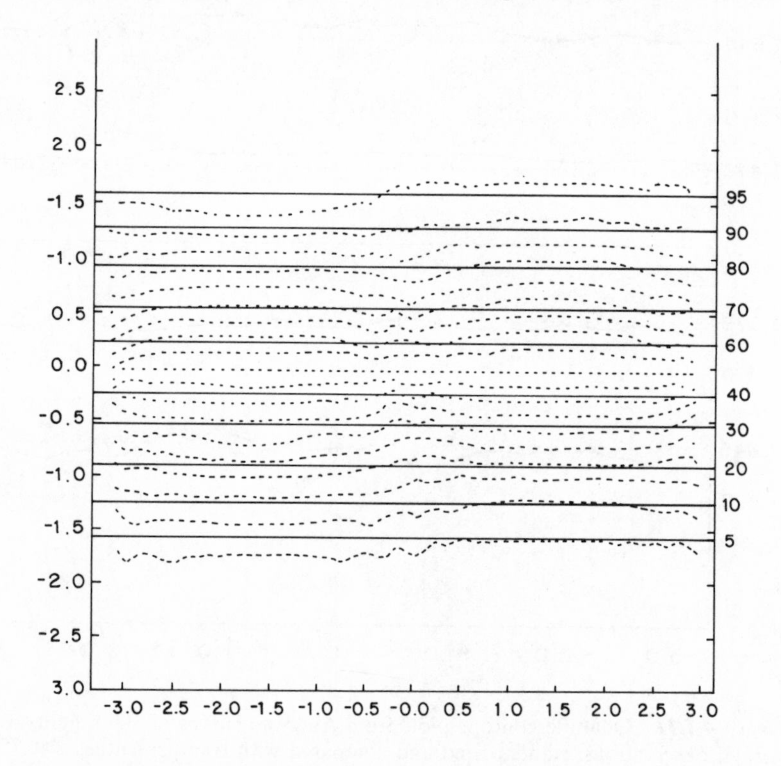

Figure 4.1.12 Quantile contour plots from Andrews curves of mixture data with some 'normal' percentiles

corresponding to jumps of 5 per cent, are drawn for a sample of 200 observations drawn from a bivariate normal mixture with

$$p(\mathbf{x}) = \frac{1}{2}\frac{1}{2\pi}\{\exp(-\tfrac{1}{2}\mathbf{x}^{\mathrm{T}}\mathbf{x}) + \exp[-\tfrac{1}{2}(\mathbf{x}-\boldsymbol{\mu}_0)^{\mathrm{T}}(\mathbf{x}-\boldsymbol{\mu}_0)]\},$$

where $\boldsymbol{\mu}_0^{\mathrm{T}} = (0, 3)$.

Here $f_{\mathbf{x}}(t) = x_1/\sqrt{2} + x_2 \sin t(-\pi < t < \pi)$. Note how the plot, on which the appropriate standard normal quantiles are superposed, shows, correctly, apparent normality at $t = 0, \pm\pi$, with increasing separation of the components towards $t = -\pi/2$.

Finally, we mention a generalization of the approaches of Tanner (1962) and Buchanan–Wollaston and Hodgson (1928) due to Postaire and Vasseur (1981). For multivariate normal mixtures, parameters are estimated by identifying regions of the sample space where the density, as estimated by the data, is concave and then by fitting elliptical contours to what are assumed to be data from the component subpopulations.

4.2 THE METHOD OF MOMENTS

4.2.1 Introduction

Suppose we have a data set of n independent observations from a population whose probability model depends on r unknown parameters, $\boldsymbol{\psi}$. Suppose $\boldsymbol{\mu}(\boldsymbol{\psi})$ denotes a vector of r functionally independent moments and that $\mathbf{m}$ denotes the corresponding set of sample moments. The method of moments estimator is the $\hat{\boldsymbol{\psi}}$ which satisfies

$$\boldsymbol{\mu}(\hat{\boldsymbol{\psi}}) = \mathbf{m}. \tag{4.2.1}$$

In general, there are a number of potential problems with moment estimators.

(a) Explicit solution of (4.2.1) may not be easy or even possible.

(b) The solution of (4.2.1) may not be unique and may not automatically lie in a feasible region of $\mathbb{R}^r$.

(c) Although consistency of $\boldsymbol{\mu}(\hat{\boldsymbol{\psi}})$ and, consequently, in typical cases, consistency of $\hat{\boldsymbol{\psi}}$ usually follows by the laws of large numbers, $\hat{\boldsymbol{\psi}}$ may not be asymptotically efficient.

(d) Exact computation of $\mathrm{cov}(\hat{\boldsymbol{\psi}})$ is not usually possible. However, a Taylor expansion argument can often be invoked to show that, approximately and for large samples,

$$\boldsymbol{\mu}(\boldsymbol{\psi}_0) + D(\boldsymbol{\psi}_0)(\hat{\boldsymbol{\psi}} - \boldsymbol{\psi}_0) = \mathbf{m}, \tag{4.2.2}$$

where D is the square matrix of derivatives of elements in $\boldsymbol{\mu}$ and $\boldsymbol{\psi}_0$ is the true value. Thus, approximately,

$$\begin{aligned} \mathrm{cov}(\hat{\boldsymbol{\psi}}) &= D(\boldsymbol{\psi}_0)^{-1}\mathrm{cov}_{\boldsymbol{\psi}_0}(\mathbf{m})[D^{\mathrm{T}}(\boldsymbol{\psi}_0)]^{-1} \\ &\approx D(\hat{\boldsymbol{\psi}})^{-1}\mathrm{cov}_{\hat{\boldsymbol{\psi}}}(\mathbf{m})[D^{\mathrm{T}}(\hat{\boldsymbol{\psi}})]^{-1}. \end{aligned} \tag{4.2.3}$$

All these potential problems occur frequently with moment estimators in mixture problems. However, there is a long history of application of such methods, partly because problem (a) typically does not occur and partly because of the computational problems associated with alternative methods such as maximum likelihood, particularly before the advent of computers.

4.2.2 Mixtures of two densities

Example 4.2.1 Mixture of two known densities

$$p(x) = \pi f_1(x) + (1 - \pi) f_2(x), \qquad 0 < \pi < 1.$$

Let $t(x)$ be such that $\mathbb{E}t(X)$ exists for each component density and denote these expected values by μ_{1t} and μ_{2t}. If we denote the sample moment from the mixture by m_t, then we can obtain an explicit moment estimator for π, based on t, from

$$m_t = \hat{\pi}_t \mu_{1t} + (1 - \hat{\pi}_t)\mu_{2t},$$

which implies that

$$\hat{\pi}_t = (m_t - \mu_{2t})/(\mu_{1t} - \mu_{2t}).$$

Note that $\hat{\pi}_t$ is unbiased for π and that its variance is

$$\begin{aligned} \operatorname{var}(\hat{\pi}_t) &= \operatorname{var}(m_t)/(\mu_{1t} - \mu_{2t})^2 \\ &= n^{-1}\operatorname{var}(t)/(\mu_{1t} - \mu_{2t})^2, \end{aligned}$$

where var (t) can be written in terms of the first two moments of t corresponding to the two component densities.

The simple solution for $\hat{\pi}_t$ results from the linearity of $p(x)$ in π. This linearity clearly extends to the mixture of more than two known densities, with an equivalent helpful spin-off (see Odell and Basu, 1976, and Tubbs and Coberly, 1976). For these cases (4.2.2) and (4.2.3) are exact.

In an important special case, $t(x)$ takes the form of an indicator function:

$$\begin{aligned} t(x) &= 1 \qquad \text{if } x < c \\ &= 0 \qquad \text{if } x \geqslant c. \end{aligned}$$

The corresponding m_t can be regarded as a 'zeroth-order' sample moment. (Here we consider only univariate x, although the multivariate version is, conceptually, no harder.)

Thus m_t is the proportion of observations lying below c and

$$\hat{\pi}_t = [m_t - F_2(c)]/[F_1(c) - F_2(c)],$$

where $F_j(\cdot)$ is the distribution function corresponding to $f_j(\cdot)$. Note that, even here, there is no guarantee that $0 \leqslant \hat{\pi}_t \leqslant 1$. We also obtain

$$\operatorname{var}(\hat{\pi}_t) = n^{-1}F(c)[1 - F(c)]/[F_1(c) - F_2(c)]^2,$$

where

$$F(c) = \pi F_1(c) + (1 - \pi)F_2(c).$$

In particular, if we can choose c to be the median of the mixture and component 1 is 'to the left' of component 2 in that $F_1(c) > \frac{1}{2}$, for the mixture with $\pi = \frac{1}{2}$ we obtain $\mathbb{E}(\hat{\pi}_t) = \frac{1}{2}$, $\text{var}(\hat{\pi}_t) = \{4n[2F_1(c) - 1]^2\}^{-1}$ and, approximately,

$$P(0 \leqslant \hat{\pi}_t \leqslant 1) = 2\Phi(|2F_1(c) - 1|\sqrt{n}) - 1, \tag{4.2.4}$$

where Φ denotes the standard normal c.d.f. Although the right-hand side of (4.2.4) tends to 1 as $n \to \infty$, it will still be small for moderate n if $F_1(c)$ is close to $\frac{1}{2}$, corresponding to poor separation.

This special case is called the *confusion matrix* method by Odell and Basu (1976) because of its relation to the 2×2 'confusion matrix' of probabilities of correct and incorrect classification into the two components based on the cut-off c; see also Johnson (1973) and James (1978), the latter considering also mixtures of more than two components. Macdonald (1975) and Ganesalingam and McLachlan (1981), for the multivariate version, chose the cut-off corresponding to minimum total misclassification probability and the latter paper builds the estimator into an estimation procedure for multivariate normal mixtures when there are some fully categorized data; see also Guseman and Walton (1977, 1978) and Walker (1980).

An 'optimal' choice of c is that which minimizes $\text{var}(\hat{\pi}_t)$. Such a c satisfies the stationarity condition

$$f(c)[1 - 2F(c)][F_1(c) - F_2(c)] = 2F(c)[1 - F(c)][f_1(c) - f_2(c)].$$

Of course, it is not easy to solve this equation and the solution clearly depends on the unknown π, but we follow up this idea with the next example.

Example 4.2.2 Mixture of two known exponentials

Suppose

$$p(x) = \pi e^{-x} + (1 - \pi)2e^{-2x}, \qquad x > 0, 0 < \pi < 1.$$

If

$$t_s(x) = x^s/\Gamma(s + 1), \qquad s \geqslant 0,$$

Then

$$\mu_{t_s} = \mathbb{E}[t_s(X)] = \pi + (1 - \pi)/2^s,$$
$$\hat{\pi}_{t_s} = (2^s m_{t_s} - 1)/(2^s - 1)$$

and

$$\begin{aligned} n \,\text{var}\, \hat{\pi}_{t_s} &= [\Gamma(s + 1)(2^s - 1)]^{-2} 2^{2s} \,\text{var}\, X^s \\ &= [\Gamma(s + 1)(1 - 2^{-s})]^{-2}\{\Gamma(2s + 1)\mu_{t_{2s}} - [\Gamma(s + 1)\mu_{t_s}]^2\}. \end{aligned} \tag{4.2.5}$$

The optimal power-moment estimator of π is thus given by the s^* which minimizes (4.2.5). Also, the relative efficiency of the simple estimator based on the sample mean ($s = 1$) can be obtained as

$$\mathrm{RE}(\pi) = \mathrm{var}(\hat{\pi}_{t_{s*}})/\mathrm{var}(\hat{\pi}_{t_1}).$$

Some values of s^* and $\mathrm{RE}(\pi)$ are given below:

π	0.1	0.3	0.5	0.7	0.9
s^*	1.45	0.90	0.66	0.50	0.36
$\mathrm{RE}(\pi)$	0.94	0.99	0.95	0.88	0.81

In practice, computation of s^* requires a preliminary estimate of π, possibly $\hat{\pi}_{t_1}$. Note that these calculations ignore the fact that $\hat{\pi}$ may not lie in $0 \leqslant \hat{\pi} \leqslant 1$.

Example 4.2.3 Mixture of two unknown exponentials (Rider, 1961, 1962)

Suppose

$$p(x|\psi) = \pi_1\theta_1^{-1}\exp(-x/\theta_1) + \pi_2\theta_2^{-1}\exp(-x/\theta_2), \qquad x > 0,$$

with π_1, θ_1 and θ_2 all unknown. In order to simplify the notation of Example 4.2.2, let us define

$$m_s = n^{-1}\sum_{i=1}^{n} x_i^s/\Gamma(s+1), \qquad s = 0, 1, \ldots,$$

where $m_0 = 1$. Given m_1, m_2, m_3, moment estimates are available from

$$\pi_1\theta_1^s + \pi_2\theta_2^s = m_s, \qquad s = 0, 1, 2, 3. \tag{4.2.6}$$

Eliminating π_1 and π_2 using the first two equations and considering the remaining two, we find that θ_1, θ_2 are two solutions of the quadratic equation

$$(m_1^2 - m_0m_2)\theta^2 + 2(m_0m_3 - m_1m_2)\theta + m_2^2 - m_1m_3 = 0.$$

There is no guarantee that these roots will be real, let alone positive, although they are consistent if the true θ_j's are unequal.

Tallis and Light (1968) show that, by using moment equations based on values of s other than 1, 2, 3, spectacular gains in efficiency can sometimes be obtained. (However, they do not attempt the difficult task of calculating an optimal set of power moments in the spirit of Example 4.2.2.) Table 4.2.1 shows the gains in relative efficiency that are possible. Efficiency is measured in terms of the determinant of the asymptotic covariance matrix, given in (4.2.3), relative to that obtained using maximum likelihood. In the table, $\eta = \theta_1/\theta_2$ and the moment equations used were for $s_1 = 1$, s_2^*, s_3^*, the latter two having been identified as optimal by searching over a grid of values with increments of $\frac{1}{4}$ in each direction. E_1 and E_2 are relative efficiencies for the 'natural' powers $(1, 2, 3)$ and the 'suboptimal' set $(1, s_2^*, s_3^*)$. The greatest improvements occur if η is very different from 1 and if fractional moments are used.

Table 4.2.1 Relative efficiencies for Example 4.2.3. (Adapted from Tallis and Light, 1968 by permission of the American Statistical Association)

		η			
π		1.5	2	5	10
0.1	E_1	0.942	0.737	0.155	0.041
	E_2	0.950	0.829	0.561	0.501
	s_2^*, s_3^*	2.25, 2.75	2.00, 2.25	0.75, 1.50	0.75, 1.25
0.5	E_1	0.784	0.475	0.052	0.008
	E_2	0.950	0.872	0.625	0.507
	s_2^*, s_3^*	1.75, 2.00	1.25, 1.50	0.50, 0.75	0.50, 0.75
0.9	E_1	0.652	0.343	0.022	0.002
	E_2	0.956	0.884	0.665	0.522
	s_2^*, s_3^*	1.50, 1.75	0.75, 1.50	0.25, 0.75	0.25, 0.50

Example 4.2.4 Mixture of two univariate normals

Let

$$p(x|\psi) = \pi\phi(x|\mu_1, \sigma_1) + (1-\pi)\phi(x|\mu_2, \sigma_2),$$

$0 < \pi < 1$, σ_1, $\sigma_2 > 0$. Estimation using the method of moments in this five-parameter mixture by Pearson (1894) is often thought of as the starting point of the analysis of mixtures. Pearson's approach was later streamlined by Charlier (1906) and Charlier and Wicksell (1924); perhaps the clearest descriptions of the calculations are set out in Holgersson and Jorner (1979), Johnson and Kotz (1970a, Section 13.7.2), and Cohen (1967). Equations based on the first five central moments are used and, after much elimination, we are left with the problem of finding a negative root for the famous 'nonic' equation. We do not derive the nonic here but rather indicate the 'reverse' procedure for calculating parameter estimates from data.

In what follows, m_s and k_s denote the sth sample central moment and sample cumulant, respectively. The nonic equation, to be solved for v, is

$$a_9v^9 + a_8v^8 + a_7v^7 + a_6v^6 + a_5v^5 + a_4v^4 + a_3v^3 + a_2v^2 + a_1v + a_0 = 0$$

with

$$\begin{aligned}
a_9 &= 24,\\
a_8 &= 0,\\
a_7 &= 84k_4,\\
a_6 &= 36m_3^2,\\
a_5 &= 90k_4^2 + 72k_5m_3,\\
a_4 &= 444k_4m_3^2 - 18k_5^2,\\
a_3 &= 288m_3^4 - 108m_3k_4k_5 + 27k_4^3,\\
a_2 &= -(63m_3^2k_4^2 + 72m_3^3k_5),
\end{aligned}$$

$$a_1 = -96m_3^4k_4,$$
$$a_0 = -24m_3^6.$$

Having obtained a negative root v, we calculate

$$\eta = (-6m_3v^3 + 2k_5v^2 + 9m_3k_4 + 6m_3^3)/(2v^3 + 3k_4v + 4m_3^2)$$

and

$$\omega = \eta - m_3.$$

Then we compute $\rho = \omega/v$ and solve the quadratic equation

$$\delta^2 - \rho\delta + v = 0, \tag{4.2.7}$$

giving roots δ_1 and δ_2, with $\delta_1 > 0 > \delta_2$, say.

We also calculate

$$\beta = (2\omega - m_3)/3v.$$

We may now express our estimates in the form

$$\hat{\sigma}_j^2 = \delta_j\beta + m_2 - \delta_j^2, \qquad j = 1, 2,$$
$$\hat{\pi} = \delta_2/(\delta_1 - \delta_2),$$
$$\hat{\mu}_j = \delta_j + \bar{x}, \qquad j = 1, 2.$$

Of course, the parameter estimates may be non-feasible. Indeed, the quadratic equation (4.2.7) may not have real roots and the nonic, as well as being awkward to solve (particularly using the techniques available to early workers in the field!), may have more than one negative root or none at all (Pearson, 1894; Martin, 1936).

In general, when there are multiple solutions to the moment equations it is to be hoped that they all produce good fits. A 'best' choice can then be based on auxiliary criteria such as the following: the relative closeness of fitted higher-order moments to the sample versions (Pearson, 1894); a similar approach using the more stable zeroth-order moments, such as that considered in Example 4.2.1 (Fukunaga and Flick, 1983); the χ^2 goodness-of-fit statistic (Everitt and Hand, 1981, p. 18); or the likelihood. However, Hawkins (1972) reports a case where different solutions to the moment equations do not fit equally well.

In general, a feasible set of estimates is obtained if the nonic yields a negative solution such that

(a) (4.2.7) has real roots;
(b) $\hat{\sigma}_1^2 > 0$, $\hat{\sigma}_2^2 > 0$, $0 \leqslant \hat{\pi} \leqslant 1$.

Bowman and Shenton (1973) try to find out how often this happens by solving the moment equations for 35 000 sets of consistent values of (m_3, k_4, k_5). Regions corresponding to two negative roots for the nonic usually have $k_4 \geqslant 0$ and, as the mixture itself becomes more 'normal', the less likely it is that a feasible set of moment estimates is found for the parameters of the mixture. A similar investigation is reported by Kanno (1975).

In special versions of this mixture the computational difficulties are not so great.

I. $\rho = \delta_1 + \delta_2$ is known (Cohen, 1967; Tan and Chang, 1972a)
In this case v satisfies a cubic equation with exactly one negative root.

II. $\sigma_1 = \sigma_2 = \sigma$ (Pearson, 1894; Charlier and Wicksell, 1924; Rao, 1948; Cohen, 1967; Tan and Chang, 1972a)
Again the problem is reduced to the solution of a cubic, namely,

$$2v^3 + k_4 v + m_3^2 = 0.$$

Unless $m_3 = 0$, this has exactly one negative root. Also, $\rho = -m_3/v$ and $\hat{\sigma}^2 = v + m_2$. From these latter equations and (4.2.7) the estimates may be calculated. (Doetsch, 1928, obtains a solution in terms of non-central moments.)

III. $\pi = \frac{1}{2}$, $\sigma_1 = \sigma_2 = \sigma$ (Pearson, 1914; Gottschalk, 1948; Cohen, 1967)
Considerable simplication occurs in this case, and we obtain

$$v = -\sqrt{(-k_4/2)},$$
$$\delta_2 = -\delta_1 = \sqrt{-v},$$
$$\hat{\sigma}^2 = v + m_2$$

and

$$\hat{\mu}_j = \delta_j + \bar{x}, \qquad j = 1, 2.$$

IV. $\mu_1 = \mu_2 = \mu$ (Gottschalk, 1948; Cohen, 1967; Gridgeman, 1970)
Here $\hat{\mu} = \bar{x}$ and, if we let $t_j = \hat{\sigma}_j^2 - m_2, j = 1, 2$, then t_1, t_2 are the solutions of the quadratic

$$t^2 - k_6 t/5k_4 + k_4/3 = 0.$$

Gridgeman (1970) points out that case IV corresponds to a symmetric, *leptokurtic* density which may be a useful model in certain geodesic and astronomical problems.

Preston (1953) provides, for case II, a computational aid in the form of a skewness-kurtosis diagram, which removes the need to solve the cubic.

Dick and Bowden (1973) point out that, if categorized data on the first component are used to estimate μ_1 and σ_1, the first three moment equations can be used to estimate the other parameters, although $\hat{\pi}_1$ and $\hat{\sigma}_2^2$ are often non-feasible.

Numerical illustration of fitting a univariate mixture of two normals using the method of moments with and without the equal variances assumption is given by Ghose (1970), using a data set of fossil dimensions.

Almost nothing has been written about mixtures of more than two normals (but see Gridgeman, 1970, and Pollard, 1934).

Fryer and Robertson (1972) use Taylor expansions of the moment equations in order to approximate the biases occurring in the five-parameter univariate normal mixture and Tan and Chang (1972a) obtain the asymptotic covariance matrix for the four-parameter (equal variances) case; see also Kanno (1975).

Tan and Chang's (1972a) results on the four-parameter normal mixture establish the lack of efficiency of moment estimators, particularly if the two-component densities are close together and especially for estimating the mixing weight, π. For example, if $\delta = |\mu_1 - \mu_2|/\sigma$ and $E(\pi)$ denotes the asymptotic efficiency of $\hat{\pi}$, then

(a) $E(\pi) \leqslant 0.01$ if $\delta \leqslant 0.5$ and $\pi \leqslant 0.1$;
(b) $E(\pi) \leqslant 0.1$ if $\delta \leqslant 0.5$ and $0.2 \leqslant \pi \leqslant 0.4$;
(c) $E(\pi) > 0.9$ if $\delta > 5$;
(d) $E(\pi) \approx 0.78$ if $\delta = 2$ and $\pi = 0.5$.

Case (d) corresponds to the critical symmetric case between unimodality and bimodality.

However, although the method of moments estimator $\hat{\pi}$ is poor, it has to be acknowledged that no method will be particularly effective in estimating the mixing weight with heavily overlapping component distributions.

Example 4.2.5 Mixture of two multivariate normals

Charlier and Wicksell (1924) consider mixtures of two circular bivariate normals, and Day (1969) and John (1970b) look at mixtures of two d-variate normals with equal covariance matrices. John follows the practice of his other papers (John, 1970a, 1970c) in parameterizing in terms of the component sample sizes. He estimates the means and variances marginally as in case II of Example 4.2.4 above and the covariances are estimated using a sample version of the identity

$$\sigma_{ij} = \operatorname{cov}(X_i, X_j) = [\operatorname{var}(X_i + cX_j) - \operatorname{var}(X_i) - c^2 \operatorname{var}(X_j)]/2c.$$

Day (1969) gives a more natural generalization of the method and points out that, unless $d = 1$, not all third- and fourth-order moments need be considered in order to define estimates. He needs only all first- and second-order central moments, all marginal third-order central moments, and a final equation which is obtained using a function of the third- and fourth-order central moments which is invariant under rotation of the sample space.

For the simple case in which the common covariance matrix takes the form $\Sigma = \sigma^2 I$, moment estimates can be found with the help of moments along the direction of the first principal component (Cooper and Cooper, 1964). More recent developments are presented in Fukunaga and Flick (1983).

No extension of the univariate analysis has been made in the case of unequal covariance matrices. For such examples the method will become more unreliable because of the potential instability of high-order sample moments (Martin, 1936; Day, 1969).

We conclude this subsection with a survey of some miscellaneous topics and references.

John (1970a) considers mixtures of two components of various discrete distributional types.

Falls (1970) combines graphical and moment methods for mixtures of two-parameter Weibulls, but appears to assume that the moment estimators are asymptotically efficient.

As usual, mixtures of uniform densities do not fit into the general pattern. Gupta and Miyawaki (1978) consider three cases of mixtures of $\mathrm{Un}(0, \theta)$ and $\mathrm{Un}(\theta, 1)$, $0 < \theta < 1$:

(a) π_1, π_2 known: first moment used;
(b) θ known: first moment used;
(c) π_1, π_2, θ unknown: first two moments used.

Explicit solution is possible and the asymptotic theory is described.

Direct use of fractional moments appears in Joffe (1964), where the following model is proposed for the grain size distribution of mine dust:

$$p(x \mid \psi) = \pi A_1 \exp(-\theta_1 \sqrt{x}) + (1-\pi) A_2 \exp(-\theta_2 \sqrt{x}), \qquad x > x_0,$$

where A_1, A_2 are normalizing constants and x_0 is the *known* minimum grain size. Moment equations for the first three half-moments are solved numerically for π, θ_1, and θ_2; for details see Johnson and Kotz (1970a, Section 18.10) and Everitt and Hand (1981, Section 3.2.1).

Cohen (1965) estimates a two-component Poisson mixture using the first two sample moments along with a third equation based on the frequency in the zero cell. He also uses estimation based on factorial moments for two specific problems: a mixture of two Poisson distributions with missing zero-cell frequencies and a mixture of a Poisson with a binomial $\mathrm{Bi}(N, \theta)$, with N known (Everitt and Hand, 1981, Section 4.4).

We have commented on the phenomenon of non-real or non-feasible parameter estimates. For two-component mixtures of the general type of Example 4.2.6 (see later), Ord (1972, Section 4.6) chooses estimators which fit the moment equations as well as possible, in a well-defined sense, subject to their lying in the parameter space. He then shows that the problem can be given a linear programming formulation.

Brownie, Habicht, and Robson (1983) consider a semi-non-parametric mixture of a normal component with an arbitrary second component, under the assumption that the probability mass associated with the latter is concentrated completely on one side or other of the normal mean. An interesting procedure for estimating the mixing weight is developed from moment equations generated by the indicator function $t(\cdot)$ introduced in Example 4.2.1.

4.2.3 Mixtures of k densities

The method of solving the moment equations (4.2.6) generalizes to a wide range of mixture problems.

Example 4.2.6 A general class of k component mixtures

Suppose

$$p(x|\psi) = \sum_{j=1}^{k} \pi_j f(x|\theta_j)$$

and $t_s(x)$ is such that $t_0(x) \equiv 1$ and

$$\mathbb{E}_\theta t_s(X) = \theta^s, \qquad s = 0, 1, \ldots, 2k-1.$$

Suppose also that

$$m_s = n^{-1} \sum_{i=1}^{n} t_s(x_i).$$

Then we have a set of moment equations in θ_j ($j = 1, \ldots, k$), given by

$$\sum_{j=1}^{k} \pi_j \theta_j^s = m_s, \qquad s = 0, 1, \ldots, 2k-1,$$

which can be expressed as

$$P\boldsymbol{\pi} = \mathbf{m}, \tag{4.2.8}$$

where $\quad P_{sj} = \theta_j^s, \qquad s = 0, \ldots, 2k-1, j = 1, \ldots, k.$

Techniques for dealing with sets of non-linear equations such as (4.2.8) are well known (Blischke, 1964; Medgyessy, 1977, pp. 172, 175). Provided the θ_j's are all different, $\boldsymbol{\pi}$ can be solved in terms of the $\{\theta_j\}$ and therefore eliminated. Furthermore, the required $\{\theta_j\}$ are the roots of

$$\theta^k + D_{k-1}\theta^{k-1} + \cdots + D_1\theta + D_0 = 0,$$

where $\mathbf{D} = (D_0, \ldots, D_{k-1})^{\mathrm{T}}$ satisfies

$$\begin{pmatrix} m_0 & m_1 & \cdots & m_{k-1} \\ m_1 & m_2 & \cdots & m_k \\ \vdots & & & \\ m_{k-1} & m_k & \cdots & m_{2k-2} \end{pmatrix} \mathbf{D} = -\begin{pmatrix} m_k \\ m_{k+1} \\ \vdots \\ m_{2k-1} \end{pmatrix}.$$

See also Rennie (1974), Lingappaiah (1975), and Cornell (1962).

Specific cases to which the above analysis applies include the following:

(a) *Mixture of exponentials*: $t_s(x) = x^s/\Gamma(s+1)$, $s \geq 0$.

(b) *Mixture of binomials*, $\mathrm{Bi}(N, \theta)$, with N fixed:

$$t_s(x) = \frac{x(x-1)\cdots(x-s+1)}{N(N-1)\cdots(N-s+1)}, \qquad s \geq 1. \tag{4.2.9}$$

See Muench (1936, 1938), Rider (1962), Blischke (1962, 1964, 1965), and Johnson and Kotz (1969, Section 3.11).

(However, remember that, for identifiability, $N \geqslant 2k - 1$.)

(c) *Mixtures of negative binomials*, $\mathrm{NeBi}(N, \psi)$, with N given and

$$f(x \mid \psi) = \binom{N + x - 1}{x}(1 - \psi)^N \psi^x, \qquad x = 0, 1, \ldots.$$

Define $\theta = \psi/(1 - \psi)$ and take $t_s(x)$ as in (4.2.9). See Rider (1962), Blischke (1965), and Johnson and Kotz (1969, Section 5.11).

(d) *Mixture of Poissons*, $\mathrm{Po}(\theta)$. Take

$$t_s(x) = x(x - 1) \cdots (x - s + 1), s = 1, 2, \ldots,$$

See Pearson (1915), Muench (1938), Schilling (1947), Rider (1962), Cohen (1960), Johnson and Kotz (1969, Section 4.10), and Krolikowska (1975). Gumbel (1940) also considers moment estimators for mixtures of two exponentials and two Poissons with known mixing weight.

(e) *Mixtures of Weibulls*, $\mathrm{We}(N, \theta)$ with N fixed and

$$f(x \mid \theta) = N\theta^{-N}x^{N-1} \exp\left[(-x/\theta)^N\right], \qquad x > 0.$$

Take

$$t_s(x) = [\Gamma(1 + s/N)]^{-1} x^s, \qquad s = 0, 1, \ldots.$$

See Rider (1962).

(f) *Mixtures of gammas*, $\mathrm{Ga}(N, \theta)$ with N fixed and

$$f(x \mid \theta) = [\theta^N \Gamma(N)]^{-1} x^{N-1} \exp(-x/\theta), \qquad x > 0.$$

Take

$$t_s(x) = x^s / N(N + 1) \cdots (N + s - 1), \qquad s = 1, 2, \ldots.$$

See John (1970c).

Equations similar to (4.2.8) also arise in the estimation procedure for mixtures of one-parameter exponential family distributions described by Kabir (1968) (see also Everitt and Hand, 1981, Section 3.2) and in Rennie's (1974) treatment of mixtures of a class of location parameter densities.

Calculation of variances and covariances for all these moment estimators tends to be restricted to the Taylor-expansion-based approximation (4.2.3); see, for example, Blischke (1964) and Rider (1962). Blischke (1962, 1964) notes that, for binomial mixtures, asymptotic efficiency (as $N \to \infty$) obtains, given identifiability ($N \geqslant 2k - 1$). If, however, the mixing weights are known, the asymptotic efficiencies are zero, not unity! For 'small' N, the moment estimators are not very efficient, except in the 'critical' case, $N = 2k - 1$.

Application of the method of moments to generalized and compound Poisson distributions is described by Beall (1940), McGuire, Brindley, and Bancroft (1957), Katti and Gurland (1961), Tucker (1963), Shenton and Bowman (1967) and Press (1968). The cases of compound binomial and multinomial distributions are treated by Skellam (1948) and Mosimann (1962).

4.3 MAXIMUM LIKELIHOOD

4.3.1 Introduction

Expressions for the likelihood function associated with data from a mixture appeared in Section 1.2.1. Given a sample of n independent observations from the mixture itself, for instance, the likelihood function is

$$L_0(\boldsymbol{\psi}) = \prod_{i=1}^{n} \left[\sum_{j=1}^{k} \pi_j f(x_i | \boldsymbol{\theta}_j) \right]. \tag{4.3.1}$$

Maximization of $L_0(\boldsymbol{\psi})$ with respect to $\boldsymbol{\psi}$, for given data $\mathbf{x}$, yields the *maximum likelihood estimate* of $\boldsymbol{\psi}$. Equivalently, and more usually, the quantity maximized is the log-likelihood $\mathscr{L}_0(\boldsymbol{\psi}) = \log_e L_0(\boldsymbol{\psi})$.

For simple parametric models the maximum likelihood approach is very popular, partly because it fits into the philosophy of likelihood-based inference, partly because of the existence of attractive asymptotic theory, and partly because the estimates are often easy to compute. Techniques for computing maximum likelihood estimates are also typically useful for calculating Bayesian posterior modes. For mixture models, however, we shall discover that the asymptotic theory and computational aspects are not always so straightforward.

Example 4.3.1 Mixture of two known densities

Suppose

$$\mathscr{L}_0(\boldsymbol{\psi}) = \mathscr{L}_0(\pi) = \sum_{i=1}^{n} \log [\pi f_1(x_i) + (1-\pi) f_2(x_i)]$$

$$= \sum_{i=1}^{n} \log [\pi (f_{i1} - f_{i2}) + f_{i2}],$$

where $f_{ij} = f_j(x_i), \quad j = 1, 2, i = 1, \ldots, n.$

If we also write $p_i = \pi f_1(x_i) + (1-\pi) f_2(x_i)$, then the likelihood equation is

$$0 = \partial \mathscr{L}_0 / \partial \pi = \sum_{i=1}^{n} \frac{f_{i1} - f_{i2}}{p_i}. \tag{4.3.2}$$

There are two worrying features about the solution of (4.3.2). The first is that (4.3.2) is equivalent to a polynomial equation of degree up to $(n-1)$ in π. However, there is at most one real root, because of the concavity of $\mathscr{L}_0$:

$$\partial^2 \mathscr{L}_0 / \partial \pi^2 = - \sum_{i=1}^{n} [(f_{i1} - f_{i2})/p_i]^2 < 0.$$

The second problem arises because the solution, $\hat{\pi}$, to (4.3.2) may not satisfy $0 \leqslant \hat{\pi} \leqslant 1$, so that the maximum likelihood estimate of π is

(a) $\hat{\pi}$ if $0 \leqslant \hat{\pi} \leqslant 1$;

(b) 0 if $\partial \mathscr{L}_0/\partial \pi|_{\pi=0} < 0$;
(c) 1 if $\partial \mathscr{L}_0/\partial \pi|_{\pi=1} > 0$.

Although it is reassuring that (4.3.2) has only one real root, explicit solution is not possible and we need to use numerical methods, such as that of Newton–Raphson or the Method of Scoring. A third iterative procedure can be contrived by substituting for f_{i2} in terms of p_i and f_{i1} in (4.3.2) and rearranging the resulting equation to give

$$\pi = n^{-1} \sum_{i=1}^{n} \pi f_{1i}/p_i = n^{-1} \sum_{i=1}^{n} w_{i1}(\pi), \text{ say,} \tag{4.3.3}$$

where $w_{i1}(\pi)$ is clearly between 0 and 1. This suggests the following successive approximations procedure:

$$\pi^{(m+1)} = n^{-1} \sum_{i=1}^{n} w_{i1}(\pi^{(m)}), \qquad m = 0, 1, \ldots. \tag{4.3.4}$$

We shall say more about the properties of this procedure later. Di Gesu and Maccarone (1984) use the finite sample space version of (4.3.4) as a numerical method for solving a problem in image reconstruction.

Example 4.3.2 Mixture of two univariate normal densities

Here,

$$\begin{aligned}\mathscr{L}_0(\psi) &= \mathscr{L}_0(\pi, \mu_1, \mu_2, \sigma_1^2, \sigma_2^2)\\ &= \sum_{i=1}^{n} \log[\pi\phi(x_i|\mu_1, \sigma_1) + (1-\pi)\phi(x_i|\mu_2, \sigma_2)].\end{aligned}$$

It does not take long to discover that the set of likelihood equations cannot be solved explicitly. However, there is worse to follow. Although this is possibly the most commonly applied mixture model, the resulting likelihood surface is littered with singularities. Suppose, for instance, we set $\mu_1 = x_1$. Then, as $\sigma_1 \to 0$, $\mathscr{L}_0(\cdot) \to \infty$. We have, therefore, a multitude of 'useless' global maxima! This feature, along with the computational difficulties of solving the likelihood equations, doubtless accounts for the comparative neglect of maximum likelihood methods for this example until fairly recently. We shall later describe how maximum likelihood is still viable, provided, roughly speaking, we 'keep the variances away from zero'. Sometimes this occurs automatically, if, for instance, we demand that $\sigma_1 = \sigma_2$ or if there are supplementary sets of fully categorized data (data structures M1 and M2) containing at least two distinct observations per subpopulation.

A feature common to both Examples 4.3.1 and 4.3.2 is that, were the data fully categorized, maximum likelihood estimation would be explicit. In Example 4.3.1 we should be estimating a binomial parameter and in Example 4.3.2 a binomial parameter together with the means and variances of two normal populations. It is intriguing and relevant to notice that the iteration defined in

(4.3.4) bears a marked resemblance to the fully categorized version of Example 4.3.1. Indeed, the latter corresponds to setting the $w_{1i}(\pi^{(m)})$ to be either zeros or ones.

Fully categorized data can in general be represented as

$$\{y_i, i = 1, \ldots, n\} = \{(x_i, \mathbf{z}_i); i = 1, \ldots, n\},$$

where each $\mathbf{z}_i = (z_{ij}, j = 1, \ldots, k)$ is an indicator vector of length k with 1 in the position corresponding to the appropriate category and zeros elsewhere. The likelihood corresponding to $(y_1, \ldots, y_n)$ can then be written in the form

$$g(y_1, \ldots, y_n | \boldsymbol{\psi}) = \prod_{i=1}^{n} \prod_{j=1}^{k} \pi_j^{z_{ij}} f_j(x_i | \theta_j)^{z_{ij}} \tag{4.3.5}$$

with logarithm

$$l_0(\boldsymbol{\psi}) = \sum_{i=1}^{n} \mathbf{z}_i^{\mathrm{T}} \mathbf{V}(\boldsymbol{\pi}) + \sum_{i=1}^{n} \mathbf{z}_i^{\mathrm{T}} \mathbf{U}_i(\boldsymbol{\theta}), \tag{4.3.6}$$

where $\mathbf{V}(\boldsymbol{\pi})$ has the jth component $\log \pi_j$ and $\mathbf{U}_i(\boldsymbol{\theta})$ has the jth component $\log f_j(x_i | \boldsymbol{\theta}_j)$. The form of the mixture likelihood $L_0(\boldsymbol{\psi})$ of (4.3.1) corresponds to the marginal density of $x_1, \ldots, x_n$ obtained by summing (4.3.5) over $\mathbf{z}_1, \ldots, \mathbf{z}_n$. This emphasizes the interpretation of mixture data as incomplete data, with the indicator vectors as missing values. Consequently, those readers who are familiar with incomplete data problems will be hardly surprised that explicit calculation of maximum likelihood estimates is usually not possible. However, it does make available the general class of iterative procedures known as EM (Expectation-Maximization) algorithms. In the next subsection we consider these and other numerical methods in some detail.

4.3.2 EM and other numerical algorithms

The definitive reference for the EM algorithm is the paper by Dempster, Laird, and Rubin (1977), but many manifestations of the EM algorithm had appeared before in the treatment of incomplete data and other authors had already gone far in noticing the general pattern (Baum *et al.*, 1970; Orchard and Woodbury, 1972; Sundberg, 1974). The general EM algorithm works as follows.

Suppose we have to find $\boldsymbol{\psi} = \hat{\boldsymbol{\psi}}$ to maximize the likelihood $L(\boldsymbol{\psi}) = f(\mathbf{x} | \boldsymbol{\psi})$, where $\mathbf{x}$ is a set of 'incomplete' data. Let $\mathbf{y}$ denote a typical 'complete' version of $\mathbf{x}$ and let $\mathscr{Y}(\mathbf{x})$ denote the set of all possible such $\mathbf{y}$. (In the mixture context of (4.3.1), $\mathscr{Y}(\mathbf{x})$ contains k^n points, corresponding to the k^n choices for $\mathbf{z}_1, \ldots, \mathbf{z}_n$.) Let the likelihood from $\mathbf{y}$ be denoted by

$$g(\mathbf{y} | \boldsymbol{\psi}).$$

The EM algorithm generates, from some initial approximation, $\boldsymbol{\psi}^{(0)}$, a sequence $\{\boldsymbol{\psi}^{(m)}\}$ of estimates. Each iteration consists of the following double step:

E step: Evaluate $\mathbb{E}[\log g(\mathbf{y} | \boldsymbol{\psi}) | \mathbf{x}, \boldsymbol{\psi}^{(m)}] = Q(\boldsymbol{\psi}, \boldsymbol{\psi}^{(m)})$, say.
M step: Find $\boldsymbol{\psi} = \boldsymbol{\psi}^{(m+1)}$ to maximize $Q(\boldsymbol{\psi}, \boldsymbol{\psi}^{(m)})$.

Very often the M step is explicit: it is typically as easy (or difficult) as implementing the 'complete-data' maximum likelihood procedure.

Also, an exceptionally simple general proof, based on Jensen's inequality, shows that

$$L(\boldsymbol{\psi}^{(m+1)}) \geqslant L(\boldsymbol{\psi}^{(m)}), \qquad r = 0, 1, \ldots,$$

so that the likelihoods *of interest* are monotonic increasing. Equality usually means that we are at a stationary point of the likelihood. The version of the EM algorithm for the finite-mixture problems is dealt with in Section 4.3 of Dempster, Laird, and Rubin (1977). Remembering that, in this case, $\mathbf{z}_1, \ldots, \mathbf{z}_n$ are the missing quantities, we have the following, from (4.3.6):

E step:
$$Q(\boldsymbol{\psi}, \boldsymbol{\psi}^{(m)}) = \sum_{i=1}^{n} \mathbf{w}_i(\boldsymbol{\psi}^{(m)})^{\mathrm{T}} V(\boldsymbol{\pi}) + \sum_{i=1}^{n} \mathbf{w}_i(\boldsymbol{\psi}^{(m)})^{\mathrm{T}} \mathbf{U}_i(\boldsymbol{\theta}),$$

where
$$\mathbf{w}_i(\boldsymbol{\psi}^{(m)}) = \mathbb{E}(\mathbf{z}_i \mid x_i, \boldsymbol{\psi}^{(m)}).$$

That is,
$$w_{ij}(\boldsymbol{\psi}^{(m)}) = [\mathbf{w}_i(\boldsymbol{\psi}^{(m)})]_j = \pi_j^{(m)} f_j(x_i \mid \boldsymbol{\theta}_j^{(m)}) / p(x_i \mid \boldsymbol{\psi}^{(m)}), \qquad \text{for each } i, j.$$

These 'weights' are therefore the probabilities of category membership for the ith observation, conditional on x_i and given that the parameter is $\boldsymbol{\psi}^{(m)}$.

Usually, the parameters $\boldsymbol{\pi}$ and $\boldsymbol{\theta}$ are distinct, with the consequence that the M step for $\boldsymbol{\pi}$ is

$$\pi_j^{(m+1)} = n^{-1} \sum_{j=1}^{n} w_{ij}(\boldsymbol{\psi}^{(m)}), \qquad j = 1, \ldots, k.$$

This reveals that, for Example 4.3.1, the iteration (4.3.4) *is* the relevant EM algorithm.

The form of the M step for $\boldsymbol{\theta}$ is more problem specific, although it corresponds, in general, to maximization of the second summation term in $Q(\boldsymbol{\psi}, \boldsymbol{\psi}^{(m)})$.

Example 4.3.3 Mixtures of exponential family densities of the same type

Suppose
$$\log f_j(x \mid \boldsymbol{\theta}_j) = b(x) + \mathbf{t}(x)^{\mathrm{T}} \boldsymbol{\theta}_j - a(\boldsymbol{\theta}_j),$$

and let
$$\boldsymbol{\phi}_j = \mathbb{E}[\mathbf{t}(X) \mid \boldsymbol{\theta}_j] = \frac{\partial}{\partial \boldsymbol{\theta}_j} [a(\boldsymbol{\theta}_j)], \qquad j = 1, \ldots, k.$$

It is easy to see that the M step for $\boldsymbol{\theta}_j$ reduces to the maximization of

$$\sum_{i=1}^{n} w_{ij}(\boldsymbol{\psi}^{(m)}) [\mathbf{t}(x_i)^{\mathrm{T}} \boldsymbol{\theta}_j - a(\boldsymbol{\theta}_j)].$$

This gives

$$\boldsymbol{\phi}_j^{(m+1)} = [n_j(\boldsymbol{\psi}^{(m)})]^{-1} \sum_{i=1}^{n} w_{ij}(\boldsymbol{\psi}^{(m)}) \mathbf{t}(x_i), \qquad j = 1, \ldots, k,$$

where $\quad n_j(\boldsymbol{\psi}^{(m)}) = \sum_{i=1}^{n} w_{ij}(\boldsymbol{\psi}^{(m)}) = n\pi_j^{(m+1)}.$

Note the 'weighted average' nature of $\boldsymbol{\phi}_j^{(m+1)}$. Note also that $n_j(\boldsymbol{\psi}^{(m)})$ corresponds to a pseudo sample size associated with the jth subpopulation, with the observations allocated, by the fractions defined by the weights, to the various subpopulations.

For this example, the likelihood equations themselves take an interesting form. We can write the equation for π_j (cf. Example 4.3.1) as

$$\pi_j = n^{-1} \sum_{i=1}^{n} w_{ij}(\boldsymbol{\psi}) = n_j(\boldsymbol{\psi})/n.$$

Differentiation with respect to $\boldsymbol{\theta}_j$ similarly leads to the stationarity condition

$$\boldsymbol{\phi}_j = [n_j(\boldsymbol{\psi})]^{-1} \sum_{i=1}^{n} w_{ij}(\boldsymbol{\psi})\mathbf{t}(x_i),$$

inevitably reminiscent of the above M step. It follows therefore that, for stationarity,

$$\sum_{j=1}^{k} \pi_j \boldsymbol{\phi}_j = n^{-1} \sum_{j=1}^{k} \sum_{i=1}^{n} w_{ij}(\boldsymbol{\psi})\mathbf{t}(x_i) = n^{-1} \sum_{i=1}^{k} \mathbf{t}(x_i)$$

Thus, for the minimal sufficient statistics $\mathbf{t}(x)$, the maximum likelihood estimator of $\mathbb{E}[\mathbf{t}(X) | \boldsymbol{\psi})]$ is given by the corresponding sample mean. Manifestations of this, with varying degrees of generality, have been noticed by Behboodian (1970a), Fryer and Holt (1970), and Wilson and Sargent (1979).

An appealing feature of the EM algorithm is that the approximations generated maintain this relationship, in that

$$\sum_{j=1}^{k} \pi_j^{(m)} \boldsymbol{\phi}_j^{(m)} = n^{-1} \sum_{i=1}^{k} \mathbf{t}(x_i),$$

for all m except possibly $m = 0$.

Having discussed the EM algorithm and stationarity conditions for the exponential family model, we can safely leave simple special cases as exercises for the reader. These include mixtures of Poissons, exponentials, binomials, geometrics, and normals, univariate or multivariate, with or without equal variances or covariance matrices. The inclusion of fully categorized data can be coped with by specifying 'degenerate' weights in the iterations. For the time being, we shall confine our treatment of special cases to normal mixtures.

Example 4.3.2 (continued) *Mixture of two univariate normals*

Straightforwardly applying the EM algorithm, we obtain, if $n_j^{(m)} = \Sigma_i w_{ij}(\boldsymbol{\psi}^{(m)})$, for $j = 1, 2$,

$$\pi_j^{(m+1)} = n_j^{(m)}/n$$

$$\mu_j^{(m+1)} = [n_j^{(m)}]^{-1} \sum_{i=1}^{n} w_{ij}^{(m)} x_i$$

$$\sigma_j^{2(m+1)} = [n_j^{(m)}]^{-1} \sum_{i=1}^{n} w_{ij}^{(m)} (x_i - \mu_j^{(m+1)})^2, \qquad (4.3.7)$$

where $w_{ij}^{(m)} = w_{ij}(\boldsymbol{\psi}^{(m)})$. Note the superscript $(m+1)$ with μ_j on the right-hand side of (4.3.7) (so that the recursions in Hosmer, 1973a, and Leytham, 1984 are, therefore, not quite EM), and note also the now-familiar analogy with the maximum-likelihood formulae for the case of fully categorized data.

Extension of both this and Example 4.3.1 to the case of mixtures of k component densities ($k > 2$) is straightforward. For the latter, Peters and Coberly (1976) establish the concavity of the log-likelihood function ; see also Kazakos (1977).

Example 4.3.4 Mixture of two multivariate normals with equal covariance matrices (Day, 1969; O'Neill, 1978; Ganesalingam and McLachlan, 1981)

Denote the two mean vectors by $\boldsymbol{\mu}_1, \boldsymbol{\mu}_2$ and the common covariance matrix by Σ. Then $p(\mathbf{x}|\boldsymbol{\psi}) = \pi_1\phi_d(\mathbf{x}|\boldsymbol{\mu}_1, \Sigma) + \pi_2\phi_d(\mathbf{x}|\boldsymbol{\mu}_2, \Sigma)$, where ϕ_d denotes the d-dimensional normal density function. The likelihood equations can be rearranged in the form (cf. Example 4.3.2)

$$\pi_l = n^{-1} \sum_{i=1}^{n} w_{il}(\boldsymbol{\psi}), \qquad l = 1, 2,$$

$$\boldsymbol{\mu}_l = n_l^{-1} \sum_{i=1}^{n} w_{il}(\boldsymbol{\psi})\mathbf{x}_i, \qquad l = 1, 2$$

$$\Sigma = n^{-1} \sum_{j=1}^{2} \sum_{i=1}^{n} w_{ij}(\boldsymbol{\psi})(\mathbf{x}_i - \boldsymbol{\mu}_j)(\mathbf{x}_i - \boldsymbol{\mu}_j)^{\mathrm{T}},$$

where

$$n_l = \sum_i w_{il}(\boldsymbol{\psi}), \qquad l = 1, 2,$$

and

$$w_{ij}(\boldsymbol{\psi}) = \pi_j\phi_d(\mathbf{x}_i|\boldsymbol{\mu}_j, \Sigma)/p(\mathbf{x}_i|\boldsymbol{\psi}), \qquad \text{for each } i, j.$$

A drawback to the direct EM algorithm, which is, effectively, a successive approximations iteration based on the above equations, is the need to calculate $(\Sigma^{(m)})^{-1}$ at each stage in order to update the $\{w_{ij}\}$. However, Day (1969) proposes an elegant way of avoiding this problem. In parallel with Example 4.3.3, it turns out that the maximum likelihood estimators for the mean $\boldsymbol{\mu}$ and covariance matrix, V, of the *mixture* density are given by

$$\hat{\boldsymbol{\mu}} = n^{-1} \sum_{i=1}^{n} \mathbf{x}_i = \bar{\mathbf{x}}$$

and

$$\hat{V} = n^{-1} \sum_{i=1}^{n} (\mathbf{x}_i - \bar{\mathbf{x}})(\mathbf{x}_i - \bar{\mathbf{x}})^{\mathrm{T}}.$$

Furthermore, for each i,

$$w_{i1}(\boldsymbol{\psi}) = [1 + \exp(\boldsymbol{\delta}^{\mathrm{T}}\mathbf{x}_i + b)]^{-1},$$

where

$$\boldsymbol{\delta} = \Sigma^{-1}(\boldsymbol{\mu}_2 - \boldsymbol{\mu}_1) = V^{-1}(\boldsymbol{\mu}_2 - \boldsymbol{\mu}_1)/[1 - \pi_1\pi_2(\boldsymbol{\mu}_1 - \boldsymbol{\mu}_2)^{\mathrm{T}}V^{-1}(\boldsymbol{\mu}_1 - \boldsymbol{\mu}_2)]$$

and

$$\begin{aligned} \beta &= \tfrac{1}{2}(\boldsymbol{\mu}_1^{\mathrm{T}}\Sigma^{-1}\boldsymbol{\mu}_1 - \boldsymbol{\mu}_2^{\mathrm{T}}\Sigma^{-1}\boldsymbol{\mu}_2) + \log(\pi_2/\pi_1) \\ &= -\tfrac{1}{2}\boldsymbol{\delta}^{\mathrm{T}}(\boldsymbol{\mu}_1 + \boldsymbol{\mu}_2) + \log(\pi_2/\pi_1). \end{aligned}$$

Given $\boldsymbol{\mu}$, V, $\boldsymbol{\delta}$, and β, all the original parameters, $\boldsymbol{\psi}$, can be evaluated. This permits an iterative procedure that requires only the initial inversion of $\hat{V}$. At stage m, given $\boldsymbol{\psi}^{(m)}$, we may compute $\boldsymbol{\delta}^{(m+1)}$, $\beta^{(m+1)}$, and an updated set of $\{w_{ij}\}$'s. Substitution of these in the original likelihood equations gives $\boldsymbol{\psi}^{(m+1)}$. Ganesalingam and McLachlan (1981) generalize this to incorporate fully categorized data. (Note that the new parameterization includes, directly, the parameters $(\beta, \boldsymbol{\delta})$ of the linear discriminant function; see also Section 5.7.)

So far we have laid great emphasis on the EM algorithm for the calculation of maximum likelihood estimates. It is important, however, to bear in mind the existence of competing numerical methods, of which the most familiar are

(a) Newton–Raphson (NR);
(b) the Method of Scoring (MS).

Suppose we write the log-likelihood of interest as $\mathscr{L}(\boldsymbol{\psi})$. Then the iterative step for the NR algorithm is defined by

$$\boldsymbol{\psi}^{(m+1)} = \boldsymbol{\psi}^{(m)} - \alpha_m[D^2\mathscr{L}(\boldsymbol{\psi}^{(m)})]^{-1}\mathbf{D}\mathscr{L}(\boldsymbol{\psi}^{(m)}), \qquad m = 0, 1, \ldots, \tag{4.3.8}$$

and for the MS algorithm is defined by

$$\boldsymbol{\psi}^{(m+1)} = \boldsymbol{\psi}^{(m)} + \alpha_m[I(\boldsymbol{\psi}^{(m)}]^{-1}\mathbf{D}\mathscr{L}(\boldsymbol{\psi}^{(m)}), \qquad m = 0, 1, \ldots. \tag{4.3.9}$$

In both cases, the non-negative constant α_m has been introduced to provide a slight increase in generality (in the usual versions, $\alpha_m = 1$). In MS, $I(\boldsymbol{\psi})$ is the Fisher information matrix and, throughout, $\mathbf{D}$ and D^2 represent differentiation, once and twice, respectively, with respect to $\boldsymbol{\psi}$.

The following comparative remarks can be made:

(a) EM is usually simple to apply and satisfies the appealing monotonic property alluded to earlier.
(b) NR and MS are more complicated, particularly in view of the matrix inversion required, and there is no guarantee of monotonicity. The inversion problems can be overcome by having recourse to one of many quasi-Newton methods studied by numerical analysts. Moreover, MS also involves an integration in evaluating $I(\boldsymbol{\psi})$ and, as we have seen in Section 3.2, this is usually an awkward numerical problem.
(c) If NR converges it is of second order (i.e. fast), whereas EM is often excruciatingly slow. However, if the separation between components is poor,

even the numerical performance of NR can be disappointing. Dick and Bowden (1973) consider simulations for M0 data from a range of normal mixtures. For $(\pi_1, \mu_1, \sigma_1, \mu_2, \sigma_2) = (0.1, 8.0, 1.0, 7.0, 1.25)$ and $n = 400$, NR 'fails to converge' in about half the simulations. They also found that the sample variances of the estimators were generally appreciably larger than the asymptotic values, even for $n = 1000$.

(d) If the familiar asymptotic theory holds, then the iterations for NR and MS incorporate approximations to the covariance matrix of $\hat{\psi}$, the maximum likelihood estimator of ψ. This is not available directly with EM, but Louis (1982) gives a method of finding $-D^2(\hat{\psi})$ within the EM framework.

(e) Unconditional convergence to $\hat{\psi}$ is not guaranteed with any of the techniques. This is not surprising and is a familiar characteristic of NR. Of course, even with the monotonic EM algorithm we would not necessarily wish for convergence to a global maximum—bearing in mind, for example, the singularity-riddled likelihood surface of Example 4.3.2.

In certain problems, the simplicity of the resulting EM algorithm reveals intriguing parallels with the form of iteration for NR and MS. For example, Titterington (1984) shows that, when the complete-data problem corresponds to a simple exponential family model, the EM recursion can be written in the form

$$\psi^{(m+1)} = \psi^{(m)} + [I_c(\psi^{(m)})]^{-1} \mathbf{D}\mathscr{L}(\psi^{(m)}), \qquad m = 0, 1, \ldots, \qquad (4.3.10)$$

where $I_c(\psi)$ denotes the Fisher information matrix corresponding to 'complete' data, and is typically very easy to evaluate and invert. For other models, (4.3.10) often represents a reasonable approximation to the EM algorithm (also see Section 6.4.4).

The (lack of) speed of EM has been commented on by many authors. Everitt and Hand (1981) quote convergence rates for a sample of 200 observations drawn from a four-component Poisson mixture. From one set of starting values EM takes 192 iterations; from another, 365! Furthermore, the two sets of final values are different, suggesting the existence of multiple maxima. The following simple illustration exemplifies the comparative speed of EM and NR.

Example 4.3.5 Mixture of two Poissons

The data shown below give the number of death notices for women aged 80 years and over, from the *Times* newspaper for each day in the three-year period 1910 to 1912:

Number of notices (i)	0	1	2	3	4	5	6	7	8	9
Frequency (n_i)	162	267	271	185	111	61	27	8	3	1

A single Poisson gives a very poor fit, with $\chi^2 = 1122.0$ on 7 degrees of freedom, the last two cells having been pooled. Hasselblad (1969) fits a mixture of two

Table 4.3.1 Calculation of maximum likelihood estimates for Example 4.3.5

Starting point			Iterations for Newton (F = fail)	Iterations for EM	
π_1	β	$l(\hat{\psi})$		Correct to two decimal places	Correct to four decimal places
0.50	0.50	− 553.1551	F	740	1740
0.50	0.75	− 537.1126	9	800	1800
0.40	0.60	− 542.3506	7	560	1565
0.40	0.75	− 540.0742	10	580	1585
0.35	0.60	− 548.5111	6	280	1285
0.35	0.75	− 543.6015	F	290	1290
Solution		− 535.3698			

Poissons, with means θ_1 and θ_2. He uses the moment estimates ($\hat{\pi}_1 = 0.2870$, $\hat{\theta}_1 = 1.1006$, $\hat{\theta}_2 = 2.5818$) as initial approximations for the EM algorithm, which he then stops at $\hat{\pi}_1 = 0.3590$ (0.0428), $\hat{\theta}_1 = 1.2546$ (0.0189), $\hat{\theta}_2 = 2.6623$ (0.0772). The estimated standard errors are given in brackets and indicate that the moment estimates and maximum likelihood estimates are not at all close.

Table 4.3.1 displays the results of application of the EM algorithm and the modified Newton algorithm available in subroutine EØ4EBF of NAG(1978). Various starting points were chosen, as indicated in the table. The starting point involved choices of π_1 and β, where $\theta_1 = \beta\bar{x}$ and $\bar{x}$ is the overall sample mean, 2.1569. The level of the log-likelihood was indicated by the value $l(\hat{\psi})$, defined by

$$l(\psi) = \sum_{i=0}^{9} n_i \log[\pi_1\theta_1^i e^{-\theta_1} + (1 - \pi_1)\theta_2^i e^{-\theta_2}].$$

As expected, the Newton algorithm either failed, by wandering too far from the solution, or converged quickly, in 8 to 11 iterations. The point of convergence was $\hat{\pi}_1 = 0.3599$, $\hat{\theta}_1 = 1.2561$, $\hat{\theta}_2 = 2.6634$, slightly different from that obtained by Hasselblad using the EM algorithm (probably because the EM algorithm was stopped 'too early'). From Table 4.3.1, the slow but comparatively sure properties of EM are only too evident. The numbers of iterations required in order that $\hat{\pi}_1$ be 'correct' to two decimal places and to four decimal places are given. The difference between the two is, in each case, about 1000 iterations! Indeed, it requires a further 740 EM steps to take Hasselblad's $\hat{\pi}_1(= 0.3590)$ up to 0.3599!

The maximum likelihood Poisson mixture fits very well, with $\chi^2 = 1.27$ on 5 degrees of freedom. The posited real-life justification of the mixture model is that there could be different patterns of deaths in winter and summer.

Example 4.3.6 Mixture of two known normals

A sample of 300 values was generated from the mixture

$$p(x) = 0.8\phi(x \mid -1, 1) + 0.2\phi(x \mid 1, 1).$$

Table 4.3.2 Results for Example 4.3.6

Sample size	$\hat{\pi}$	Estimated standard error	$1/\hat{I}$	Iterations required EM	NR
25	0.775	0.116	0.311	14	4
50	0.774	0.088	0.398	16	4
100	0.813	0.061	0.378	19	4
300	0.821	0.032	0.310	16	3

The EM and NR algorithms were used to find the maximum likelihood estimate of the mixing weight, assuming that all other parameters were known. Both algorithms were started off with an initial estimate of 0.5, and samples of sizes 25, 50, 100, and 300 were extracted from the total set.

Table 4.3.2 displays the results based on a stopping rule $|\pi_1^{(m+1)} - \pi_1^{(m)}| < 10^{-5}$, where $\pi_1^{(m)}$ and $\pi_1^{(m+1)}$ are consecutive iterates. Also shown are the estimated standard errors of the estimates and the corresponding estimate of I^{-1}, where I is the Fisher information per observation. This should be compared with the theoretical value of $I^{-1} = 0.327$, evaluated by the methods of Section 3.2.

Redner and Walker (1984) apply the EM algorithm to normal mixtures. They discover that, although 'convergence' typically takes very many iterations, it does not take long to get 'high up' on the likelihood surface. Usually, 95 per cent of the change in the log-likelihood is achieved in the first five iterations. This suggests that a composite algorithm, in which a few EM iterations are followed by one or two further iterations using MS or NR, would be worth investigation.

The idea of speeding up EM itself was raised by Peters and Walker (1978a, 1978b). Suppose we write the EM iteration as

$$\psi^{(m+1)} = G(\psi^{(m)})$$

and we define

$$\tilde{\psi}^{(m+1)} = \tilde{\psi}^{(m)} + \alpha[G(\tilde{\psi}^{(m)}) - \tilde{\psi}^{(m)}],$$

reminiscent of (4.3.8) and (4.3.9), with $\alpha = 1$ corresponding to EM. Peters and Walker look at versions of Examples 4.3.1 and 4.3.2. They find that strongly consistent maximum likelihood estimators, $\hat{\psi}$, for ψ exist and that, provided initial estimates are chosen close enough to $\hat{\psi}$, 'local' convergence occurs for $0 < \alpha < 2$. Asymptotically optimal rates of convergence occur for some α^* with $1 < \alpha^* < 2$. Typically, if the component densities are well separated, α^* is near 1 and convergence is fast. Otherwise, α^* is near 2 and convergence is slow.

Incorporation of a crude one-dimensional search technique into the EM algorithm for exponential mixtures is described by Wilson and Sargent (1979).

4.3.3 Theoretical considerations

Two important questions are raised by the above discussion.

(a) Do maximum likelihood estimators exist with the familiar, desirable, asymptotic properties?
(b) What convergence properties can be established for the various computational methods available, particularly the recently proposed EM algorithm?

The reader may well be surprised to have just been informed that the answer to (a) seems to be 'yes' for the normal mixture of Example 4.3.2, in spite of all its singularity problems. However, the singularities occur at certain points on the boundary of the parameter space as the variances tend to zero. In most examples, provided we keep away from these boundaries we seem to find a sensible local maximum on the likelihood surface. The following useful summary of the theoretical results available is abstracted from Redner and Walker (1984), to whom the reader should refer for full regularity conditions.

Suppose ψ' denotes the parameters $(\pi_1, \ldots, \pi_{k-1}, \boldsymbol{\theta})$, with the redundant π_k omitted, and suppose all $\pi_j > 0$, $j = 1, \ldots, k$. Suppose also that the class of mixtures is identifiable. The likelihood equations are, therefore,

$$D_{\psi'} \mathscr{L}(\psi') = O. \tag{4.3.11}$$

Theorem 4.3.1 Given the existence of, and certain boundedness conditions on, derivatives of $f(x \mid \psi')$, of orders up to 3, and given that $I(\psi'_0)$, the Fisher information matrix evaluated at the true ψ'_0 is well defined and positive definite, there is a unique solution $\hat{\psi}'_n$ of (4.3.11) in a certain small neighbourhood of ψ'_0 for all sufficiently large sample sizes, n. This $\hat{\psi}'_n$ locally maximizes the likelihood and

$$\sqrt{n}(\hat{\psi}'_n - \psi'_0) \to N(0, I(\psi'_0)^{-1}),$$

is distribution, as $n \to \infty$.

Theorem 4.3.2 Given slightly stronger conditions than in Theorem 4.3.1, $\hat{\psi}'_n$ is the unique strongly consistent maximizer of the likelihood.

These results give considerable theoretical reassurance about maximum likelihood estimation in that the regularity conditions are satisfied by many familiar examples. In addition to Redner and Walker (1984), there are other papers devoted to this aspect; see Sundberg (1972, 1974), Peters and Walker (1978a, 1978b), Kiefer (1978a), and Example 5.6 of Lehmann (1983). In passing, we should note an obvious lack of uniqueness in some examples. In Example 4.3.2, for instance, the likelihoods evaluated at $(\pi, \mu_1, \mu_2, \sigma_1, \sigma_2)$ and $(1 - \pi, \mu_2, \mu_1, \sigma_2, \sigma_1)$ are necessarily the same. This phenomenon, called 'label switching' by Redner and Walker (1984), can be eliminated by restricting the parameter space, to $\pi \leqslant \frac{1}{2}$, or $\sigma_1 \leqslant \sigma_2$, for instance. Theorems 4.3.1 and 4.3.2 discuss *local* consistency of maximum likelihood estimators. Redner (1981) shows that, if attention can be restricted to a compact subset, Θ^*, of the parameter space, which includes the true ψ_0, then the global maximum likelihood estimator within Θ^* is strongly consistent. Hathaway (1983) demonstrates, for k-component normal mixtures, that such a compact subset exists, provided some preliminary constraints are placed on the parameters. The constraints, of the form

$$\min_{i \neq j}(\sigma_i/\sigma_j) \geqslant c > 0,$$

cut out all the singularities, as well as reducing the chance of spurious local maxima. In later work, Hathaway (1985) shows that the existence of a strongly consistent, global maximum can be estalished without having to 'compactify'. Provided the data consist of at least $(k+1)$ distinct points, the result can be established using a device from Section 6 of Kiefer and Wolfowitz (1956); see also Perlman (1970). The essence of the method is that it is enough to verify the appropriate regularity conditions in terms of the joint density of $(k+1)$ observations. It turns out that this can indeed be done, whereas it is not possible in terms of the density of a single observation. Hathaway (1983) also constructs a corresponding constrained version of the EM algorithm which produces improved convergence behaviour, particularly if poor initial estimates are chosen.

For the normal mixture of Example 4.3.2, many reports of computational experience back up the results contained in the above theory and point to the existence of a satisfactory local maximum. Generally, it is not necessary to eliminate the singularities by assuming equal variances, by grouping the data (Hasselblad, 1966, Cohen, 1966b), by insisting on there being some fully categorized data (Hosmer, 1973a), or by conditioning on there being at least two distinct x's from each subpopulation (Policello, 1981).

Example 4.3.7 A special two-component normal mixture

Suppose

$$p(x|\mu,\sigma) = \tfrac{1}{2}[\phi(x-\mu) + \sigma^{-1}\phi((x-\mu)/\sigma)]$$

and that independent data $x_1, \ldots, x_n$ are available. This example was used by Kiefer and Wolfowitz (1956) specifically to illustrate that, if $\mu = x_i$, for any i, the log-likelihood, as a function of σ, tends to ∞ as $\sigma \to 0$ from above.

While this is irrefutable, in practice the singularities are irrelevant because they make their presence felt only for extremely small σ, well away from σ values of interest. To illustrate this, consider Figure 4.3.1, which displays log-likelihoods from thirty independent observations from the above mixture, with $\mu = 0$ and $\sigma = 2$. The figure shows the section of the log-likelihood $\mathscr{L}_0(\mu, \sigma)$ for μ equal to an observation whose value happened to be about -0.414. Figure 4.3.1(a) plots $\mathscr{L}_0(\mu, \sigma)$ for σ between 0.1 and 3.0. Note that there is a slight dimple near $\sigma = 0.6$ and a well-defined maximum near $\sigma = 1.4$. Figure 4.3.1(b) then plots $\mathscr{L}_0(\mu, \sigma)$ against $\log_{10}\sigma$ to show that there really is a singularity, although hardly one with practical impact. As $\sigma \to 0$, the dominating quantity in the log-likelihood is $\log(2\sigma)^{-1}$. Note that Figure 4.3.1(b) shows that, for $r \geqslant 3$, $\mathscr{L}_0(\mu, 10^{-(r+1)}) - \mathscr{L}_0(\mu, 10^{-r}) \approx \log_e 10$, as it should.

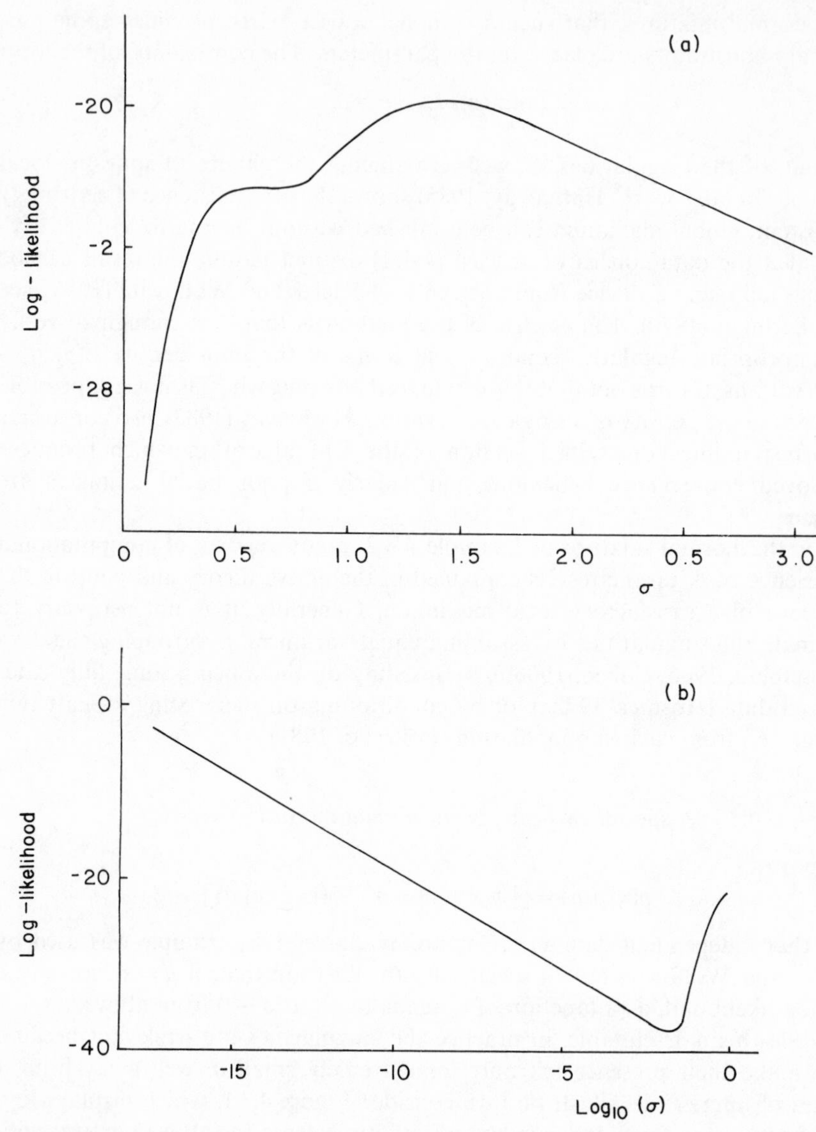

Figure 4.3.1 Graphs '$\mathscr{L}_0(\mu, \sigma)$ + constant' at $\mu = -0.414$ for data from Example 4.3.7

It is possible, however, for singularities to be picked out in numerical investigations, particularly in data with outliers, or if unlucky initial approximations are chosen. Everitt and Hand (1981) report the phenomenon for the well-known Iris setosa data, having started the EM algorithm well away from the satisfactory local maximum (see also Everitt, 1984).

The second question, (b), of convergence of the algorithm, is of obvious importance. From the MS formulation of (4.3.9), the degree of conditioning of $I(\psi'_0)$ will be an important factor in the stability of MS and NR. Redner and Walker (1984) illustrate how this conditioning is determined by the mutual separation of the component densities: the greater the separation, the better the conditioning.

Unless, as in Example 4.3.1, for instance, the log-likelihood is concave, only local convergence is likely to be provable, whatever algorithm is used.

For the EM algorithm, the convergence theory is aided by the nice monotonicity property of EM. Dempster, Laird, and Rubin (1977) do discuss convergence but their argument contains a flaw, recognition of which had led to more rigorous investigations by Boyles (1983) and Wu (1983). One particularly reassuring result of Wu (1983) can be summarized as follows. If $\mathscr{L}(\psi)$ is unimodal, with a unique maximizer, and if $D_{\psi_1}Q(\psi_1, \psi_2)$ is continuous in both sets of arguments, then the sequence $\{\psi^{(m)}\}$ converges to the unique maximizer of $\mathscr{L}(\psi)$. Often, of course, $\mathscr{L}(\psi)$ will be multimodal, but the locally restricted version of the above result is still very useful. Redner and Walker (1984) confirm that, for many mixture problems, regularity conditions sufficient for these local convergence results do hold. One of their most useful results is the following.

Example 4.3.3 (continued) *Mixture of exponential family densities*

If all the true mixing weights are positive, if the Fisher information matrix at ψ_0 is positive definite, if $\{\psi^{(m)}\}$ are generated by the EM algorithm, and if $\psi^{(0)}$ is close enough to ψ_0, then

(a) for large enough n, the unique strongly consistent solution $\hat{\psi}_n$ of the likelihood equations is well defined;
(b) $\psi^{(m)} \to \hat{\psi}_n$, as $m \to \infty$, at a linear rate.

In terms of practical performance the main findings from various simulation studies can be summarized as follows.

(a) Unless the component densities are well separated and the mixing weights not too extreme, estimation is poor, in real terms, at least for some of the parameters (Dick and Bowden, 1973; Hosmer, 1973b; Wilson and Sargent, 1979).
(b) Generally, maximum likelihood is more efficient than the method of moments, thus reflecting the asymptotic theory (Fryer and Robertson, 1972; Tan and Chang, 1972a; Hosmer, 1973a, b, 1978b).
(c) The inclusion of even a small proportion of fully categorized data (10 to 20 per cent) increases the efficiency of estimation substantially (Hosmer, 1973a, and cf. Section 3.2). Table 4.3.3, from Hosmer (1973a), illustrates this for a poorly separated normal mixture.
(d) Speed of convergence of the numerical procedures is greatly enhanced by

Table 4.3.3 Means, variances, and mean-squared errors of M0, M1, and M2 estimates computed from ten samples of size 300; $\pi_1 = 0.3$ $\mu_1 = 0$, $\sigma_1 = 1$, $\mu_2 = 3$, $\sigma_2 = 1.5$.
For M1 samples, the number of categorized observations for each component is 150K.
For M2 samples, the number of categorized observations altogether is 300K.
Reproduced from: D. W. Hosmer, Jr., 'A Comparison of Iterative Maximum Likelihood Estimates of the Parameters of a Mixture of Two Normal Distributions Under Three Different Types of Samples', *Biometrics*, **29**, 761–770 (1973), with permission from the Biometric Society

		π		μ_1		σ_1^2		μ_2		σ_2^2	
K		M1	M2	M1	M2	M1	M2	M1	M2	M1	M2
	Mean	0.396		0.036		1.001		3.120		1.943	
0	Var	0.008		0.100		0.066		0.052		0.183	
	MSE	0.011		0.101		0.066		0.067		0.277	
	Mean	0.317	0.315	−0.009	−0.076	0.975	0.913	3.059	2.997	2.019	2.111
0.1	Var	0.003	0.001	0.021	0.017	0.019	0.008	0.027	0.019	0.080	0.056
	MSE	0.004	0.002	0.021	0.023	0.019	0.016	0.030	0.019	0.133	0.076
	Mean	0.302	0.316	0.016	−0.060	1.015	0.942	3.069	2.993	2.016	2.124
0.2	Var	0.001	0.001	0.015	0.009	0.020	0.012	0.012	0.010	0.054	0.033
	MSE	0.001	0.001	0.016	0.012	0.020	0.015	0.017	0.010	0.109	0.049
	Mean	0.262	0.315	−0.010	−0.054	0.971	0.944	3.048	2.984	2.037	2.147
0.3	Var	0.000	0.000	0.009	0.010	0.010	0.014	0.008	0.007	0.036	0.044
	MSE	0.002	0.000	0.009	0.013	0.011	0.017	0.011	0.008	0.082	0.055

each of the following factors: more fully categorized data; larger sample size; better separation (Sundberg, 1976).

These empirical findings are almost all gleaned from the examination of mixtures of two univariate normal mixtures.

In a small but informative comparative study, Everitt (1984) compares the performance of the EM algorithm, two forms of NR, the conjugate gradient method of Fletcher and Reeves (1964), and the simplex method of Nelder and Mead (1965). Examples of two-component, five-parameter normal mixtures were considered with ten samples of size 50 from each. The simplex and Fletcher–Reeves methods were more liable to find inferior local maxima and were sometimes very slow; the Fletcher–Reeves method sometimes 'stopped' at a non-stationarity point; EM was sometimes very slow; the simplex method sometimes converged to a singularity, NR did so only once, and EM not at all. In a simulation study on the use of an EM-type algorithm on a mixture of the above type, Leytham (1984) does report convergence to singularities, mainly in examples with the very small sample size of 30—this in spite of the fact that the true parameter values were used to start the recursion.

4.3.4 Further examples

Maximum likelihood analysis has been very widely applied, as detailed in Table 4.3.4. As might be expected, in the case of mixtures of uniform densities (Gupta and Miyawaki, 1978), points of discontinuity in the likelihood yield the maximum likelihood estimates.

We now look in some detail at several particularly interesting examples.

Example 4.3.8 Mixture of multinomials with extra, fully categorized data, according to data structure M2

If the multinomial has L cells, the mixture data can be written as frequencies $\{n_{0l}: l = 1, \ldots, L\}$ and the fully categorized data as $\{n_{jl}, j = 1, \ldots, k, l = 1, \ldots, L\}$. Thus the data can be regarded as a set of data $\{n_{jl}\}$ from a two-way contingency table along with further data $\{n_{0l}\}$ on one of the margins. In this example, a certain amount of fully categorized data is required to ensure identifiability. Suppose the cell probabilities associated with the two-way table are $\{\theta_{jl}\}$, summing to 1. They are related to the parameters in the mixture model by

$$\theta_{jl} = \pi_j f_{jl},$$

where $f_{jl} = P(\text{in cell } l \mid \text{in subpopulation } j)$.

If, instead, we use the equivalent parameterization

$$\theta_{jl} = \eta_{jl} \psi_l,$$

where $\psi_l = P(\text{in cell } l)$

$\eta_{jl} = P(\text{in subpopulation } j \mid \text{in cell } l)$,

Table 4.3.4 Maximum likelihood estimation with mixture data

Example	References	Computational method, if characterizable
Binomials	John (1970a)	Cluster analysis approach
	Hasselblad (1966, 1969)	EM
Geometrics	Harris (1983)	Gradient method
Negative binomials	John (1970a)	Cluster analysis
Hypergeometrics	John (1970a)	Cluster analysis
Gammas	John (1970c)	Cluster analysis
Poissons	Samaniego (1976)	
	Hasselblad (1969)	EM
	John (1970a)	Cluster analysis
Compound gamma	Hill, Saunders, and Land (1980)	Newton
Generalized Poisson	Samaniego (1976)	
Compound Poisson	Simar (1976)	Steepest ascent
Uniforms	Gupta and Miyawaki (1978)	Successive approximations
Exponentials	Weiner (1962)	
	Tallis and Light (1968)	
	Hasselblad (1969)	EM
	Suchindran and Lachenbruch (1974)	EM
	Wilson and Sargent (1979)	EM
Exponentials with data grouped	Suchindran and Lachenbruch (1974)	Scoring
Exponentials with truncation	Mendenhall and Hader (1958) (see also Johnson and Kotz, 1970b, Section 18.10, and Everitt and Hand, 1981, Section 3.2)	
Two univariate normals	Molenaar (1965)	
	Odell and Basu (1976)	
	Hosmer (1973a, 1978a)	EM
	Dick and Bowden (1973)	Newton
	Tan and Chang (1972a, 1972b)	
	Cohen (1967)	Newton
	Bryant and Williamson (1978)	Cluster analysis
	Covey-Crump (1970)	Search
	Kumar, Nicklin, and Paulson (1979)	Search
	Aitkin and Tunnicliffe Wilson (1980)	EM
	Quandt and Ramsey (1978)	Davidson–Fletcher–Powell
	Newcomb (1886)	EM
	Sundberg (1976)	EM

Table 4.3.4 (*contd.*)

Example	References	Computational method, if characterizable
	Leytham (1984)	EM
	Everitt (1984)	Fletcher–Reeves, EM, NR, simplex
Univariate normals (grouped data)	Rao (1948)	Newton
	Hasselblad (1966)	EM, Newton
	Cohen (1966b)	
k Univariate normals	Odell and Basu (1976)	EM
	Aitkin and Tunnicliffe Wilson (1980)	EM
Multivariate normals	Blåfield (1980)	
	Wolfe (1965, 1970, 1971)	EM, scoring
	John (1970b)	Cluster analysis
	O'Neill (1978)	EM
	Chang (1976, 1979)	
	Marriott (1975)	
	Day (1969)	EM
	Little (1978)	EM
	Ganesalingam and McLachlan (1978, 1979a, 1981)	EM
	Do and McLachlan (1984)	EM
MVNs with missing values	Mill (1983)	EM
von Mises	Jones and James (1972)	Steepest ascent,
	Mardia and Sutton (1975)	Newton
Switching regressions	Quandt (1972)	Conjugate gradients
	Hosmer (1974)	EM
	Goldfeld and Quandt (1973, 1976)	
	Quandt and Ramsey (1978)	Davidson–Fletcher–Powell
	Hartley (1978)	EM
Latent class	Skene (1978)	EM
	Dawid and Skene (1979)	EM
Logistic model	Anderson (1979)	Quasi-Newton
Estimating mixing distributions	Macdonald (1971)	Newton
	Odell and Basu (1976)	
	Peters and Coberly (1976)	EM
	Tubbs and Coberly (1976)	
(Empirical Bayes)	{ Laird (1978a)	EM
	Robbins (1964) }	
	Lindsay (1983a, 1983b)	EM
	Fabi and Rossi (1983)	EM
	Di Gesu and Maccarone (1984)	EM
Quantal response (probit)	Ashford and Walker (1972)	Search
Truncated Poisson + spike	Cohen (1960), Umbach (1981)	
Truncated negative binomial + spike	Cohen (1966a)	
Robustified normals	Campbell (1984)	EM

then the log-likelihood is

$$\mathscr{L}(\boldsymbol{\eta}, \boldsymbol{\psi}) = \sum_j \sum_l n_{jl} \log(\eta_{jl}\psi_l) + \sum_l n_{0l} \log \psi_l.$$

This yields maximum likelihood estimates

$$\hat{\psi}_l = (n_{.l} + n_{0l})/N, \qquad \hat{\eta}_{jl} = n_{jl}/n_{.l},$$

where $n_{.l} = \sum_j n_{jl}$ and N is the total number of observations. In terms of the parameters of interest, we have

$$\hat{\pi}_j = \sum_l \frac{n_{jl}}{n_{.l}} \frac{n_{.l} + n_{0l}}{N}, \qquad j = 1, \ldots, k,$$

and

$$\hat{f}_{jl} = \hat{\eta}_{jl}\hat{\psi}_l / \hat{\pi}_j, \qquad \text{for each } j, l.$$

Thus, at last, we have an example in which *explicit* maximum likelihood estimates exist! See Brown (1976) and Hall and Titterington (1984) for further treatment of this example.

Example 4.3.9 Grouped data

Very often, data are given in the form of a histogram, with cell boundaries $a_0, \ldots, a_M$, such that the ith cell corresponds to the interval (a_{i-1}, a_i). If n_i denotes the number of observations falling in the ith cell, then the log-likelihood is

$$\mathscr{L}_0(\boldsymbol{\psi}) = \sum_{i=1}^{M} n_i \log P_i(\boldsymbol{\psi}),$$

where $$P_i(\boldsymbol{\psi}) = \int_{a_{i-1}}^{a_i} p(x \mid \boldsymbol{\psi})\, \mathrm{d}x, \qquad i = 1, \ldots, M.$$

If the cells are each of a small constant width h, then $P_i(\boldsymbol{\psi}) \approx hp(\bar{x}_i \mid \boldsymbol{\psi})$, where $\bar{x}_i = \frac{1}{2}(a_{i-1} + a_i)$. If now this approximation is used in $\mathscr{L}_0(\boldsymbol{\psi})$, we obtain a log-likelihood which corresponds to all the n_i observations in cell i being treated as $\bar{x}_i$ (Hasselblad, 1966).

If this approximation is thought to be too crude, it is possible to write down an EM algorithm from which we can obtain maximum likelihood estimates for the original $\mathscr{L}_0(\boldsymbol{\psi})$. Grouped data can themselves be thought of as incomplete (Dempster, Laird and Rubin, 1977, Section 4.2) and an E step can be devised to account simultaneously for the grouping and mixing incompleteness—an exercise for the reader!

Example 4.3.10 Markov chain mixtures

Suppose that the underlying indicator vectors $\mathbf{z}_1, \ldots, \mathbf{z}_n$ are not independent, but follow a Markov chain, with stationary transition probabilities

$$\{\pi_{jl}: j, l = 1, \ldots, k\} = \boldsymbol{\pi}, \text{ say.}$$

Suppose, for convenience, that $\mathbf{z}_1$ is fixed. Then the 'complete-data' log-likelihood is

$$\log g(\mathbf{y}|\boldsymbol{\psi}) = \sum_{i=2}^{n} \mathbf{z}_{i-1}^{\mathrm{T}} V(\boldsymbol{\pi})\mathbf{z}_i + \sum_{i=1}^{n} \mathbf{z}_i^{\mathrm{T}} \mathbf{U}_i(\boldsymbol{\theta}),$$

where now

$$[V(\boldsymbol{\pi})]_{jl} = \log \pi_{jl}.$$

The *E step* of the EM algorithm requires, therefore, the calculation of

$$Z^{(m)} = \sum_{i=2}^{n} \mathbb{E}(\mathbf{z}_i \mathbf{z}_{i-1}^{\mathrm{T}} | \mathbf{x}, \boldsymbol{\psi}^{(m)})$$

and

$$\mathbb{E}(\mathbf{z}_i | \mathbf{x}, \boldsymbol{\psi}^{(m)}), \qquad i = 2, \ldots, n.$$

The *M step* for $\boldsymbol{\theta}$ is much as before. For the transition probabilities we obtain, for each j, l,

$$\pi_{jl}^{(m+1)} = z_{jl}^{(m)} / \sum_s z_{js}^{(m)}.$$

For details, see Baum and Eagon (1967), Goldfeld and Quandt (1973, 1976, Chapter 1) and, particularly, Baum *et al.* (1970) and Lindgren (1978). Baum *et al.* (1970) develop the EM principle to a degree of generality approaching that of Dempster, Laird, and Rubin (1977).

Example 4.3.11 Non-parametric estimation of mixtures with M2 data

We assume here that the k component densities are unknown and do not necessarily have parametric forms. For any possibility of identifiability, therefore, there has to be some fully categorized data from each subpopulation. Example 4.3.8 represents the discrete version of this problem. In this more general setting, we mention two methods, each based on a particular approach to non-parametric density estimation.

(a) Kernel method

The formula for a kernel-based density estimate was described in Section 2.2 and involves an arbitrary smoothing parameter, h. The following *ad hoc* procedure, which performs well in practice, was proposed by Murray and Titterington (1978).

(i) From the fully categorized data from each subpopulation, obtain a kernel-based estimate, $\hat{f}_j$, for the component density. Estimate the mixing weights by the relative frequencies occurring in the fully categorized data.

(ii) For each of the n uncategorized observations, evaluate weights

$$w_{ij} = \hat{\pi}_j \hat{f}_j(x_i) / \sum_l \hat{\pi}_l \hat{f}_l(x_i), \qquad i = 1, \ldots, n, j = 1, \ldots, k,$$

and allocate the observations, by these fractions, to the subpopulations, subsequently reestimating the mixing weights.

(iii) Repeat (ii) until convergence occurs.

The algorithm is of EM type and 'monotonicity' of the likelihood can be established.

(b) Maximum penalized likelihood

Given a univariate sample $x_1, \ldots, x_n$ from a population, a completely non-parametric approach to maximum likelihood is to choose a density function $p(\cdot)$ to maximize

$$\sum_{i=1}^{n} \log p(x_i).$$

The solution to this problem is the excessively rough 'density' whose integral is the empirical distribution function from the data. To obtain a more satisfactory solution, a penalized log-likelihood,

$$\sum_{i=1}^{n} \log p(x_i) - \alpha H(p),$$

is maximized, where $H(p)$ is a roughness penalty function and $\alpha > 0$ is an adjustable parameter which determines the degree of smoothing. If

$$H(p) = \int \{[p'(x)]^2/p(x)\}\, dx,$$

then the solution is an exponential spline (see Tapia and Thompson, 1978, Chapter 4; Anderson and Blair, 1982).

For M2 data, the corresponding penalized log-likelihood is

$$L(\boldsymbol{\pi}, \mathbf{f}) = \sum_{j=1}^{k} \sum_{i=1}^{n_j} \log[\pi_j f_j(x_{ji})] + \sum_{i=1}^{n} \log\left[\sum_{j=1}^{k} \pi_j f_j(x_i)\right] - \sum_{j=1}^{k} \alpha_j H(f_j).$$

The corresponding 'complete-data' log-likelihood is

$$l(\boldsymbol{\pi}, \mathbf{f}) = \sum_{j=1}^{k} \sum_{i=1}^{n_j} \log[\pi_j f_j(x_{ji})] + \sum_{i=1}^{n} \sum_{j=1}^{k} z_{ij} \log[\pi_j f_j(x_i)] - \sum_{j=1}^{k} \alpha_j H(f_j)$$

and a monotonic EM algorithm can be defined as follows.

E step: Evaluate $\mathbb{E}l(\boldsymbol{\pi}, \mathbf{f} \mid \text{data}; \boldsymbol{\pi}^{(m)}, \mathbf{f}^{(m)}) = Q(\boldsymbol{\pi}, \mathbf{f} \mid \boldsymbol{\pi}^{(m)}, \mathbf{f}^{(m)})$.

M step: Choose $(\boldsymbol{\pi}, \mathbf{f}) = (\boldsymbol{\pi}^{(m+1)}, \mathbf{f}^{(m+1)})$ to maximize Q.

The practical drawback is that each *M step* involves a fairly complicated calculation.

Example 4.3.12 General parametric mixtures

Suppose, as in Equation (1.1.4), that

$$p(x|G) = \int_{\Theta} f(x|\boldsymbol{\theta})\mathrm{d}G(\boldsymbol{\theta}),$$

where f represents a parametric density and $G(\cdot)$ is some distribution over the parameter space, $\boldsymbol{\Theta}$. Although we professed, in Chapter 1, to have little interest in this case, we shall see that, so far as maximum likelihood is concerned, we need only deal with finite mixtures. By the nature of $p(\cdot|G)$, our results will have application also in the context of empirical Bayes estimation (Robbins, 1964; Laird, 1978a; see, also, Example 2.2.6).

If $G = \hat{G}_n$ is the maximizer of

$$\mathscr{L}(G) = \sum_{i=1}^{n} \log p(x_i|G),$$

then, given identifiability and certain other mild conditions, $\hat{G}_n$ is consistent for G.

Now suppose

$$\mathbf{f}_{\boldsymbol{\theta}} = [f(y_1|\boldsymbol{\theta}), \ldots, f(y_m|\boldsymbol{\theta})]^{\mathrm{T}},$$

where $y_1, \ldots, y_m$ $(m \leqslant n)$ are the set of *distinct* values among the sample observations $x_1, \ldots, x_n$, and let

$$\mathscr{T} = \{\mathbf{f}_{\theta}: \boldsymbol{\theta} \in \boldsymbol{\Theta}\}.$$

Then we have the following:

(a) If $c(\mathscr{T})$ denotes the convex hull of $\mathscr{T}$, any member of $c(\mathscr{T})$ can be expressed as a convex combination of at most $(m+1)$ members of $\mathscr{T}$ and, for a boundary point of $c(\mathscr{T})$, this number can be reduced to m (Carathéodory's theorem).
(b) For any G, $\mathscr{L}(G)$ is a strictly concave function of the coordinates $p(y_i|G)$ on $c(\mathscr{T})$.
(c) Consequently, if $\mathscr{T}$, and therefore $c(\mathscr{T})$, are closed and bounded, $\mathscr{L}(G)$ attains its maximum at a unique point, $\hat{\mathbf{f}}$, on the boundary of $c(\mathscr{T})$.
(d) From (a) and (c), this $\hat{\mathbf{f}}$ can be written in terms of a finite mixture of at most m components.

These results are given by Lindsay (1981, 1983a) in the specific context of the mixture problem, although they are essentially well-known general theorems in optimization theory. Equivalent results in optimal design theory are summarized by Silvey (1980). A further common feature of these two fields of application of the general theorems is a set of necessary and sufficient conditions for the maximizing $\hat{\mathbf{f}}$. For the mixture problem, the result is as follows.

Let $D(G_1, G_2)$ denote the directional derivative of $\mathscr{L}(G)$ at G_1 in the direction of G_2, defined by

$$D(G_1, G_2) = \lim_{\varepsilon \downarrow 0} \varepsilon^{-1}\{\mathscr{L}[(1-\varepsilon)G_1 + \varepsilon G_2] - \mathscr{L}(G_1)\}.$$

Let G_θ denote the degenerate probability measure which assigns unit mass to $\boldsymbol{\theta}$. Then

(a) $\hat{G}_n$ maximizes $\mathscr{L}(G)$ if and only if $D(\hat{G}_n, G_\theta) \leqslant 0$ for all $\boldsymbol{\theta} \in \boldsymbol{\Theta}$.
(b) $\boldsymbol{\theta}$ is in the support of $\hat{G}_n$ only if

$$D(\hat{G}_n, G_\theta) = 0. \tag{4.3.12}$$

Result (a) simply means that no direction of view from the top of a concave hill can be upwards. It provides a check of optimality for a trial solution $\hat{G}_n$. Result (b) provides equations from which the mixing weights can be calculated, once the optimal set of $\{\boldsymbol{\theta}_j : j = 1, \ldots, k\}$ has been found. For our problem, equation (4.3.12) turns out to be

$$\sum_{i=1}^{n} \frac{f(x_i \mid \boldsymbol{\theta}_l)}{\sum_{j=1}^{k} \pi_j f(x_i \mid \boldsymbol{\theta}_j)} = n, \qquad l = 1, \ldots, k, \tag{4.3.13}$$

where $\{\pi_j\}$ are the mixing weights. This just takes us back to (4.3.3)!

Calculation of $\hat{G}_n$ in practice requires that we find an optimal support, $\boldsymbol{\theta}_1, \ldots, \boldsymbol{\theta}_k$ and then, from (4.3.13), the corresponding optimal weights. If, at some stage, we have calculated a trial $\hat{G}_n$ that is not the maximum likelihood estimator then result (a) will not hold for at least some $\boldsymbol{\theta}$, and this will indicate a direction of ascent from the current position. Many useful algorithms have been proposed in the context of optimal design (see Silvey, 1980, Chapter 4). Although Lindsay (1981, 1983a) seems to have been the first to discuss the mixture problem in this general setting, previous papers have looked at special cases and have advocated particular computational methods. Laird (1978a) adapts the EM algorithm; Simar (1976) proposes, for the problem of a compound Poisson mixture, an algorithm whose design analogue is set out by Silvey and Titterington (1973) and whose convergence is established by Bohning (1982); Hill, Saunders, and Land (1980) give a detailed treatment of the compound gamma problem, with a countable parameter space, and use a Newton-type method in computations. Both they and Lindsay (1983a) consider constraining the number of components in the mixture to be at most k, for specified k. Jewell (1982) looks at exponential mixtures and Lindsay (1983b) considers some very special cases. Tierney and Lambert (1984) develop asymptotic theory for estimators of functionals of mixing distributions, comparing non-parametric estimators with estimators induced by the above maximum likelihood procedure. The case of mixed Poisson distributions is dealt with in more detail by Lambert and Tierney (1984).

Finally, in this section, we mention a maximum likelihood related approach which might be called the *clustering method*. Given data from a certain parametric k-component mixture, all possible clusterings of the data into k groups are considered. For each clustering, the resulting likelihood is maximized with respect to the parameters. Originally (Scott and Symons, 1971), it was thought that this method would

(a) represent a reasonable method for cluster analysis;
(b) lead to useful parameter estimates.

Aim (a) is reasonably well satisfied, but not so (b). The method, mentioned also by John (1970a, b), Rayment (1972), and Kazakos (1977), corresponds to choosing indicators $\{\mathbf{z}_i\}$ and parameters $\{\boldsymbol{\theta}\}$ to maximize

$$\prod_{i=1}^{n} \prod_{j=1}^{k} f_j(x_i \mid \boldsymbol{\theta}_j)^{z_{ij}}. \tag{4.3.14}$$

The failure in respect of aim (b) was pointed out by Marriott (1975), who showed the lack of consistency of the estimators of $\boldsymbol{\theta}$. For instance, for data from a mixture of two univariate normal densities, each with the same variance, the optimal clustering is based on a cut-off value x_0: all x_i's below x_0 go into cluster 1, all above x_0 into cluster 2. Although this greatly cuts down the number of groupings we need to examine (from 2^n to $0(n)$), it is clear that treating the clustered data as if they were two normal samples leads to biases. The features of overlap and effective truncation are ignored, with the result that the common variance is estimated with negative bias and the difference between the means with positive bias (see Figure 4.1.7). Woodward *et al.* (1984), however, show that robust estimators of location and scale parameters based on the clustered data prove remarkably efficient.

Marriott's (1975) results are followed up by Bryant and Williamson (1978), who show that, even if the component densities are well separated, the biases may be appreciable.

The multivariate normal version of the problem (equal covariance matrices) leads to choosing the optimal clustering to minimize

$$n \log |W|,$$

where W is the within-groups sample dispersion matrix. Symons (1981) comments on the tendency of the method to produce roughly equal clusters, in contrast to those obtained by minimizing

$$n \log |W| - \sum_{j=1}^{k} n_j \log n_j.$$

This latter procedure corresponds to choosing $\{\mathbf{z}_i\}$, $\{\boldsymbol{\theta}\}$, *and* $\{\boldsymbol{\pi}\}$ to maximize

$$\prod_{i=1}^{n} \prod_{j=1}^{k} \pi_j^{z_{ij}} f_j(x_i \mid \boldsymbol{\theta}_j)^{z_{ij}}, \tag{4.3.15}$$

the complete-data likelihood of (4.3.5)!

Expression (4.3.14) is the 'equal π_j's' version of this, thus explaining the 'equal-clusters' phenomenon. So far as parameter estimation is concerned, biases will still be present. Symons (1981) does, however, recommend the use of (4.3.15) for generating initial estimates for iterative procedures to find the 'proper' maximum likelihood estimates. A basic flaw in this *clustering method* for parameter estimation is the treatment of the $\{\mathbf{z}_i\}$ as if they were parameters, rather than treating them as missing random variables. For other illustrations of this phenomenon, see Little and Rubin (1983).

Sclove (1977, 1983) provides further illustrations of this method, which he calls the conditional mixture model ('conditional' on a prescribed clustering); see also Bezdek and Dunn (1975).

Bayesian versions of the method are described by Symons (1981) and Binder (1978a); see, also, Section 4.4.

4.4 BAYESIAN METHODS

4.4.1 Introduction

Given a model with an unknown parameter vector $\boldsymbol{\psi} \in \boldsymbol{\Psi}$, the Bayesian inference paradigm is very easily described. If $L(\boldsymbol{\psi})$ denotes the likelihood for $\boldsymbol{\psi}$ given sample data $\mathbf{x} = (x_1, \ldots, x_n)$, Bayes theorem provides the mechanism whereby beliefs about $\boldsymbol{\psi}$ *prior* to observing $\mathbf{x}$, expressed as a density $p(\boldsymbol{\psi})$, are updated into beliefs about $\boldsymbol{\psi}$ *posterior* to observing $\mathbf{x}$, denoted by $p(\boldsymbol{\psi} \mid \mathbf{x})$ and given by

$$p(\boldsymbol{\psi} \mid \mathbf{x}) = \frac{L(\boldsymbol{\psi})p(\boldsymbol{\psi})}{\int_{\boldsymbol{\Psi}} L(\boldsymbol{\psi})p(\boldsymbol{\psi})\,\mathrm{d}\boldsymbol{\psi}}. \tag{4.4.1}$$

In the context of finite mixture models, $L(\boldsymbol{\psi})$ could take any of the forms $L_0(\boldsymbol{\psi})$, $L_1(\boldsymbol{\psi})$, $L_2(\boldsymbol{\psi})$ described by equations (1.2.1), (1.2.2), (1.2.3), with obvious modifications if some of the component parameters are known. There is little that can be said in general terms about the specification of $p(\boldsymbol{\psi})$, since, within the Bayesian philosophical framework, (4.4.1) is a tool for exploring actual or potential prior to posterior belief mappings in particular applications (see Smith, 1984, Section 2.3 (vii)). We note, however, that a simple analytic treatment of (4.4.1), such as results when $L(\boldsymbol{\psi})$ corresponds to an exponential family model and $p(\boldsymbol{\psi})$ is chosen to be a member of the corresponding conjugate family (cf. the discussion in Example 2.2.5), will not be possible when $L(\boldsymbol{\psi})$ derives from a finite mixture model.

It follows that the implementation of the Bayesian paradigm for finite mixture models will not be at all straightforward, unless $\boldsymbol{\psi}$ consists of just one or two unknown parameters. In this latter case, plots or summaries of joint or marginal posterior densities can easily be obtained from (4.4.1) at little computational expense by simply evaluating $p(\boldsymbol{\psi} \mid \mathbf{x})$ over suitable grids of points. When $\boldsymbol{\psi}$ consists of three or more unknown parameters, however, we encounter the general problem of carrying out efficient numerical integration in several dimensions (in order to obtain the denominator of (4.4.1), as well as marginal inferences on lower dimensional functions of $\boldsymbol{\psi}$). A general discussion of this problem, together with details of some proposed strategies, is given in Naylor and Smith (1982) and Smith *et al.* (1985). We shall not enter into detail here, since, once we invoke a general numerical integration strategy, a finite mixture model is simply a special case of a multiparameter Bayesian inference problem and the implementation strategy takes little note of the mixture structure.

One situation in which analytic progress is possible occurs when the sample information about $\boldsymbol{\psi}$ may be thought of as overwhelming the prior information as specified in $p(\boldsymbol{\psi})$. To all intents and purposes, $p(\boldsymbol{\psi}|\mathbf{x})$ may then be regarded as a normalized version of $L(\boldsymbol{\psi})$, so that the Bayesian asymptotic theory closely parallels that reviewed for the maximum likelihood approach in Section 4.3. Subject to appropriate regularity conditions (see, for example, Heyde and Johnstone, 1979), the asymptotic posterior distribution of $\boldsymbol{\psi}$ is $N(\hat{\boldsymbol{\psi}}, \hat{\Sigma})$, where $\hat{\boldsymbol{\psi}}$ is the maximum likelihood estimate and

$$(\hat{\Sigma}^{-1})_{ij} = -\left.\frac{\partial^2 L(\boldsymbol{\psi})}{\partial\psi_i \partial\psi_j}\right|_{\boldsymbol{\psi}=\hat{\boldsymbol{\psi}}},$$

the usual estimated inverse information matrix. It follows that some of the computational algorithms discussed in Section 4.3 are equally useful for the large-sample Bayesian approach (or, in the small-sample situation, for finding the mode of the posterior distribution derived from an arbitrary $p(\boldsymbol{\psi})$).

Another situation in which limited analytic insight is available arises when only the mixing weights $\boldsymbol{\pi}$ are unknown and are assigned a Dirichlet prior distribution, which has density proportional to

$$\prod_{j=1}^{k} \pi_j^{\alpha_j - 1}, \qquad \alpha_j > 0, j = 1, \ldots, k. \tag{4.4.2}$$

If the data consist of n independent observations $x_1, \ldots, x_n$ from the mixture

$$p(x|\boldsymbol{\pi}) = \pi_1 f_1(x) + \cdots + \pi_k f_k(x),$$

then

$$L(\boldsymbol{\psi}) = \sum_{i=1}^{n} \left[\sum_{j=1}^{k} \pi_j f_j(x_i) \right] \tag{4.4.3}$$

$$= \sum_{r_1 + \cdots + r_k = n} C(\mathbf{x}; r_1, \ldots, r_k) \prod_{j=1}^{k} \pi_j^{r_j}, \tag{4.4.4}$$

where $C(\mathbf{x}; r_1, \ldots, r_k)$ is an easily identified function of $\{f_j(x_i), i = 1, \ldots, n, j = 1, \ldots, k\}$. Since the posterior density for $\boldsymbol{\pi}$ is proportional to the product of (4.4.2) and (4.4.4), it follows that $p(\boldsymbol{\pi}|\mathbf{x})$ has the form of a mixture of k^n Dirichlet densities. The latter correspond to the k^n possible posterior densities derivable from (4.4.2) assuming a particular identification of the n individual observations with the k possible 'source distributions'. If $J_1, \ldots, J_{k^n}$ denote the possible identifications, we have

$$p(\boldsymbol{\pi}|\mathbf{x}) = \sum_{s=1}^{k^n} p(\boldsymbol{\pi}|\mathbf{x}, J_s) p(J_s|\mathbf{x}),$$

so that the weights in the posterior mixture reflect the relative plausibility of the identification of observations with 'sources'. We shall not consider such forms further here but, instead, we defer discussion until Chapter 6, where the general problem is considered in a sequential setting.

A special feature of this example, in which $\boldsymbol{\pi}$ alone is unknown, is that the Dirichlet factors $\{p(\boldsymbol{\pi}|\mathbf{x}, J_s)\}$ in $p(\boldsymbol{\pi}|\mathbf{x})$ are independent of $\mathbf{x}$. As a result, the number of *distinct* component Dirichlet densities in $p(\boldsymbol{\pi}|\mathbf{x})$ is much less than k^n. For the simplest case of $k = 2$, for instance, the number of terms in $p(\boldsymbol{\pi}|\mathbf{x})$ can be reduced from 2^n to $(n+1)$, each corresponding to a distinct beta component. Behboodian (1972b) claims that, for general k, the same reduction can be made, from k^n to $(n+1)$, but this appears to be incorrect.

As we have indicated, the general Bayesian inference strategy for a finite mixture model in the non-sequential case differs little, if at all, from the approach adopted for any multiparameter problem. In what follows, therefore, we shall concentrate our discussion on just two special topics where the Bayesian approach specifically exploits the mixture structure.

4.4.2 Bayesian approaches to outlier models

Most Bayesian approaches to modelling situations in which 'aberrant' observations may occur have taken the view that an *outlier* is 'any observation that has not been generated by the mechanism that generated the majority of observations in the data set' (Freeman, 1981).

In the one-sample location-scale situation, for example, it is often assumed that most of the observations are from a normal distribution, with density $\phi(x|\mu, \sigma)$, but that a small proportion of observations are from an alternative (contaminant) normal density $\phi_\lambda(x|\mu, \sigma)$, the latter taken to be either of the form

$$\phi_\lambda(x|\mu, \sigma) = \phi(x|\mu + \lambda, \sigma), \qquad \text{the } \textit{location-shift} \text{ model,}$$

or

$$\phi_\lambda(x|\mu, \sigma) = \phi(x|\mu, \lambda\sigma), \qquad \lambda > 1, \qquad \text{the } \textit{inflated-variance} \text{ model.}$$

The data are then modelled as a sample of n independent observations from the contaminated normal mixture model

$$p(x|\boldsymbol{\psi}) = \pi\phi(x|\mu, \sigma) + (1 - \pi)\phi_\lambda(x|\mu, \sigma), \tag{4.4.5}$$

where $\boldsymbol{\psi}^{\mathrm{T}} = (\mu, \sigma, \lambda)$, assuming, for simplicity, that π is known. See, also, the discussion of Example 2.2.1.

Variants of this approach have been considered by a number of authors. In particular, modified forms of mixture model can be arrived at by allowing λ to vary from one aberrant observation to another and, in the location-shift case, by allowing the different λ's to have possibly different signs (so that one might distinguish lower and upper outliers). Various such elaborations, together with extensions to linear model and multivariate normal problems, are considered in detail by Box and Tiao (1968), Abraham and Box (1978), Guttman, Dutter, and Freeman (1978), Freeman (1981) and Pettit and Smith (1984, 1985). A unified discussion of these various outlier models is greatly facilitated by noting that, for appropriate choices of vectors and matrices, they can be formulated within the structure

$$\mathbf{x} \sim N(A_1\boldsymbol{\theta}_1, C_1), \qquad \boldsymbol{\theta}_1 \sim N(A_2\boldsymbol{\theta}_2, C_2). \tag{4.4.6}$$

Thus, for example, for location-scale models, with λ_i denoting the shift if x_i is an aberrant observation under the location-shift assumption, we have the following possibilities for specifying that $x_1, \ldots, x_{r_1}$ are lower outliers and $x_{n-r_2+1}, \ldots, x_n$ are upper outliers, with $r_1 + r_2 = r$.

(a) Location-shift model; λ_i's different

We consider, as an example, the model of Guttman, Dutter, and Freeman (1978), with the additional assumption that, conditional on σ^2, the λ_i's are $N(\lambda_0, \eta^2\sigma^2)$ in the case of upper outliers and $N(-\lambda_0, \eta^2\sigma^2)$ in the case of lower outliers. If the location μ is assigned, conditional on σ^2, an $N(\mu_0, \kappa^2\sigma^2)$ distribution, then in the notation of (4.4.6) we have

$$A_1 = \left[\mathbf{1}_n \begin{array}{cc} I_{r_1} & 0 \\ 0 & 0 \\ 0 & I_{r_2} \end{array} \right], \quad C_1 = \sigma^2 I_n, \quad \boldsymbol{\theta}_1^{\mathrm{T}} = (\mu, \lambda_1, \ldots, \lambda_{r_1}, \lambda_{n-r_2+1}, \ldots, \lambda_n),$$

$$A_2 = \begin{bmatrix} 1 & 0 \\ 0 & -\mathbf{1}_{r_1} \\ 0 & \mathbf{1}_{r_2} \end{bmatrix}, \quad C_2 = \sigma^2 \begin{bmatrix} \kappa^2 & 0 \\ 0 & \eta^2 I_r \end{bmatrix}, \boldsymbol{\theta}_2 = \begin{pmatrix} \mu_0 \\ \lambda_0 \end{pmatrix}.$$

(b) Location-shift model; $|\lambda_i|$'s equal

Here, we consider, as an example, the model of Abraham and Box (1978) extended to include upper and lower outliers, with the same prior specification for μ as in the previous model and with the assumption that each outlier is normally distributed with positive mean value λ_0 and variance $\eta^2\sigma^2$. In the notation of (4.4.6), we then have

$$A_1 = \left[I_n \begin{array}{c} -\mathbf{1}_{r_1} \\ 0 \\ \mathbf{1}_{r_2} \end{array} \right], \qquad C_1 = \sigma^2 I_n, \qquad \boldsymbol{\theta}_1^{\mathrm{T}} = (\mu, \lambda)$$

$$A_2 = I_2, \qquad C_2 = \sigma^2 \begin{bmatrix} \kappa^2 & 0 \\ 0 & \eta^2 \end{bmatrix}, \qquad \boldsymbol{\theta}_2 = \begin{bmatrix} \mu_0 \\ \lambda_0 \end{bmatrix}.$$

(c) Inflated-variance model

With the same prior specification for μ as in the previous models, the inflated-variance model of Box and Tiao (1968) corresponds, in the notation of (4.4.6), to

$$A_1 = 1_n, \qquad C_1 = \sigma^2 \begin{bmatrix} \lambda^2 I_{r_1} & 0 & 0 \\ 0 & I_{n-r} & 0 \\ 0 & 0 & \lambda^2 I_{n-r_2} \end{bmatrix}, \qquad \theta_1 = \mu$$

$$A_2 = 1, \qquad C_2 = \kappa^2 \sigma^2, \qquad \theta_2 = \mu_0.$$

If we impose no constraints on the possible number of outliers in models such as these the number of possible identifications of subsets of $x_1, \ldots, x_n$ as outliers may prove computationally prohibitive. In practice, however, it might be more realistic to allow for up to only 10 or 20 per cent of the sample to be outliers and therefore to limit the number of possible identifications.

If we denote the class of possible identifications of subsets of observations as outliers that we wish to consider by J_s, $s = 1, \ldots, M < 2^n$, then overall inference for unknown parameters is based on

$$p(\psi|\mathbf{x}) = \sum_{s=1}^{M} p(\psi|\mathbf{x}, J_s) p(J_s|\mathbf{x}). \tag{4.4.7}$$

In this mixture posterior form, $p(\psi|\mathbf{x}, J_s)$ represents the inference to be made about ψ, were it to be assumed that a particular subset of the observations (as defined by J_s) are outliers; the weight factor $p(J_s|\mathbf{x})$ represents the posterior plausibility of the assumed identification J_s. If k_r denotes the set of J_s which specify a total of r outliers, then

$$p(r \text{ outliers}|\mathbf{x}) = \sum_{J_s \in k_r} p(J_s|\mathbf{x}), \qquad r = 0, 1, \ldots, R, \tag{4.4.8}$$

forms the basis for a Bayesian assessment of the existence and number of outliers (assuming an upper bound of R). Plots of joint or marginal densities based on $p(\psi|\mathbf{x}, J_s)$ for various s provide insights into the sensitivity of conclusions to assumptions about numbers of outliers.

The detailed forms of $p(\psi|\mathbf{x}, J_s)$ and $p(J_s|\mathbf{x})$ are easily derived using general results for (4.4.6) given by Lindley and Smith (1972). If C_1, C_2 are known, the posterior probability for an J_s specified in terms of appropriate choices of A_i, $\boldsymbol{\theta}_i$, C_i, $i = 1, 2$, is proportional to the prior probability multiplied by

$$|C_1 + A_1 C_2 A_1^{\mathrm{T}}|^{-1/2} \exp[-\tfrac{1}{2}(\mathbf{x} - A_1 A_2 \boldsymbol{\theta}_2)^{\mathrm{T}} (C_1 + A_1 C_2 A_1^{\mathrm{T}})^{-1} (\mathbf{x} - A_1 A_2 \boldsymbol{\theta}_2)]. \tag{4.4.9}$$

Alternatively, if, conditional on σ^2, $C_1 = \sigma^2 I_n$, $C_2 = \sigma^2 V$, with V known, and the prior for the unknown σ^2 is specified in the form $\nu\lambda/\sigma^2 \sim \chi^2_\nu$, (4.4.9) is replaced by

$$|I_n + A_1 V A_1^{\mathrm{T}}|^{-1/2} [\nu\lambda + (\mathbf{x} - A_1 A_2 \boldsymbol{\theta}_2)^{\mathrm{T}} (I_n + A_1 V A_1^{\mathrm{T}})^{-1} (\mathbf{x} - A_1 A_2 \boldsymbol{\theta}_2)]^{-(n+\nu)/2}. \tag{4.4.10}$$

If interest centres on inference for $\boldsymbol{\theta}_1$, we note (see, for example, Lindley and Smith, 1972) that the distribution of $\boldsymbol{\theta}_1$, given $\mathbf{x}$, C_1, C_2, is $N(Bb, B)$, where

$$B^{-1} = A_1^{\mathrm{T}} C_1^{-1} A_1 + C_2^{-1}, \qquad \mathbf{b} = A_1^{\mathrm{T}} C_1^{-1} \mathbf{x} + C_2^{-1} A_2 \boldsymbol{\theta}_2.$$

Under the alternative specification for unknown σ^2 given above, the posterior distribution for $\boldsymbol{\theta}_1$ is Student-t with degrees of freedom $n+v$ and mean and dispersion matrix given by

$$(A_1^{\mathrm{T}}A_1+V^{-1})^{-1}(A_1\mathbf{x}+V^{-1}A_2\boldsymbol{\theta}_2) \qquad \text{and} \qquad (A_1^{\mathrm{T}}A_1+V^{-1})^{-1},$$

respectively.

Overall inference for $\boldsymbol{\theta}_1$ thus has the form of a weighted average of such normal or t-densities, with weights given by (4.4.9) or (4.4.10).

Example 4.4.1 Outlier analysis of Darwin's data

Abraham and Box (1978) present an analysis of Darwin's data on the differences in heights between fifteen pairs of self-fertilized and cross-fertilized plants grown in the same conditions. The ordered values are:

$$-67,\ -48,\ 6,\ 8,\ 14,\ 16,\ 23,\ 24,\ 28,\ 29,\ 41,\ 49,\ 56,\ 60,\ 75.$$

Using a simple location-shift model and assuming that $\pi=0.95$ in (4.4.5), Abraham and Box calculate $p(J_s|\mathbf{x})$ for the set of n identifications which specify for each observation in turn that it is the only outlier. They thus obtain the set of posterior weights

$$w_i = p(x_i \text{ is the outlier}|\mathbf{x}, 1 \text{ outlier}),$$

as shown in Figure 4.4.1.

Figure 4.4.2 shows plots of various posterior densities for μ (the location parameter for the differenced data) corresponding to different assumptions about values of π and numbers of outliers. Curve A is the density obtained when all the observations are assumed to be from the normal component $\phi(x|\mu,\sigma)$; curve B is the density obtained by forming a suitable weighted combination of the $p(\mu|\mathbf{x},J_s)$ densities assuming exactly two outliers, so that

$$p_{\mathrm{B}}(\mu|\mathbf{x}) = \sum_{J_s\in k_2} p(\mu|\mathbf{x},J_s)p(J_s|\mathbf{x}) \Big/ \sum_{J_s\in k_2} p(J_s|\mathbf{x}).$$

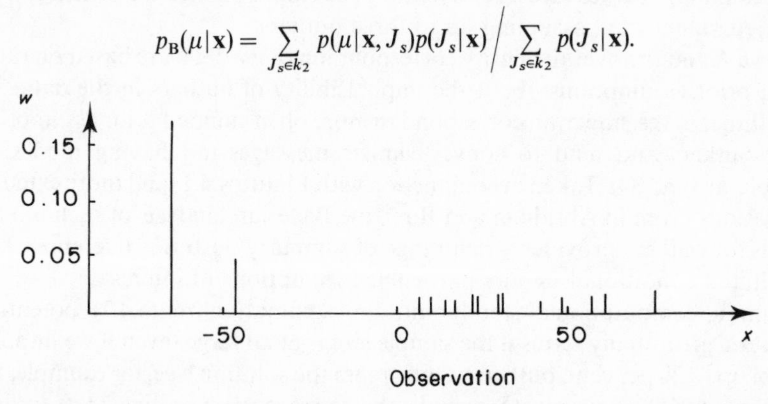

Figure 4.4.1 Weights w for the plant heights data corresponding to $\pi=0.95$. Reproduced by permission of the Royal Statistical Society from Abraham and Box (1978)

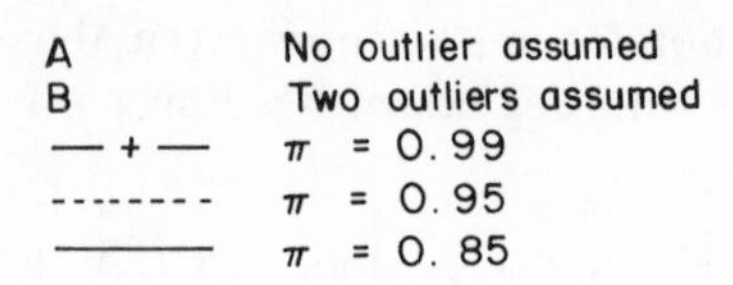

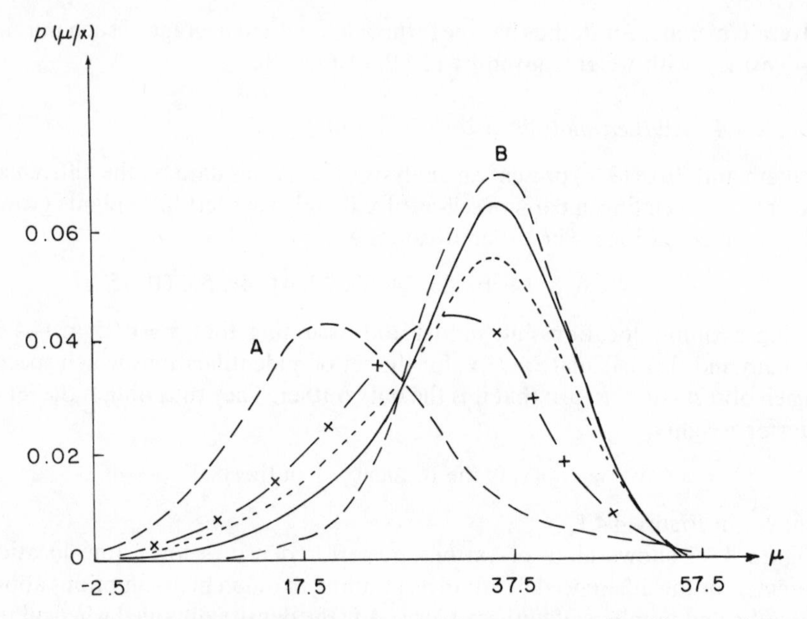

Figure 4.4.2 Fitted densities for plant height data. Reproduced by permission of the Royal Statistical Society from Abraham and Box (1978)

The remaining curves are the overall posterior densities obtained for the indicated values of π, assuming up to three outliers.

Curve A and the overall density corresponding to $\pi = 0.99$ are based on rather strong prior assumptions about the implausibility of outliers in the data. The other three curves, however, correspond to more open-minded prior assumptions about outliers and tend to convey similar messages (all having modes, for example, around 34). Taken in conjunction with Figure 4.4.1 (and more extensive calculations given in Abraham and Box), the Bayesian analysis of such mixture models for outliers provides a rich range of summary posterior inferences, both overall and conditional on any particular assumptions of interest.

From a computational standpoint, the summation in (4.4.7) potentially involves a great many terms if the sample size is at all large (even if we impose a limit of up to 20 per cent. outliers). One pragmatic solution (see, for example, Box and Tiao, 1968) is simply to include the terms corresponding to 0 outliers, 1 outlier, ..., etc., up to the point where it is computationally convenient to stop. Pettit and Smith (1984, 1985) have argued that such an approach could be made more efficient by considering only those J_s which identify as 'outliers' the observations which actually do appear to 'outlie'. They discuss a possible justification for this approximation in terms of the form of prior density assigned to λ in (4.4.5).

4.4.3 Bayesian cluster analysis

In Section 4.3 we mentioned the use of mixture models in cluster analysis, which seeks to partition a set of n observations, $\mathbf{x} = (x_1, \ldots, x_n)$, into a set of k mutually exclusive groups, where k may or may not be prescribed. Bayesian methods have also been applied in this context.

Wolfe (1970) bases his approach to cluster analysis on the mixture density itself.

Binder (1978a) constructs the following general formulation in which a prior density is specified jointly for the number of clusters, k, the grouping of the data, and the parameters, $\boldsymbol{\theta}$, of the component densities of individual observations within the k categories. If we represent any particular grouping of the data by an appropriate choice of the corresponding set of n indicator vectors $\{\mathbf{z}_i\} = Z_n$, the prior specification is of the form

$$p(k, Z_n, \boldsymbol{\theta}) = p(k)p(Z_n|k)p(\boldsymbol{\theta}|k, Z_n).$$

This combines with the likelihood

$$\prod_{i=1}^{n} \prod_{j=1}^{k} [f_j(x_i|\boldsymbol{\theta}_j)]^{z_{ij}}$$

to give the posterior density

$$p(k, Z_n, \boldsymbol{\theta}|\mathbf{x}),$$

which will, of course, be zero unless Z_n corresponds to a grouping into k clusters.

The marginal posterior probabilities of particular groupings,

$$p(Z_n|\mathbf{x}) = \sum_k \int p(k, Z_n, \boldsymbol{\theta}|\mathbf{x})\, \mathrm{d}\boldsymbol{\theta}, \tag{4.4.11}$$

then provide the basis for choosing a 'best' clustering of the data. Binder (1978a) discusses the solution of this decision problem for various loss functions based on different measures of the 'closeness' of the chosen grouping to the true grouping. The simplest such loss function takes the value zero or one, depending on whether the true and chosen groupings match perfectly or not. The resulting choice of clustering corresponds to that Z_n which gives the mode of (4.4.11). In some special cases this reproduces familiar clustering criteria. Suppose, for example, that k is specified, so that the summation disappears from (4.4.11), and that the component densities are multivariate normal with the same covariance matrix. If suitable priors are chosen for $\boldsymbol{\theta}$ and Z_n, the optimal clustering turns out to be that for which $|W|$ is minimized, where, as in Section 4.3, W is the within-groups sample dispersion matrix. The 'suitable prior' for Z_n is

$$p(Z_n) \propto \prod_{j=1}^{k} n_j^{d/2},$$

where d is the dimensionality of the individual data points and n_j is the number of data points put into cluster j, according to Z_n.

For details of alternative, more complicated, loss functions, see Binder (1978a).

4.5 MINIMUM DISTANCE ESTIMATION BASED ON DISTRIBUTION FUNCTIONS

4.5.1 Introduction to distance measures

A variety of procedures for estimating a probability density function or a distribution function can be subsumed under the heading of 'minimum distance estimators'. Suppose $F(\cdot\,|\,\psi)$ is the distribution function of interest, with ψ the unknown parameters, and let $F_n(\cdot)$ denote the empirical distribution function obtained from a sample of n independent observations, $\mathbf{x}$. Also, let

$$\delta(G, F)$$

be a measure of distance between two distribution functions, F and G.

Then a minimum distance estimator, $\hat{\psi}$, for ψ is a value of ψ which minimizes

$$\delta[F_n(\cdot), F(\cdot\,|\,\psi)].$$

If this distance is denoted by $\delta(\psi)$ for convenience, typically $\hat{\psi}$ will be a stationary point of $\delta(\psi)$ satisfying

$$\mathbf{D}_{\psi}\delta(\hat{\psi}) = \mathbf{0}, \tag{4.5.1}$$

This is, of course, reminiscent of the maximum likelihood solution and we shall be able to obtain maximum likelihood as a special case of the distance measure approach. As in Section 4.3, the key to deriving the asymptotic properties of $\hat{\psi}$ will be a linear Taylor expansion of (4.5.1) about the true ψ_0, subject, of course, to the usual kinds of regularity conditions. Approximately, for $\hat{\psi}$ 'near' ψ_0,

$$\mathbf{D}_{\psi_0}\delta(\psi_0) + D^2_{\psi_0}\delta(\psi_0)(\hat{\psi} - \psi_0) = 0. \tag{4.5.2}$$

Often this will lead to the asymptotic result that

$$\hat{\psi} \to N(\psi_0, V(\psi_0)),$$

in distribution, as $n \to \infty$, where

$$V(\psi_0) = [\mathbb{E}D^2_{\psi_0}\delta(\psi_0)]^{-1} \operatorname{cov}[D_{\psi_0}\delta(\psi_0)][\mathbb{E}D^2_{\psi_0}\delta(\psi_0)]^{-1}.$$

Furthermore, as we shall see later, this motivates invoking versions of Newton–Raphson (NR) or the Method of Scoring (MS) algorithms for calculating $\hat{\psi}$ when numerical methods are required.

Although much of our introductory discussion will be in terms of continuous univariate distributions on the real line, multivariate and discrete distributions may also be dealt with. In the latter case, F represents a discrete probability measure.

There is clearly a wide range of estimators that can be derived by this method, depending on the choice of the distance measure $\delta(\cdot,\cdot)$; see Parr (1981) for a recent bibliography. The word 'distance' will be used rather loosely in that $\delta(\cdot,\cdot)$ may or may not satisfy the formal properties of a metric. The triangle inequality is not important in the present context, nor is it vital that $\delta(\cdot,\cdot)$ be symmetric in its two arguments.

We do demand, however, that

$$\delta(G, F) \geqslant \delta(G, G), \qquad \text{for any } F, G,$$

with equality only if $F(x) = G(x)$ almost everywhere. Usually, $\delta(G, G) = 0$ for all G.

Under these circumstances, consistency of the estimator is assured provided certain regularity conditions hold and

(a) the family $F(\cdot \mid \psi)$ is identifiable;
(b) $\{F_n(\cdot)\}$ is consistent for the true distribution function.

This latter is generally true when, as suggested above, $F_n(\cdot)$ is the empirical distribution function, as is the case in most of the published work on this topic. There is, however, considerable scope for investigating other consistent non-parametric estimators, such as kernel-based density estimators, in the role of $F_n(\cdot)$.

There have been a number of applications of minimum distance estimators to mixtures and Table 4.5.1 displays some of the distance measures that have been suggested. Versions are given for continuous and discrete sample spaces, where appropriate, and the subscript nomenclature is an attempt to provide a useful *aide-memoire*. Several points should be emphasized:

(a) Although some of the measures are metrics, this is not true, for example, of the Kullback–Leibler (KL), Levy (L), chi-squared (C), modified chi-squared (MC), and averaged L_2-norm measures.
(b) Some of the measures can be regarded as special cases of others: for instance, δ_C is a special version of δ_{WL}.
(c) $\delta_C(F, G) = \delta_{MC}(G, F)$.
(d) $\delta_{KL}\{F_n(\cdot), F(\cdot \mid \psi)\}$ is equivalent to the maximum likelihood criterion, so far as the implied estimate of ψ is concerned.
(e) In the 'weighted' measures, non-negative weight functions $w(\cdot)$ are used.
(f) In the versions for discrete distributions, $(\mathbf{p}, \mathbf{q})$ are the sets of probabilities associated with the measures (F, G). In such cases we shall write, for instance, $\delta(\mathbf{p}, \mathbf{q})$. The vector $\mathbf{r}$ will denote the corresponding quantity for $F_n(\cdot)$; i.e. the vector of relative frequencies.
(g) In the discrete versions of measures such as δ_{LA}, δ_{WLA}, and δ_S, based on

Table 4.5.1 Some distance measures

	Continuous sample space	Discrete sample space
L_2-norm with distribution functions		
$\delta_{LA}(F, G)$	$\int [F(x) - G(x)]^2 \, dx$	$\sum_{i=1} \left(\sum_{j=1}^{i} p_j - \sum_{j=1}^{i} q_j \right)^2$
L_2-norm with densities		
$\delta_{LB}(F, G)$	$\int [f(x) - g(x)]^2 \, dx$	$\sum_{i=1} (p_i - q_i)^2$
Weighted L_2-norms		
$\delta_{WLA}(F, G)$	$\int [F(x) - G(x)]^2 w(x) \, dx$	$\sum_{i=1} \left[\sum_{j=1}^{i} (p_j - q_j) \right]^2 w_i$
$\delta_{WLB}(F, G)$	$\int [f(x) - g(x)]^2 w(x) \, dx$	$\sum_{i=1} (p_i - q_i)^2 w_i$
Averaged L_2-norms		
$\delta_{ALA}(F, G)$	$\int [F(x) - G(x)]^2 dF(x)$	$\sum_{i=1} \left[\sum_{j=1}^{i} (p_j - q_j) \right]^2 p_i$
$\delta_{ALB}(F, G)$	$\int [f(x) - g(x)]^2 \, dF(x)$	$\sum_{i=1} (p_i - q_i)^2 p_i$
Chi-squared		
$\delta_C(F, G)$	$\int \frac{[f(x) - g(x)]^2}{f(x)} \, dx$	$\sum_{i=1} (p_i - q_i)^2 / p_i$
Modified chi-squared		
$\delta_{MC}(F, G)$	$\int \frac{[f(x) - g(x)]^2}{g(x)} \, dx$	$\sum_{i=1} (p_i - q_i)^2 / q_i$
Sup-norm		
$\delta_S(F, G)$	$\sup_x \lvert F(x) - G(x) \rvert$	$\sup_i \left\lvert \sum_{j=1}^{i} (p_j - q_j) \right\rvert$
Wolfowitz distance		
$\delta_W(F, G)$	$\int \lvert F(x) - G(x) \rvert \, dx$	$\sum_i \left\lvert \sum_{j=1}^{i} (p_j - q_j) \right\rvert$
Hellinger distance (Beran, 1977)		
$\delta_H(F, G)$	$\int [\sqrt{f(x)} - \sqrt{g(x)}]^2 \, dx$	$\sum_{i=1} (\sqrt{p_i} - \sqrt{q_i})^2$
Levy distance		
$\delta_L(F, G) = \inf_x \{\varepsilon: F(x - \varepsilon) - \varepsilon \leqslant G(x) \leqslant F(x + \varepsilon) + \varepsilon\}$		
Kullback–Leibler (Kullback and Leibler, 1951)		
$\delta_{KL}(F, G)$	$\int \log [dF(x)/dG(x)] \, dF(x)$	$\sum_{i=1} p_i \log (p_i/q_i)$

distribution functions, there is an implicit ordering of the points in the sample space. In many applications this will be quite natural, particularly if the data consist of grouped univariate data from a continuous sample space. If no such ordering is natural, however, it may not make sense to use these measures.

(h) Problems arise with $\delta_{KL}(\mathbf{p}, \mathbf{q})$ if, for example, $p_i \neq 0$ but $q_i = 0$. The same is true for δ_C and δ_{MC}. Preliminary smoothing of the relative frequencies is helpful in this context.

(i) Formally, the versions of *density*-based distance measures in examples with continuous sample spaces may not be valid because of the lack of differentiability of $F_n(\cdot)$. Again, preliminary smoothing of $F_n(\cdot)$ gets round the problem.

Choice of distance measure may be quite important. In view of its relation with maximum likelihood, δ_{KL} is clearly the preferred option for likelihood adherents. For problems where it is important to do well in the tails, δ_C and δ_{MC} are favoured. From an asymptotic point of view, the different measures tend to have the same characteristic as δ_{KL} in that they produce consistent and asymptotically normal estimators; see, for instance, Wolfowitz (1957), Rao (1965), Choi (1969a), and Macdonald and Pitcher (1979). In practical terms, the small-sample behaviour of the estimator and the degree of difficulty of the associated computations may be the most important considerations in the choice of measure. Since small-sample behaviour will vary from application to application, computational feasibility is perhaps the dominant question. For instance, this seems to rule out almost entirely the use of the Levy measure, δ_L, although Yakowitz (1969) has suggested a possible algorithm based on this approach for obtaining consistent estimators of the mixing weights of a finite mixture (see also Fisher and Yakowitz, 1970).

At the opposite end of the scale of difficulty is the estimation of mixing weights using *quadratic* distance measures. The simplicity of the measures, in conjunction with the linearity, in $\boldsymbol{\pi}$, of $p(x \mid \boldsymbol{\psi})$ and $F(x \mid \boldsymbol{\psi})$ accounts for the comparative ease of solution, as indicated in the next subsection.

4.5.2 Estimation of mixing weights based on quadratic distances

Example 4.5.1 Mixtures of known component densities

As an example, consider the distance measure adopted by Choi and Bulgren (1968):

$$\delta_{ALA}[F_n(\cdot), F(\cdot \mid \boldsymbol{\psi})] = n^{-1} \sum_{i=1}^{n} [F(x_i \mid \boldsymbol{\psi}) - i/n]^2$$

$$= n^{-1} \sum_{i=1}^{n} \left[\sum_{j=1}^{k} \pi_j F_j(x_i) - i/n \right]^2. \tag{4.5.3}$$

There are two general ways of writing (4.5.3):

(a)
$$n^{-1}(B\boldsymbol{\pi} - \mathbf{d})^{\mathrm{T}} V (B\boldsymbol{\pi} - \mathbf{d}), \tag{4.5.4}$$

where, without loss of generality, $x_1 \leqslant \cdots \leqslant x_n$,

$$B_{ij} = F_j(x_i), \qquad i = 1, \ldots, n, j = 1, \ldots, k,$$

$$V = \text{the } n \times n \text{ identity matrix,}$$

and
$$d_i = i/n, i = 1, \ldots, n.$$

(b)
$$n^{-1}(\boldsymbol{\pi}^{\mathrm{T}} A \boldsymbol{\pi} - 2\boldsymbol{\pi}^{\mathrm{T}}\mathbf{b} + c), \tag{4.5.5}$$

where $A = B^{\mathrm{T}}VB$, $\mathbf{b} = B^{\mathrm{T}}V\mathbf{d}$, and $c = \mathbf{d}^{\mathrm{T}}V\mathbf{d}$.

It can be shown that A is positive definite (Hardy, Littlewood, and Polya, 1952, p. 16), so that the unique unconstrained minimum occurs at

$$\tilde{\boldsymbol{\pi}} = A^{-1}\mathbf{b}.$$

The resulting function $\sum_{j=1}^{k} \tilde{\pi}_j F_j(x)$ $(x \in \mathscr{X})$ will, subject to sufficient regularity conditions, provide a (consistent, as $n \to \infty$) approximation to $F(\cdot|\boldsymbol{\psi})$, but it may not itself, for any given sample, be a distribution function.

To guarantee the latter, we need to minimize (4.5.4) subject to

$$\pi_1 + \cdots + \pi_k = 1,$$
$$\pi_1 \geqslant 0, \ldots, \pi_k \geqslant 0. \tag{4.5.6}$$

This represent a *quadratic programming* problem and terminating simplex algorithms are available for its solution (see, for instance, Walsh, 1975, Chapter 2). If we are prepared to compromise and risk the possibility of negative π_j's, while insisting on (4.5.6), we may again write down the solution explicitly as

$$\hat{\boldsymbol{\pi}} = A^{-1}\mathbf{b} + A^{-1}\mathbf{1}(1 - \mathbf{1}^{\mathrm{T}}A^{-1}\mathbf{b})/\mathbf{1}^{\mathrm{T}}A^{-1}\mathbf{1}, \tag{4.5.7}$$

where $\mathbf{1}^{\mathrm{T}}$ is a vector of ones and (4.5.6) has been incorporated using a Lagrange multiplier.

This solution does represent a constrained minimum (see, for instance, Fletcher, 1971, Macdonald, 1975, and Macdonald and Pitcher, 1979). Alternatively, π_k may be eliminated using (4.5.6) and the resulting quadratic function of $(\pi_1, \ldots, \pi_{k-1}) = \boldsymbol{\eta}^{\mathrm{T}}$ minimized, without further constraints, at

$$\hat{\boldsymbol{\eta}} = A^{*-1}\mathbf{b}^*,$$

where
$$(A^*)_{jl} = \sum_i [F_j(x_i) - F_k(x_i)][F_l(x_i) - F_k(x_i)], \qquad j, l = 1, \ldots, k-1,$$

$$b_j^* = \sum_i [F_j(x_i) - F_k(x_i)][i/n - F_k(x_i)], \qquad j = 1, \ldots, k-1.$$

It is easy to check that this solution is equivalent to (4.5.7).

Several other distance measures lead to similar quadratic criteria. For instance, $\delta_{\mathrm{WLB}}(F_n, F)$ can be written in the form

$$\delta_{\mathrm{WLB}}(F_n, F) = \int \frac{[\mathrm{d}F_n(x) - \mathrm{d}F(x|\boldsymbol{\psi})]^2}{\mathrm{d}H(x)};$$

see Bartlett and Macdonald (1968).

Although, as remarked by Macdonald (1975), the right-hand side is infinite if $H(\cdot)$ is differentiable, formal minimization shows that $\hat{\boldsymbol{\psi}}$ would satisfy the vector equation

$$n^{-1}\sum_{i=1}^{n} w(x_i)\frac{\mathrm{d}p(x_i\,|\,\boldsymbol{\psi})}{\mathrm{d}\boldsymbol{\psi}} = \int w(x)\frac{\mathrm{d}p(x\,|\,\boldsymbol{\psi})}{\mathrm{d}\boldsymbol{\psi}}\,\mathrm{d}x, \tag{4.5.8}$$

where $w(x) = [\mathrm{d}H(x)/\mathrm{d}x]^{-1}$.

In our example, with $\boldsymbol{\psi} = \boldsymbol{\eta}$, we have

$$\frac{\partial p(x\,|\,\boldsymbol{\eta})}{\partial \eta_j} = f_j(x) - f_k(x), \qquad j = 1, \ldots, k-1, \text{ for all } x.$$

Equations (4.5.8) may therefore be written

$$\mathbf{T} = J\boldsymbol{\eta} + \mathbf{R},$$

where

$$T_j = n^{-1}\sum_{i=1}^{n} w(x_i)[f_j(x_i) - f_k(x_i)], \qquad j = 1, \ldots, k-1,$$

$$J_{jl} = \int [f_j(x) - f_k(x)][f_l(x) - f_k(x)]w(x)\mathrm{d}x, \qquad j, l = 1, \ldots, k-1,$$

and

$$R_j = \int [f_j(x) - f_k(x)]f_k(x)w(x)\,\mathrm{d}x, \qquad j = 1, \ldots, k-1.$$

Thus

$$\hat{\boldsymbol{\eta}} = J^{-1}(\mathbf{T} - \mathbf{R}).$$

It is easily shown (Macdonald, 1975) that $\hat{\boldsymbol{\eta}}$ is unbiased and

$$\mathrm{cov}\,(\hat{\boldsymbol{\eta}}) = n^{-1}[J^{-1}HJ^{-1} - (\boldsymbol{\eta} + J^{-1}\mathbf{R})(\boldsymbol{\eta} + J^{-1}\mathbf{R})^{\mathrm{T}}],$$

where

$$H_{jl} = \int [f_j(x) - f_k(x)][f_l(x) - f_k(x)]w^2(x)p(x\,|\,\boldsymbol{\eta})\mathrm{d}x, \qquad j, l = 1, \ldots, k-1.$$

The discrete version of this distance measure is better defined in the form

$$\delta_{\mathrm{WLB}}(\mathbf{p}(\boldsymbol{\psi}), \mathbf{r}),$$

where $\quad p_i(\boldsymbol{\psi}) = \sum_{j=1}^{k} \pi_j f_{ij}(\boldsymbol{\theta}_j), \qquad i = 1, 2, \ldots,$

say. If only $\boldsymbol{\pi}$ is unknown, the criterion takes the form (4.5.4) with

$$B_{ij} = f_{ij}, \qquad i = 1, 2, \ldots, j = 1, \ldots, k,$$
$$V = \mathrm{diag}\,(w_1, w_2, \ldots)$$

and

$$d_j = r_i, \qquad i = 1, 2, \ldots;$$

see Macdonald, 1975.

In practice, the sample space will be made finite, possibly as a result of grouping, before evaluating relative frequencies and weights.

The attraction of being able to obtain explicit results might lead one in some applications to prefer the *modified chi-squared* criterion $\delta_{MC}(\mathbf{p}(\boldsymbol{\pi}), \mathbf{r})$ to the *chi-squared* criterion $\delta_C(\mathbf{p}(\boldsymbol{\pi}), \mathbf{r})$, particularly in view of the asymptotic equivalence of the two resulting estimators (Rao, 1965, Chapter 5).

Whereas the distance measures discussed above led to quadratic programming problems, if we use the sup-norm $\delta_S(\cdot,\cdot)$ associated with the Kolmogorov–Smirnov test we are led to a *linear programming* problem.

Suppose the sample space, $\mathscr{X}$, is continuous. Then, for any n, $\sup_x |F(x|\boldsymbol{\pi}) - F_n(x)|$ is attained at one of the data points $x_1, \ldots, x_n$. As before, we assume $x_1 \leqslant \cdots \leqslant x_n$. Our objective is, therefore, to minimize π_0 subject to

$$\left| \sum_{j=1}^{k} \pi_j F_j(x_i) - i/n \right| \leqslant \pi_0, \qquad i = 1, \ldots, n,$$

$$\left| \sum_{j=1}^{k} \pi_j F(x_i) - (i-1)/n \right| \leqslant \pi_0, \qquad i = 1, \ldots, n,$$

$$\pi_1 + \cdots + \pi_k = 1, \quad \pi_0, \pi_1, \ldots, \pi_k \geqslant 0.$$

Since all the above constraints can be written as ordinary inequalities, linear in $\pi_0, \pi_1, \ldots, \pi_k$, the linear programming interpretation is clear. Using a generalization of this, involving a sequence of such linear programmes, Deely and Kruse (1968) developed a method for estimating a general mixing distribution. Application of their approach is extended by Blum and Susarla (1977).

The following modification of the procedures considered so far in this section, which is due to Hall (1981), actually *requires* data from all the component distributions as well as form the mixture. The purpose is to estimate the mixing weights alone, with no interest in the component distributions and without the need to specify parametric models for them.

Example 4.5.2 Mixture of k unknown distributions

Suppose data sets of sizes $n, n_1, \ldots, n_k$ are available from the mixture and the k components, respectively, yielding empirical distributions $F_n(\cdot)$, $F_{n_1}^{(1)}(\cdot), \ldots, F_{n_k}^{(k)}(\cdot)$. Then the mixing weights, $\boldsymbol{\pi}$, are estimated by minimizing

$$\delta\left(F_n(\cdot), \sum_{j=1}^{k} \pi_j F_{n_j}^{(j)}(\cdot) \right),$$

for some distance measure $\delta(\cdot,\cdot)$. Hall (1981) considers the case of univariate continuous data with the quadratic-based distance measures δ_{LA} and δ_{WLA}. As a result, the calculation of the optimal $\boldsymbol{\pi}$ can, of course, be carried out exactly. If, as earlier, we let

$$\boldsymbol{\eta}^T = (\pi_1, \ldots, \pi_{k-1}),$$

if

$$J_{jl} = \int [F_{n_j}(x) - F_{n_k}(x)][F_{n_l}(x) - F_{n_k}(x)]w(x)\,dx, \qquad j, l = 1, \ldots, k-1,$$

and if

$$e_j = \int [F_n(x) - F_{n_k}(x)][F_{n_j}(x) - F_{n_k}(x)]w(x)\,dx, \qquad j = 1, \ldots, k-1,$$

then

$$J\hat{\boldsymbol{\eta}} = \mathbf{e}.$$

If the mixture is identifiable then, asymptotically, J^{-1} exists so $\hat{\boldsymbol{\eta}}$ is uniquely defined. For the case of uniform $w(\cdot)$, and assuming data in the form of Hosmer's (1973a) model M2, Hall (1981) proves consistency and asymptotic normality and derives the asymptotic covariance structure for $\boldsymbol{\eta}$.

If empirical versions of density functions are to be used, as in δ_{WLB}, for instance, and the data are not discrete, smoothed density estimates such as those based on kernel functions will be required. Titterington (1983) considers the use of δ_{LB}, δ_{WLB}, and δ_{ALB} with unsmoothed and smoothed discrete data and smoothed continuous data.

4.5.3 Problems with non-explicit estimators

The success of the treatments of Example 4.5.1 discussed above was founded on the *linearity* of the mixture density or distribution function in terms of the unknown parameters, and the linear or quadratic nature of the criterion to be minimized. However, most finite mixtures give rise to densities which are highly non-linear in the unknown parameters, so that explicit results of any kind are rarely available and numerical optimization will normally be essential. Usually, the Newton–Raphson method or the Method of Scoring has been used to solve the associated stationarity equations. The case of $\delta_{\mathrm{KL}}[F_n(\cdot), F(\cdot\,|\,\boldsymbol{\psi})]$ has effectively already been dealt with in our discussion of maximum likelihood in Section 4.3. As in (4.5.2), write

$$\mathbf{D}\delta(\boldsymbol{\psi}) = \frac{\partial}{\partial \boldsymbol{\psi}} \delta[F_n(\cdot), F(\cdot\,|\,\boldsymbol{\psi})]$$

and let $D^2\delta(\boldsymbol{\psi})$ denote the matrix of second derivatives. Then, having chosen initial approximations, $\boldsymbol{\psi}^{(0)}$, we generate $\{\boldsymbol{\psi}^{(r)}\}$ according to

$$\boldsymbol{\psi}^{(r+1)} = \boldsymbol{\psi}^{(r)} - \alpha_r[D^2\delta(\boldsymbol{\psi}^{(r)})]^{-1}\mathbf{D}\delta(\boldsymbol{\psi}^{(r)}), \qquad r = 0, 1, \ldots, \tag{4.5.9}$$

where α_r *may* be taken to be 1, for all r.

By analogy with the Method of Scoring, we might replace the matrix of second derivatives by its expected value, conditional on $\boldsymbol{\psi} = \boldsymbol{\psi}^{(r)}$.

A particular version of (4.5.9) is the program ROKE described by I. Clark (1977), in which grouped data from a mixture of univariate normals, possibly with unequal variances, is assessed on the basis of the discrete version of δ_{LA}, that is

$\delta_{LA}(p(\psi), \mathbf{r})$. For the same type of mixture, a modified Newton–Raphson method is used by Mundry (1972) to minimize a slightly different criterion,

$$\delta_{M}(\mathbf{p}, \mathbf{r}) = \sum_{i=1} \left[\Phi^{-1}\left(\sum_{j=1}^{i} p_j \right) - \Phi^{-1}\left(\sum_{j=1}^{i} r_j \right) \right]^2,$$

where $\Phi^{-1}(\cdot)$ is the inverse of the standard normal distribution function.

For an extensive treatment of the practical aspects of decomposing normal mixtures using quadratic distance measures, see Macdonald and Pitcher (1979).

The distance measure

$$\delta_{ALA}[F_n(\cdot), F(\cdot \mid \psi)] = n^{-1} \sum_{i=1}^{n} [F(x_i \mid \psi) - i/n]^2, \tag{4.5.10}$$

with $x_1 \leqslant \cdots \leqslant x_n$, which was used in Example 4.5.1, was investigated in detail from a theoretical point of view by Choi (1969a). He established, under the usual sort of regularity conditions, several of the asymptotic properties that might be hoped for.

(a) With probability one, there is a neighbourhood of ψ_0, the true value of ψ, such that, for large enough n, the criterion has a unique minimizing point therein; $\hat{\psi}_n$, say.
(b) With probability one, $\hat{\psi}_n \rightarrow \psi_0$ as $n \rightarrow \infty$.
(c) Asymptotically, $\hat{\psi}_n$ is multivariate normal.

Furthermore, the asymptotic covariance matrix is derived. The method is illustrated by Choi (1969b) and Choi and Bulgren (1968) in the contexts of empirical Bayes and more general mixtures.

If the roles of $F_n(\cdot)$ and $F(\cdot \mid \psi)$ are reversed as arguments in the measure (4.5.10), then the criterion may be identified as the Cramér–von Mises statistic

$$n^{-1} \sum_{i=1}^{n} [F(x_i \mid \psi) - (i - \tfrac{1}{2})/n]^2 + (12n^2)^{-1}.$$

The only essential difference is the replacement of the term i/n in (4.5.10) by $(i - \frac{1}{2})/n$. This change does not affect the asymptotic results, nor the quadratic programming nature of the problem when only the mixing weights are unknown. As might be expected, however, it leads to reduction in the bias for small-sample cases in a similar way to the analogous modification of probability-plotting procedures (see, for instance, Barnett, 1975). Empirical evidence for this is indicated by Macdonald (1971); see also Macdonald (1969).

Woodward *et al.* (1984) match Choi's asymptotic theory, but for the Cramér–von Mises distance measure. They also compare the performance of this particular minimum distance method with that of maximum likelihood estimation in a simulation study, based on the assumption of a five-parameter two-component normal mixture. Special emphasis is placed on estimation of the mixing weight. When the assumed model is correct, maximum likelihood estimation is, not surprisingly, more efficient, but the minimum distance method is shown to be more robust when the actual underlying models are mixtures of heavy-tailed densities such as exponentials or student-t's. Woodward *et al.* (1984) make many other interesting numerical points.

The use of *minimum chi-squared* estimators is fairly common for the treatment of discrete or grouped data. For a given discretization of the sample space, the procedure leads to estimators that are asymptotically equivalent to maximum likelihood estimators for the discretized problem. These, of course, minimize

$$\delta_{KL}[\mathbf{r}, \mathbf{p}(\boldsymbol{\psi})] = \sum_{i=1} r_i \log [r_i/p_i(\boldsymbol{\psi})] = \text{constant} - \sum_{i=1} r_i \log p_i(\boldsymbol{\psi}). \quad (4.5.11)$$

If continuous data are grouped, the approach obviously leads to a loss of information. However, the asymptotic theory is more secure and the 'awkwardness' of singularities in the likelihood surface, which are a feature of, say, the mixture of two normal densities with unequal variances (Section 4.3), is done away with.

If, for convenience, we introduce the notation

$$X^2(\boldsymbol{\psi}) = \delta_C(\mathbf{p}(\boldsymbol{\psi}), \mathbf{r}),$$

then, as indicated above, in equation (4.5.9), the minimum chi-squared estimates may be calculated, from initial values $\boldsymbol{\psi}^{(0)}$, using

$$\boldsymbol{\psi}^{(r+1)} = \boldsymbol{\psi}^{(r)} - [D^2X^2(\boldsymbol{\psi}^{(r)})]^{-1}\mathbf{D}X^2(\boldsymbol{\psi}^{(r)}), \qquad r = 0, 1, \ldots, \quad (4.5.12)$$

where, in this case, we have set the step-length parameter to unity. The Taylor expansion of $X^2(\boldsymbol{\psi})$ is also the key to studying the asymptotic properties and, by this means, Fryer and Robertson (1972) derive the dominant terms of the biases and mean-squared errors of the estimators. In terms of mean-squared error, the method of minimum chi-squared and the associated 'grouped-data' maximum likelihood approach based on (4.5.11) are generally superior to the method of moments, at least for the data sets which Fryer and Robertson treat empirically. For these data sets, a mixture of two univariate normal densities with unequal variances is a suitable model. However, the pattern of superiority is not uniform, nor do these authors address the question of choosing the *grouping* of the data in an optimal manner. The method is also investigated, for this type of mixture, by Hasselblad (1966), and a modified version is applied to Poisson mixtures by Saleh (1981).

In multiparameter problems, of course, the use of the Newton–Raphson algorithm in its basic form is computationally very expensive, as mentioned in Section 4.3. The NR algorithm, however, produces second-order convergence. If, therefore $\boldsymbol{\psi}^{(0)}$, the initial approximation, is itself a reasonably good estimate, one might consider stopping the procedure after a *single* iteration and using $\boldsymbol{\psi}^{(1)}$ as the estimate. For instance, Blischke (1964) points out that, if (4.5.12) is used for the case of a mixture of binomials $\{\text{Bi}(N, \theta_j), j = 1, \ldots, k\}$ with N and k known, and if the moment estimator is used for $\boldsymbol{\psi}^{(0)}$, then $\boldsymbol{\psi}^{(1)}$ is a best asymptotically normal (BAN) estimator.

The same principle of a one-stage Newton–Raphson procedure is implicit in the work of Boes (1966, 1967), but with the log-likelihood or, equivalently, the Kullback–Leibler directed divergence measure, taking the place of $X^2(\psi)$. In fact, Boes suggests a one-stage method of scoring for estimating the mixing weights when the component densities are known. Suppose, for brevity, we write

$$p(x_i|\boldsymbol{\pi}) = \boldsymbol{\pi}^{\mathrm{T}}\mathbf{f}_i,$$

where $\mathbf{f}_i = [f_1(x_i), \ldots, f_k(x_i)]^{\mathrm{T}}, \quad i = 1, \ldots, n.$

As before, let

$$\boldsymbol{\eta}^{\mathrm{T}} = (\pi_1, \ldots, \pi_{k-1}),$$

and let

$$\{I(\boldsymbol{\eta})\}_{jl} = \int \frac{[f_j(x) - f_k(x)][f_l(x) - f_k(x)]}{p(x|\boldsymbol{\pi})} \mathrm{d}x$$

$$j = 1, \ldots, k-1, \; l = 1, \ldots, k-1$$

$$\{\mathbf{d}(\boldsymbol{\eta})\}_j = \sum_{i=1}^{n} \frac{f_j(x_i) - f_k(x_i)}{p(x_i|\boldsymbol{\eta})}, \qquad j = 1, \ldots, k-1.$$

Then $I(\boldsymbol{\eta})$ is the Fisher information matrix for one observation and the Method of Scoring gives

$$\boldsymbol{\eta}^{(1)} = \boldsymbol{\eta}^{(0)} + [nI(\boldsymbol{\eta}^{(0)}]^{-1}\mathbf{d}(\boldsymbol{\eta}^{(0)}).$$

There is considerable similarity between this estimator and (4.5.8), at least in its formal expression, the only essential difference being the replacement of

$$w(x) \quad \text{by} \quad [p(x|\boldsymbol{\eta}^{(0)})]^{-1}.$$

It is easy to show that

$$\mathbb{E}[\mathbf{d}(\boldsymbol{\eta}^{(0)})|\boldsymbol{\eta}] = nI(\boldsymbol{\eta}^{(0)})(\boldsymbol{\eta} - \boldsymbol{\eta}^{(0)}),$$

so that, whatever the true $\boldsymbol{\eta}$, $\boldsymbol{\eta}^{(1)}$ is unbiased for $\boldsymbol{\eta}$.

Furthermore, if the true value happens to be $\boldsymbol{\eta}^{(0)}$, the initial approximation, then

$$\operatorname{cov}(\boldsymbol{\eta}^{(1)}|\boldsymbol{\eta}^{(0)}) = [nI(\boldsymbol{\eta}^{(0)})]^{-1},$$

the multivariate Cramér–Rao lower bound. This is the motivation of Boes' terminology, that $\boldsymbol{\eta}^{(1)}$ or, equivalently, $\boldsymbol{\pi}^{(1)}$ is the $\boldsymbol{\pi}^{(0)}$*-efficient estimator* of $\boldsymbol{\pi}$. Boes (1966) looks in detail at the simple one-parameter case corresponding to $k = 2$.

In practice, of course, there is the problem of calculating the integrals implicit in

$$I(\boldsymbol{\eta}^{(0)}),$$

a difficulty that arises also in the use of (4.5.8). The natural suggestion is to use a sample-based version, such as

$$[\tilde{I}(\boldsymbol{\eta}^{(0)})]_{jl} = n^{-1} \sum_{i=1}^{n} \frac{[f_j(x_i) - f_k(x_i)][f_l(x_i) - f_k(x_i)]}{p^2(x_i \mid \boldsymbol{\eta}^{(0)})},$$
$$j, l = 1, \ldots, k-1.$$

Alternatively, numerical integration may be used; see Section 3.2.

The π_0-efficient estimators can be related to minimax unbiased estimators (Boes, 1967).

4.5.4 What to do with extra categorized data?

We briefly comment on the possibility of dealing with supplementary confirmed samples from the various components; i.e. with data models M1 and M2 of Chapter 1. Odell and Basu (1976, Section 4.2) treat a version of the least-squares criterion

$$\delta_{\mathrm{ALA}}(\cdot, \cdot)$$

for the multivariate case with M1 data. They use the confirmed data to provide estimates for the parameters $\boldsymbol{\theta}$. These are then substituted into the stationarity equations for the least-squares solutions from the mixture data from which estimates of $\boldsymbol{\pi}$ are then obtained. This does at least provide a best asymptotically normal procedure. In order to define a multivariate version of

$$\delta_{\mathrm{ALA}}[F_n(\cdot), F(\cdot(\boldsymbol{\psi}))],$$

it is necessary to impose an ordering on the multivariate sample space. Odell and Basu (1976) mention the possibility of doing this using the technique of *statistically equivalent blocks* (Anderson, 1966) but, for their application, they adopt the simpler, if less satisfactory, approach of evaluating δ_{ALA} for each component and minimizing the sum of these criteria. This has the attractive feature, again, of leading to stationarity equations which are linear in $\boldsymbol{\pi}$. The problem of ordering multivariate data suggests that distances based on densities themselves, or on transforms, are easier to generalize directly to the multivariate case.

A systematic approach to the incorporation of confirmed data has not yet been investigated thoroughly, apart, that is, from the case of the Kullback–Leibler directed divergence, which reduces, essentially, to maximum likelihood. In the latter case, the procedure is, of course, to *add* the criteria from the samples of the various types and optimize the resulting quantity. It is, perhaps, not obvious that this is the best procedure to adopt in all cases, no matter which distance measure is chosen. One might, alternatively, take a weighted rather than a simple sum, or even choose a different distance measure for the different samples, although this would undoubtedly be unnecessarily complicated. The simple additive procedure will be feasible for M1 data and, by the usual arguments, asymptotic theory can be developed. For M2 data, in which the confirmed data add information about the mixing weights $\boldsymbol{\pi}$ as well as $\boldsymbol{\theta}$, the generalization is not quite so straightforward, although methods can be devised based on the joint distribution, or a transform thereof, of the categorical and feature variables for a given observation.

4.6 MINIMUM DISTANCE ESTIMATORS BASED ON TRANSFORMS

4.6.1 Introduction

The approach of Section 4.5 was to choose parameter estimates so that the estimated mixture distribution was as close as possible to the data, using a distance measure between distribution functions or density functions. In this section, we use distance measure between theoretical and empirical *transforms* of the distribution function. Suppose, for some auxiliary variable $t \in \mathscr{T}$,

$$G(t|\boldsymbol{\psi}) = \mathbb{E}g(t, X) = \int g(t, x)\,\mathrm{d}F(x|\boldsymbol{\psi}),$$

provided the integral exists. If the sample space is discrete, we interpret the integral as a sum and, if x is multivariate, t may have to be vector valued. If, also, $\mathbb{E}g^2(t, X)$ is finite, for all $t \in \mathscr{T}$, if $X_1, \ldots, X_n$ represents a random sample from $F(\cdot|\boldsymbol{\psi})$ and if we define

$$\bar{g}_n(t) = n^{-1} \sum_{i=1}^{n} g(t, X_i),$$

then, by the law of large numbers,

$$\bar{g}_n(t) \to G(t|\boldsymbol{\psi}),$$

in an appropriate sense as $n \to \infty$, for each $t \in \mathscr{T}$. It may even be possible sometimes to guarantee uniform convergence. A natural source of estimators for $\boldsymbol{\psi}$, therefore, is the minimization of some measure of distance between $\bar{g}_n(\cdot)$ and $G(\cdot|\boldsymbol{\psi})$,

$$\delta[\bar{g}_n(\cdot), G(\cdot|\boldsymbol{\psi})], \text{ say.}$$

If this criterion is denoted, for convenience, by $\delta(\boldsymbol{\psi})$, then the same formal equations (4.5.1) and (4.5.2) can often be used to derive the estimator $\hat{\boldsymbol{\psi}}$ and the associated asymptotic theory.

Estimators based on the distribution function (Section 4.5) correspond to the special case in which $g(t, X)$ is the indicator function

$$\left.\begin{aligned} g(t, x) &= 1, && \text{if } x < t, \\ &= 0, && \text{otherwise.} \end{aligned}\right\} \tag{4.6.1}$$

The range of choice of possible δ is much as it was in Section 4.5. However, just as in that section we restricted our attention, largely for computational reasons, to 'quadratic' distance measures, here—for basically the same reasons—we shall confine our discussion to the weighted L_2-norm (weighted least-squares distance)

$$\delta\{\bar{g}_n(\cdot), G(\cdot \mid \psi)\} = \int_{\mathscr{L}} |G(t \mid \psi) - \bar{g}_n(t)|^2 \, dW(t), \tag{4.6.2}$$

where $W(\cdot)$ is a positive weighting measure on $\mathscr{L}$.

With $\mathscr{X} = \mathscr{L} = \mathbb{R}$ and $g(t, x)$ given by (4.6.1), (4.6.2) is the same as $\delta_{\mathrm{WLA}}[F_n(\cdot), F(\cdot \mid \psi)]$, as given in Table 4.5.1.

For univariate data, some obvious candidates for $G(t \mid \psi)$ and their empirical versions are as follows.

(a) *Characteristic function* (*Fourier transform*) (Paulson, Holcomb, and Leitch, 1975; Thornton and Paulson, 1977; Heathcote, 1977; Bryant and Paulson, 1983):

$$G_C(t \mid \psi) = \mathbb{E}(e^{itX}); \qquad \bar{g}_n(t) = n^{-1} \sum_{r=1}^{n} \exp(itx_r), \; t \in \mathbb{R}, \qquad \text{where } i = \sqrt{(-1)}.$$

(b) *Moment generating function* (*Laplace transform*) (Quandt and Ramsey, 1978):

$$G_M(t \mid \psi) = \mathbb{E}(e^{tX}); \qquad \bar{g}_n(t) = n^{-1} \sum_{r=1}^{n} \exp(tx_r).$$

Here G_M will only exist for a certain range of t.

(c) *Probability generating function* (discrete data):

$$G_P(t \mid \psi) = \mathbb{E}(t^X); \qquad \bar{g}_n(t) = n^{-1} \sum_{r=1}^{n} t^{x_r}.$$

Note that use of the modulus notation in (4.6.2) is necessary if we wish to include the characteristic function (CF) case.

The weight function $W(\cdot)$ is open to choice. As we shall see later, present practice is a combination of common-sense and convenience. The estimators that result will be functions of $W(\cdot)$ and, in principle, it may be possible to define an optimal $W(\cdot)$ that leads to estimators with a 'minimal' asymptotic covariance matrix in some sense, but the complicated dependence on $W(\cdot)$ suggests that this ideal solution is not practicable.

The most straightforward type of weight function is one concentrated on a finite number of points.

Example 4.6.1 Mixture of two univariate normals

Quandt and Ramsey (1978) applied the moment generating function (MGF) method to the five-parameter normal mixture, using distance measure (4.6.2) and a uniform five-point measure on $\mathscr{L} = \mathbb{R}$. Given support points

$t_1, \ldots, t_5$, the parameter estimates are obtained by minimizing

$$\sum_{j=1}^{5} [\pi \exp(\mu_1 t_j + \tfrac{1}{2}\sigma_1^2 t_j^2) + (1-\pi)\exp(\mu_2 t_j + \tfrac{1}{2}\sigma_2^2 t_j^2) - n^{-1} \sum_{r=1}^{n} \exp(t_j x_r)]^2.$$

The support points should be chosen so that they are:

(a) not too large, otherwise instability may be caused, as a result of large sampling variability;
(b) not too small, otherwise the surface defined by $\delta(\cdot, \cdot)$ will be very flat and any minimum will be poorly defined.

Of course, point (a) is reminiscent of a caveat that must be borne in mind when using the method of moments and, indeed, Johnson (1978) points out that the above use of the MGF method with a normal mixture is equivalent to the use of the method of moments with, possibly, fractional moments, on a mixture of lognormals! More general weight functions have been used from the start with the CF method; see the above references and, also, Kiefer (1978b), Fowlkes (1979), Clarke and Heathcote (1978), and Kumar, Nicklin, and Paulson (1979).

Consideration of point (a) also favours the CF method because numerical problems with large exponentials will not occur (Binder, 1978b), the practical range of choice of $W(\cdot)$ will be wider (Clarke and Heathcote, 1978), and, generally, numerical results show greater stability (Kumar, Nicklin, and Paulson, 1979). All these features result from the uniform boundedness of the characteristic function. The MGF criterion is easier to deal with algebraically because it does not involve imaginary numbers. However, qualitatively similar asymptotic results may be proved for both.

4.6.2 Theoretical aspects of the MGF and CF methods

As indicated earlier, the principal asymptotic results are that, under appropriate regularity conditions, the estimators obtained are, asymptotically, unbiased and multivariate normal in distribution, with a computable covariance matrix. These results do not, of course, depend on the underlying density being a mixture. For the MGF method, results of this type are given by Kiefer (1978b) and Quandt and Ramsey (1978).

In the MGF method of Quandt and Ramsey (1978), one chooses a set of Q distinct values $t_1, \ldots, t_Q$ and finds $\hat{\psi}$ to minimize

$$[\mathbf{G}_M(\psi) - \mathbf{g}_n]^{\mathrm{T}}[\mathbf{G}_M(\psi) - \mathbf{g}_n], \tag{4.6.3}$$

where $[\mathbf{G}_M(\psi)]_i = G_M(t_i | \psi)$, $(\mathbf{g}_n)_i = \bar{g}_n(t_i)$, $i = 1, \ldots, Q$. We are thus dealing with a non-linear least-squares problem, for which the stationarity equations are

$$A(\psi)^{\mathrm{T}}(\mathbf{G}_M(\psi) - \mathbf{g}_n) = \mathbf{0},$$

where $$[A(\boldsymbol{\psi})]_{ij} = \frac{\partial G_M(t_i \mid \boldsymbol{\psi})}{\partial \psi_j}.$$

Asymptotically, as $n \to \infty$,

$$\sqrt{n}(\hat{\boldsymbol{\psi}} - \boldsymbol{\psi}_0) \to N(\mathbf{0}, V_1)$$

in distribution,

where $V_1 = (A^{\mathrm{T}}A)^{-1}A^{\mathrm{T}}\Omega_0 A(A^{\mathrm{T}}A)^{-1}$
and

$$\Omega_0 = \Omega(\boldsymbol{\psi}_0) = \operatorname{cov}(\mathbf{g}),$$

where $(\mathbf{g})_i = \exp(t_i X)$.

Schmidt (1982) points out that a more efficient estimator, $\tilde{\boldsymbol{\psi}}$, can be found by minimizing, instead of (4.6.3),

$$[\mathbf{G}_M(\boldsymbol{\psi}) - \mathbf{g}_n]^{\mathrm{T}}\Omega(\boldsymbol{\psi}_0)[\mathbf{G}_M(\boldsymbol{\psi}) - \mathbf{g}_n]. \tag{4.6.4}$$

Asymptotically, as $n \to \infty$,

$$\sqrt{n}(\tilde{\boldsymbol{\psi}} - \boldsymbol{\psi}_0) \to N[\mathbf{0}, (A^{\mathrm{T}}\Omega_0^{-1}A)^{-1}].$$

Here we have replaced an ordinary least-squares criterion (4.6.3) by a generalized least-squares criterion (4.6.4). The same refinement could be used in other minimum distance contexts as well, although the dependence of (4.6.4) on $\boldsymbol{\psi}_0$ would than have to be dealt with in some way. When Q is the same as s, the number of parameters, the two methods are equivalent. (This is the case in Quandt and Ramsey's, 1978, approach to Example 4.6.1, or in the usual method of moments, for that matter.) Schmidt (1982) points out, however, that by increasing Q and thereby increasing the flexibility of the method, one can, not surprisingly, increase the efficiency. Given Q, an optimal set $t_1, \ldots, t_Q$ can in principle be found. For a five-parameter normal mixture, Schmidt (1982) shows that the optimal minimal ($Q = 5$) set of t_i are often very close together (within a range of 0.03). Unfortunately, this theoretical analysis ignores the practical complications which arise from the facts that the empirical m.g.f.'s eventually used suffer from random variability and the optimal t_i's depend on the *true* $\boldsymbol{\psi}_0$ (cf. Examples 4.2.2 and 4.2.3).

Very limited empirical comparison of the performance of the MGF method with maximum likelihood and the method of moments is reported by Hosmer (1978a) using a few normal mixtures and switching regressions. In spite of the asymptotic theory, the MGF method performed best and the method of moments worst. Kumar, Nicklin, and Paulson (1979) strongly support the superiority of maximum likelihood. They are backed up by simulations by Everitt and Hand (1981, pp. 54–56), who point out the sensitivity of the NR solution of the MGF method to the iteration starting point.

The complex variables version of the asymptotic theory needed for the CF method is given by Thornton and Paulson (1977) and by Bryant and Paulson (1979).

For real-valued transforms, such as the MGF method, the stationarity equations for (4.6.2) are

$$\int [G(t|\boldsymbol{\psi}) - \bar{g}_n(t)] \frac{\partial G(t|\boldsymbol{\psi})}{\partial \psi_j} \, \mathrm{d}W(t) = 0, \qquad j = 1, \ldots, s, \tag{4.6.5}$$

where s denotes the number of parameters in $\boldsymbol{\psi}$, provided, as will usually be the case, that we can differentiate under the integral sign. This is equivalent to

$$n^{-1} \sum_{r=1}^{n} \int [G(t|\boldsymbol{\psi}) - g(t, x_r)] \frac{\partial G(t|\boldsymbol{\psi})}{\partial \psi_j} \, \mathrm{d}W(t) = 0, \qquad j = 1, \ldots, s,$$

so that, as far as stationary points are concerned, (4.6.2) is equivalent to

$$n^{-1} \sum_{r=1}^{n} \int [G(t|\boldsymbol{\psi}) - g(t, x_r)]^2 \, \mathrm{d}W(t),$$

that is,

$$n^{-1} \sum_{r=1}^{n} \delta[g(\cdot, x_r), G(\cdot|\boldsymbol{\psi})] \tag{4.6.5}$$

which can be written as

$$\int \delta[g(\cdot|x), G(\cdot|\boldsymbol{\psi})] \, \mathrm{d}F_n(x).$$

Expression (4.6.5), in particular, emphasizes the link with traditional least-squares inference based on a sample $x_1, \ldots, x_n$, and has direct association with M-estimators (Paulson and Nicklin, 1983). As a result of this and the quadratic nature of δ in (4.6.2), we again have, as we show in the next section, the possibility of a straightforward analysis for the case of a mixture with known component densities.

4.6.3 Illustrations based on the estimation of mixing weights

Example 4.6.2 Mixture of known component distributions

Suppose

$$P(x) = \sum_{j=1}^{k} \pi_j F_j(x)$$

and that $G_j(\cdot)$ is the transform of $F_j(\cdot)$. Then

$$G(t) = \sum_{j=1}^{k} \pi_j G_j(t) = \boldsymbol{\pi}^{\mathrm{T}} \mathbf{G}(t), \text{ say.}$$

Then, we have to minimize, with respect to $\boldsymbol{\pi}$,

$$n^{-1}[\boldsymbol{\pi}^{\mathrm{T}} A \boldsymbol{\pi} - 2\boldsymbol{\pi}^{\mathrm{T}} \mathbf{b} + c],$$

where $$A_{jl} = \int G_j(t) G_l(t) \mathrm{d}W(t)$$

and $$b_j = n^{-1} \sum_{r=1}^{n} \int g(x_r, t) G_j(t) \mathrm{d}W(t).$$

The estimator $\hat{\boldsymbol{\pi}}$ is then obtained in a parallel way to that used in Section 4.5 for this example. As in the previous discussion, the constraint $\boldsymbol{\pi}^{\mathrm{T}}\mathbf{1} = 1$ may be incorporated either using a Lagrange multiplier or by eliminating π_k, say, before minimization.

The latter approach is used by Bryant and Paulson (1983) who treat this problem using the CF method. To bring out the differences resulting from the presence of complex variables we quote their result here.

Denote by $\bar{y}$ the complex conjugate of y and define

$$b_j^* = n^{-1} \sum_{r=1}^{n} \int \mathrm{Re}\{[g(x_r, t) - G_k(t)][\overline{G_j(t) - G_k(t)}]\} \, \mathrm{d}W(t)$$

$$(A^*)_{jl} = \int \mathrm{Re}\{[G_j(t) - G_k(t)][\overline{G_l(t) - G_k(t)}]\} \, \mathrm{d}W(t),$$

$$j = 1, \ldots, k-1, l = 1, \ldots, k-1,$$

where $\mathrm{Re}(\cdot)$ means 'real part of $\cdot$'. Then, the estimator of the first $(k-1)$ components of $\boldsymbol{\pi}$ is

$$A^{*-1}\mathbf{b}^*.$$

As with all these methods, there is no guarantee that, for the resulting $\hat{\boldsymbol{\pi}}$, $\hat{\boldsymbol{\pi}} \geqslant \mathbf{0}$.

Bryant and Paulson (1983) show that, as in some of the cases we have looked at earlier,

(a) A^* is positive definite so that we do obtain the unique minimum, provided the mixture is identifiable.
(b) $\hat{\boldsymbol{\pi}}$ is unbiased.
(c) The covariance matrix of the first $(k-1)$ coefficients of $\hat{\boldsymbol{\pi}}$ can be written down.

In practice, of course, to work out $\hat{\boldsymbol{\pi}}$, or its covariance matrix, or even the stationarity equations (4.6.5) or the criterion function (4.6.2) itself, we have to be able to do the necessary integration, and it is this which largely dictates the choice of weight function.

Example 4.6.3 Mixture of two known normals (Bryant and Paulson, 1983)

Suppose we mix $N(0, 1)$ and $N(\mu, \sigma^2)$, with μ and σ^2 given, so that

$$G_{\mathrm{C}}(t \mid \pi) = \pi \exp(-t^2/2) + (1 - \pi) \exp(\mathrm{i}\mu t - \sigma^2 t^2/2), \qquad t \in \mathbb{R}.$$

A convenient differentiable weight measure is

$$dW(t) = \exp(-\alpha^2 t^2/2)dt, \qquad \text{for some } \alpha > 0.$$

The integrals involved may be evaluated explicitly and the resulting estimator is

$$\hat{\pi} = (qn)^{-1}\left\{c_1 \sum_{r=1}^{n} \exp[-x_r^2/(2+4\alpha^2)] - c_2 \sum_{r=1}^{n} \exp\left[\frac{-(x_r-\mu)^2}{2\sigma^2+4\alpha^2}\right] + nc_3\right\},$$

where $\quad q = (1+\alpha^2)^{-1/2} + 2c_3 - (\sigma^2+\alpha^2)^{-1/2},$

$$c_1 = \sqrt{[2/(1+2\alpha^2)]},$$
$$c_2 = \sqrt{[2/(\sigma^2+2\alpha^2)]},$$
$$c_3 = \sqrt{[2/(1+\sigma^2+2\alpha^2)]}\exp[-\mu^2/2(1+\sigma^2+2\alpha^2)] + (\sigma^2+\alpha^2)^{-1/2}.$$

Since we are dealing with a single unknown parameter, we can choose the constant, α, which characterizes the weight function, so as to minimize var $(\hat{\pi})$, where

$$\text{var}(\hat{\pi}) = (-\pi^2 + \pi d_1 + d_0)/n,$$

in which d_0 and d_1 depend on α in an explicit but complicated way. In general, therefore, as in Section 4.2, the optimal choice depends on π, the true value of the parameter, so it will be necessary to substitute either some crude estimate or, failing that, an 'arbitrary' value such as $\pi = \frac{1}{2}$. When the two normal variances are equal, however, $d_1 = 0$, so that the same minimizing α (from d_0) will be appropriate for every π.

Bryant and Paulson (1983) show that, for $0.3 \leqslant \pi \leqslant 0.7$, for $\mu = 1, 2, 3$, and $\sigma = 1, 2, 3$, the efficiency of $\hat{\pi}$ is hardly ever below 95 per cent. In a comparison with the estimator of Boes (1966) for (μ, σ) combinations (0, 2), (1, 1), (1, 2), and (3, 3), they find that the variances of the two estimators are almost the same in the first two cases and agree to within a few per cent in the other two, for π varying between 0.1 and 0.9. (The value $\pi^{(0)} = \frac{1}{2}$ was chosen in Boes' procedure and the α-value optimal for $\pi = \frac{1}{2}$ was used.) The choice of α is not a particularly sensitive one and, in a simulation study based on a sample size of $n = 30$, they show that the ML, Boes, and CF estimators perform similarly, and better than the minimum Cramér–von Mises estimator, based on $\delta_{ALA}[F(\cdot|\psi), F_n(\cdot)]$, which was used by Macdonald (1971). This latter estimator is, itself, shown by Macdonald (1971) to perform better than that of Choi and Bulgren (1968), based on $\delta_{ALA}[F_n(\cdot), F(\cdot|\psi)]$.

Example 4.6.4. Mixture of known normals (Example 4.3.6 revisited)

Example 4.3.6 described the calculation of maximum likelihood estimates from data simulated from the mixture

$$p(x) = 0.8\phi(x|-1, 1) + 0.2\phi(x|1, 1).$$

Table 4.6.1 Performance of MGF and moment estimators for Example 4.6.4

Sample size (n)	Estimates of π_1 (A)	(B)	(C)	(D)
25	0.851	0.764	0.794	0.871
50	0.815	0.734	0.762	0.804
100	0.822	0.734	0.810	0.818
300	0.824	0.803	0.826	0.817
$\text{var}(\sqrt{n}\hat{\pi}_1)$	0.3638	0.4464	0.3971	0.4100

The mixing weight was regarded as the only unknown parameter. If the MGF method is used, based on a single value of the auxiliary variable, t, then it is easy to show that

$$\text{var}(\hat{\pi}_1) = \frac{\pi_1 \exp[2t(t-1)] + (1-\pi_1)\exp[2t(t+1)] - (\pi_1 \exp[t(\tfrac{1}{2}t-1)] + (1-\pi_1)\exp[t(\tfrac{1}{2}t+1)])}{\{\exp[t(\tfrac{1}{2}t-1)] - \exp[t(\tfrac{1}{2}t+1)]\}^2}$$

The calculations are similar to those in Examples 4.2.1 and 4.2.2. Numerical minimization of this gives the optimum t as 0.318, for $\pi_1 = 0.8$. Table 4.6.1 provides (column A) the corresponding MGF estimates of π_1 from the samples used in Example 4.3.6. For comparison, the following moment estimates are also given.

(B) Zeroth moment as in Example 4.2.1, with $c = 0$.
(C) Zeroth moment with optimum c for $\pi_1 = 0.8 (c = 0.481)$.
(D) First moment.

Remember that, in practice, π_1 will not be known, so some preliminary estimate would be required in order to use approximations to (A) and (C).

Table 4.6.1 also gives the values of $\text{var}(\sqrt{n}\hat{\pi}_1)$, for comparison with 0.3274, the inverse of the Fisher information for one observation.

In more complicated problems, numerical minimization procedures are necessary for implementing the methods of this section (Quandt and Ramsey, 1978; Kumar, Nicklin, and Paulson, 1979). This suggests that the more stable CF approach is to be preferred.

4.7 NUMERICAL DECOMPOSITION OF MIXTURES

4.7.1 Some introductory methods

The techniques to be discussed here are, as mentioned in Section 1.2, not so much statistical as numerical. The example of electrophoresis described in Section 2.1 is typical. A density or concentration *curve* is given and the objective is to decompose it into a (possibly known) number of unimodal components. The components are usually assumed to be symmetrical probability density functions and are often, in practice, taken to be normal or Cauchy in shape. In most applications the data take the form of a trace, or a curve displayed on a cathode-ray tube. In the example of electrophoresis, the curve represents the concentration of protein, say, at a certain distance from the initial boundary with the buffer solution. Further illustrations are given by Fraser and Suzuki (1966), one of which provides the trace, shown in Figure 4.7.1, of absorbance level against volume, which resulted from chromatographic analysis of a mixture of aspartic acid, methionine sulfone, threonine, and serine. Figure 4.7.1 also displays the decomposition of the curve into four normal components, one associated with each constituent of the mixture. The principle used for generating the components in this case was, in fact, one of those considered in Section 4.5; namely, that

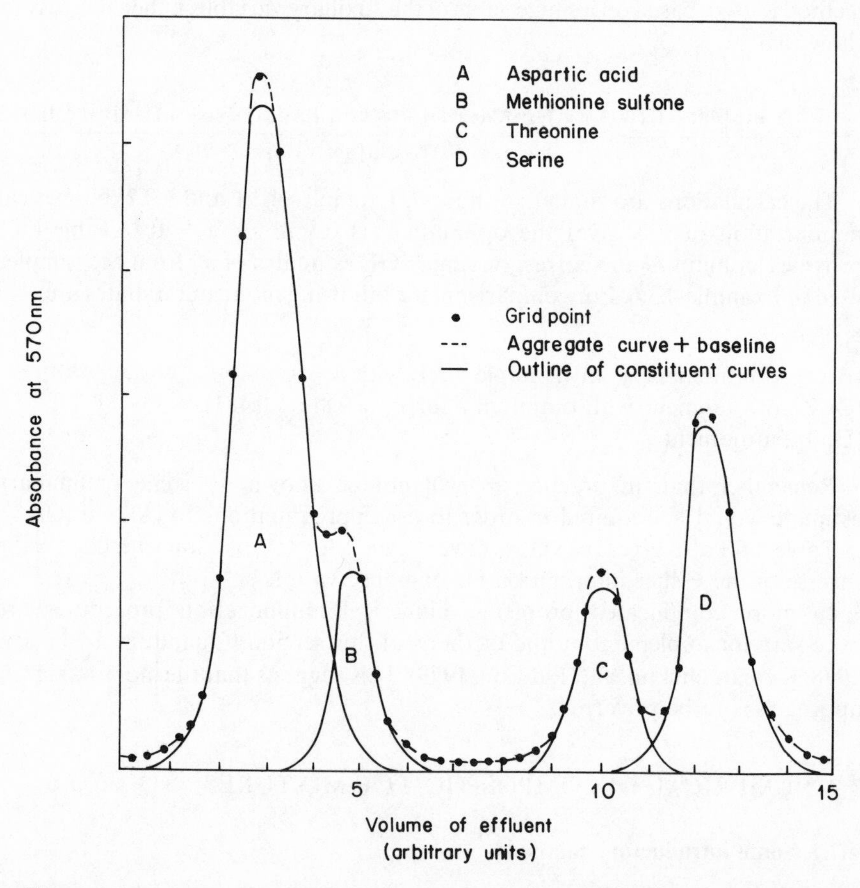

Figure 4.7.1 Output from amino acid analyser. Reprinted with permission from Fraser and Suzuki (1966), *Anal. Chem.*, **38**, 1770–1773. Copyright (1966) American Chemistry Society

of minimizing

$$\delta[F_0(\cdot), F(\cdot|\psi)] = \int [f_0(x) - p(x|\psi)]^2 \, dW(x),$$

where, in this case, $f_0(\cdot)$ denotes the datum *curve*. Thus $F_0(\cdot)$ is not an empirical distribution function as such. The weight function is generally taken to be a point measure with finite support, so that ψ is chosen to minimize

$$\sum_{i=1}^{n} [f_0(x_i) - p(x_i|\psi)]^2, \tag{4.7.1}$$

for some $x_1, \ldots, x_n$. It must be emphasized that the $x_1, \ldots, x_n$ are no longer datum points but merely represent a grid of values on which the least-squares analysis is based. Although most reported accounts of this method are based on (4.7.1), with equally spaced grid points in a univariate sample space, there is clearly scope for more subtly chosen grid points, non-uniformly weighted least-squares, or extension to multivariate sample spaces. This last extension is facilitated by the use of *density* functions in (4.7.1) as opposed to *cumulative distribution* functions.

In some applications, the data are of the more usual sample data or histogram form. Least-squares analysis of the above type has been applied to such cases by taking $f_0(\cdot)$ to be the standardized histogram (Allen and McMeeking, 1978). An altenative approach, about which there does not seem to be much literature as yet, would be to smooth the empirical distribution function or the histogram and let $f_0(\cdot)$ be the resulting smooth density curve. This could be achieved using kernel-based density estimation (see Section 2.2) or the methods of Boneva, Kendall, and Stefanov (1971), based on splines, and van Ryzin (1973).

Minimization of (4.7.1) with respect to ψ will require numerical methods. Relevant software and, sometimes, hardware are described by French *et al.* (1954), Poulik and Pinteric (1955), Noble, Hayes, and Eden (1959), van Andel (1973), Zlokazov (1978), and Allen and McMeeking (1978). The system of Noble, Hayes, and Eden (1959), for instance, allows display of the datum curve and the fitted approximation, which can be a mixture of up to ten normal or Cauchy components. The parameters can be adjusted interactively so as to improve the fit, based on integrated squared error. Dean and Jett (1974) describe an interesting application involving a mixture of two normal components linked by a third which is modelled as a polynomial.

As in Sections 4.5 and 4.6, explicit minimization is possible if a quadratic measure of fit is used and only the mixing weights are unknown.

In the early applied literature, the procedures for decomposition displayed varying degrees of adhockery, ingenuity, and formality. In electrophoresis, in particular, a major emphasis is placed on the mixing weights, which indicate the relative total concentrations of the various proteins in the mixture. Tiselius and Kabat (1939) estimate these parameters by the crude procedure of calculating the areas under the curve between successive minima; see also Section 4.1, where the accuracy of this technique is discussed in some detail. Inaccuracy is caused by the overlap with the adjacent and, to a lesser extent, the more distant components.

A more sophisticated technique is that of Pedersen (Svedberg and Pedersen, 1940, Chapter B, Section 3) for decomposing a mixture of symmetric curves. First, an attempt is made to construct the extreme left-hand or right-hand component. It is assumed that, in the former case, for example, the left-hand half is pure and not overlapped by any other curve. The median is located and a mirror-image right-hand half is drawn. The total curve is subtracted from the mixture and this process is repeated until all components have been extracted. If the later components do not look satisfactory a slightly different choice is made for the initial median and a new decomposition is constructed. Figure 4.7.2 (Figure 119 of Svedberg and Pedersen, 1940) shows how, in time, a mixture of serum albumin and γ-globulin is separated by the ultracentrifuge. Once the decomposition has been achieved satisfactorily, the relative concentrations are calculated by integration of the component curves. This method is obviously much more complicated than that of Tiselius and Kabat (1939) and it often does not perform much better (Longsworth, 1942). It can, however, be used in problems with greater overlap than can be dealt with by the crude method. This method is similar to that described by Usinger (1975) for decomposing pollen mixtures; see also Gordon and Prentice (1977, p. 363).

An even more complicated method, based on the same principle, is described by Berry and Chanutin (1955) and applied to a trace known to involve the aggregation of seventeen components! If the components can be assumed to be normal, construction of the components can be made more systematic, but the method is still essentially graphical. Suppose

$$y = y(x) = \frac{a}{\sqrt{(2\pi)}\sigma} \exp\left[-\frac{1}{2\sigma^2}(x-\mu)^2 \right]$$

and that μ is treated as known. Then two points, $(x_1, y_1), (x_2, y_2)$, on the curve give

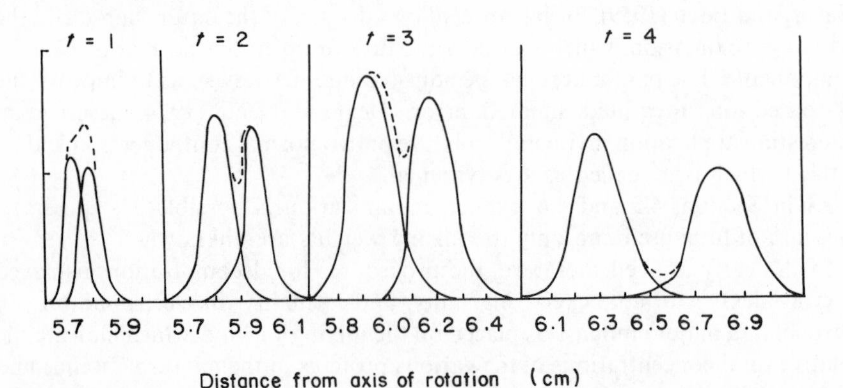

Figure 4.7.2 Ultracentrifuge resolution of the sedimentation curve for a mixture of serum albumin and γ-globulin, where t = time after reaching full speed. Reproduced by permission of Oxford University Press from Svedberg and Pedersen (1940), *The Ultracentrifuge*

an estimate of σ from

$$\hat{\sigma} = \sqrt{\left\{0.127\left[\frac{(x_2-\mu)^2-(x_1-\mu)^2}{\log y_2 - \log y_1}\right]\right\}}.$$

We then obtain an estimate of the area, a, in the form

$$\hat{a} = \sqrt{(2\pi)}\hat{\sigma}y(\mu),$$

where $y(\mu)$ is the modal height. These results form the basis of a procedure for decomposing a normal mixture suggested by Hoxter, Wajchenberg, and Mungioli (1957).

4.7.2 Formal methods for mixtures of exponentials

Example 4.7.1 Mixtures of exponentials

Before looking at the decomposition of density curves, we consider the problem of identifying the parameters in the function

$$y(x) = \pi_1 e^{-\theta_1 x} + \pi_2 e^{-\theta_2 x}, \qquad x > 0, \theta_1 > \theta_2 > 0.$$

If (π_1, π_2) are probabilities, this gives a mixture of exponential survival functions. Were π_1 or π_2 zero, the plot of y on x using semilog paper would show a straight line with slope $-\theta_2$ or $-\theta_1$. For general (π_1, π_2) the plot is not linear but, for large values of x, the contribution of the component $\pi_1 e^{-\theta_1 x}$ will be negligible and

$$\log y \approx \log \pi_2 - \theta_2 x.$$

Thus, estimates $(\hat{\theta}_2, \hat{\pi}_2)$ may be made for θ_2 and π_2 from the linear part of the semilog plot, and a new function

$$y'(x) = y(x) - \hat{\pi}_2 e^{-\hat{\theta}_2 x}$$

may be drawn from which $\hat{\theta}_1$ and $\hat{\pi}_1$ can be estimated, again using semilog paper. This treatment may be extended to deal with curves with more than two components. The method is described by Defares, Sneddon, and Wise (1973), who do not impose the constraint that $\pi_1 + \pi_2 = 1$, but consider a more general linear combination with positive coefficients.

Functions of this type are often associated with decay curves. The above method obviously relies on accurate information about the curve at large decay times. This makes the technique lose its attraction for application to an empirical plot from sample data, unless the sample size is large; see also Section 4.1. The following technique manages to avoid this sort of drawback.

Suppose

$$F(x|\boldsymbol{\psi}) = \sum_{j=1}^{k} \pi_j e^{-\theta_j x},$$

$$0 < \theta_1 < \cdots < \theta_k, \pi_j > 0, j = 1, \ldots, k, \tag{4.7.2}$$

where k itself is to be determined.

Let

$$
\begin{aligned}
x_i &= x_0 + (i-1)d, & i &= 1, \ldots, n, \\
y_j &= \exp(-\theta_j d), & j &= 1, \ldots, k, & (4.7.3) \\
\phi_j &= \pi_j \exp(-\theta_j x_0), & j &= 1, \ldots, k, & (4.7.4) \\
c_i &= F(x_i \mid \psi), & i &= 1, \ldots, n.
\end{aligned}
$$

The procedure is, for each k, to equate the observed $\{c_i\}$ to the fitted values. We must solve, therefore,

$$\phi_1 y_1^{i-1} + \cdots + \phi_k y_k^{i-1} = c_i, \qquad i = 1, \ldots, n, \tag{4.7.5}$$

a system of equations similar to (4.2.6) and (4.2.7) which arose in the method of moments. Details are given by Parsons (1968), who also shows how k can be identified, barring round-off errors in computation. Further discussion, including that of the choice of d, is given in Parsons (1970); see also Bellman (1960), Cornell (1962), and Ruhe (1980).

Alternative approaches based on transforms were developed, by Brownell and Callaghan (1963) and Gardner (1963), for dealing with radioactive tracer studies. Let

$$
\begin{aligned}
G(s \mid \psi) &= \int_0^\infty sxF(x \mid \psi) \sin(sx)\, \mathrm{d}x \\
&= \sum_{j=1}^{k} \frac{2\pi_j \theta_j s^2}{(s^2 + \theta_j^2)^2},
\end{aligned}
$$

where $F(x \mid \psi)$ is defined by (4.7.2).

This curve often shows peaks at $s = \theta_j$, particularly when π_j/θ_j is constant, for each j. An alternative method of decomposition is provided by the inverse Laplace transform $g(s \mid \psi)$, defined by

$$F(x \mid \psi) = \int_0^\infty \mathrm{e}^{-sx} g(s \mid \psi)\, \mathrm{d}s.$$

The exact $g(s \mid \psi)$ is a sequence of delta functions at $\theta_1, \ldots, \theta_k$. In practice, $g(s \mid \psi)$ will be obtained by numerical Fourier analysis and a smoother curve will be obtained which, nevertheless, should often show clear separation. We may hope to identify $k, \theta_1, \ldots, \theta_k$, and, from the heights of the peaks, $\{\pi_j/\theta_j : j = 1, \ldots, k\}$; see Gardner (1963). Interesting practical examples of this and similar techniques, facilitated by the use of fast Fourier transforms, are described by Smith, Cohn-Sfetcu, and Buckmaster (1976).

More examples along these lines are given by Medgyessy (1977, Chapter III, Section 3). However, the main approach of his book is somewhat different, as described below.

4.7.3 Medgyessy's method

Motivation is provided by the following *ad hoc* procedure.

Example 4.7.2 Mixture of two normals

Suppose

$$p(x) = p(x\,|\,\psi) = q_1(x) + q_2(x),$$

where $\quad q_j(x) = \pi_j \phi(x\,|\,\mu_j, \sigma_j), \qquad j = 1, 2.$

A plot of the curve

$$\begin{aligned} p_1(x) &= [p(x)]^2 - 2q_1(x)q_2(x) \\ &= [q_1(x)]^2 + [q_2(x)]^2 \end{aligned} \tag{4.7.6}$$

would give a composition of normal curves with the original means and half the original variances. This can be iterated, through

$$p_2(x) = [p_1(x)]^2 - 2[q_1(x)q_2(x)]^2, \qquad \text{and so on.}$$

The idea is that the two components are more clearly separated and, consequently, easier to identify, in $p_1(x)$ and $p_2(x)$ than in $p(x)$. Ageno and Frontali (1963) put this into practice by using initial estimates of ψ to estimate $2q_1(x)q_2(x)$ in (4.7.6) and hence to construct the curve $p_1(x)$. Repetition of this procedure then gives $p_2(x)$, if necessary.

This particular procedure is not very sophisticated but it illustrates a fact on which a large part of the book by Medgyessy (1977) is based; namely, that if the parameter(s) dictating the narrowness of the component densities can be adjusted to sharpen the peaks then the separation of the peaks is greatly improved. As a result, the number of components involved may be identified, along with their locations, and the way is then open to estimate the other parameters. Medgyessy (1977) considers four general cases in detail: continuous data from mixtures where the component densities are known to have distinct modes; continuous data where the component densities all have the same mode; and the corresponding two cases for discrete data. Examples of the four cases are as follows:

(a) normal mixtures with different component means;
(b) normal mixtures with the same mean and exponential mixtures;
(c) binomial mixtures and Poisson mixtures with component means which are not too similar;
(d) geometric mixtures and Poisson mixtures with means all less than unity.

In the discrete case, the difference between (c) and (d) is that, in (c), the largest probability for each component must occur at a different value of x.

The general approach for cases (a) and (c) is motivated by the following.

Suppose the true curve is a mixture of two normals, with

$$\begin{aligned} p(x) = p(x\,|\,\psi) &= \sum_{j=1}^{2} \pi_j \phi(x\,|\,\mu_j, \sigma_j) \\ &= \sum_{j=1}^{2} \pi_j f(x - \mu_j, \sigma_j), \text{ say.} \end{aligned}$$

Then, if $0 < \lambda < \min_j(\sigma_j)$, the curve

$$p_\lambda(x) = \sum_{j=1}^{2} \pi_j f(x - \mu_j, \sigma_j - \lambda), \tag{4.7.7}$$

also representing a normal mixture density, should show more clearly the peaks at μ_1 and μ_2, particularly if λ is close to the smaller of σ_1 and σ_2. A curve such as $p_\lambda(x)$ is called a *test function*, from which at least one of the components should be discernible as a result of the increased separation.

For cases (b) and (d) the same principle is used after a preliminary transformation has been made from $p(x)$ to a curve which does have different component modes; see later.

The difficulty in practice is to construct the test function from knowledge of just the curve $p(x)$ and not its constitution. Use of $p_\lambda(x)$ directly from (4.7.7), for instance, is not a practical proposition!

Example 4.7.3 A general mixture of k components

Suppose

$$p(x) = \sum_{j=1}^{k} \pi_j f(x - \mu_j, \sigma_j), \tag{4.7.8}$$

where $f(x - \mu, \sigma)$ denotes a p.d.f., $\mu_1, \ldots, \mu_k$ are distinct, and $0 < \sigma_1 \leqslant \cdots \leqslant \sigma_k$. Suppose also that the characteristic function of $f(z, \sigma)$ takes the known form

$$[G(s)]^\sigma,$$

where $G(s)$ is a characteristic function. Then it is indeed possible to use

$$p_\lambda(x) = \sum_{j=1}^{k} \pi_j f(x - \mu_j; \sigma_j - \lambda) \tag{4.7.9}$$

as a test function.

This follows because the characteristic functions from (4.7.8) and (4.7.9), $G_0(s)$ and $G_\lambda(s)$, are related by

$$G_\lambda(s) = [G(s)]^{-\lambda} G_0(s), \qquad \text{for all } s.$$

Thus

$$G_0(s) = [G(s)]^{\lambda} G_\lambda(s), \qquad \text{for all } s,$$

and the right-hand side is a product of characteristic functions. As a result of the convolution nature of this relationship,

$$p(x) = \int f(x - y, \lambda) p_\lambda(y) \mathrm{d}y, \qquad \text{for all } x. \tag{4.7.10}$$

The relationship (4.7.10) does not involve any of the unknown parameters and identifies $p_\lambda(y)$ as the solution of a Fredholm integral equation of the first kind, for which numerical solution, at least, is possible (see Medgyessy, 1977, Chapter V).

Suitable experimentation with λ ('formant changing' is Medgyessy's terminology) will reveal the number of components in the mixture.

Specific cases that conform to this pattern include mixtures of stable-law densities.

In the case of a normal mixture we have to solve

$$p(x) = (2\pi\lambda)^{-1/2} \int_{-\infty}^{\infty} \exp[-(x-y)^2/2\lambda] p_\lambda(y)\, dy.$$

Here there is an explicit solution, given by

$$p_\lambda(x) = \sum_{r=0}^{\infty} \frac{(-\frac{1}{2}\lambda)^r}{r} p^{(2r)}(x),$$

where $p^{(r)}(x)$ denotes the rth derivative of $p(x)$. Thus $p_\lambda(x)$ could be drawn quite easily, using suitable truncation of the series and numerical estimation of the derivatives (see Medgyessy, 1977, Chapter III, Section 1.1.1.1 and pp. 133–136, where the pioneering work of Sen, 1922, and Doetsch, 1928, 1936, is acknowledged).

Practical application is described by Gregor (1969), who fits a normal mixture to histogram data. He subtracts one component after another from the mixture, as they are identified, until the fit is satisfactory. Suppose it is proposed that

$$p(x) = \sum_{j=1}^{k} \pi_j \phi(x \mid \mu_j, \sigma_j).$$

The method of Example 4.7.3, using numerical calculation of Fourier transforms and their inverses, identifies the 'narrowest' component and its mean, μ_1, say. The values of π_1 and σ_1 are then found by considering the areas of two small strips s_1, s_2 (Figure 4.7.3) round the mean, μ_1.

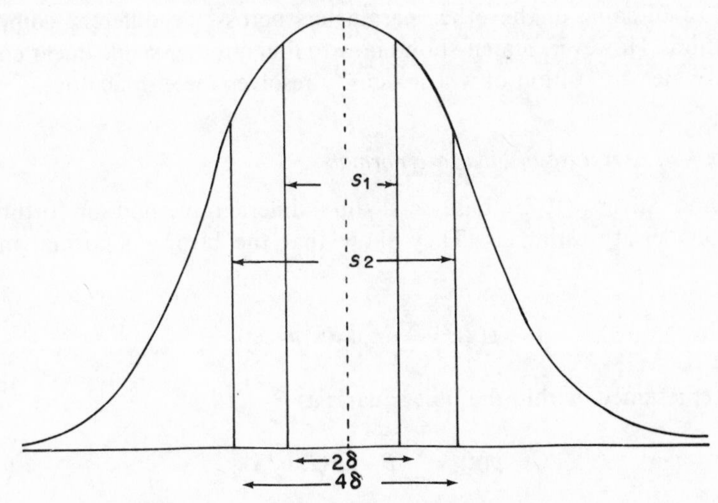

Figure 4.7.3 Strips used in Gregor's method

It is assumed that the strips are narrow enough to ensure that there is little overlap with the other components. Analytical approximations of these areas in terms of π_1 and σ_1 are equated to numerical approximations obtained from the data by Simpson's rule. Once these equations are solved for π_1 and σ_1, the component

$$\pi_1 \phi(x|\mu_1, \sigma_1)$$

is 'subtracted' from the data and the procedure repeated.

Further data-based examples are given in Sammon (1968).

4.7.4 Further examples

Sammon (1968) and Stanat (1968) both extend the method of Example 4.7.3 to the multivariate case. Stanat (1968) deals with multivariate normal and multivariate Bernoulli mixtures. The normal case is described briefly below.

Example 4.7.4 Mixture of multivariate normals

Suppose

$$p(\mathbf{x}) = \sum_{j=1}^{k} \pi_j \phi_d(\mathbf{x}|\boldsymbol{\mu}_j, \Sigma_j),$$

where ϕ_d denotes the d-dimensional multivariate normal density.

The method is based on decomposition of univariate projections of this density. The mixing weights, means, and variances can be obtained from the marginal densities and then the covariances $\{(\Sigma_j)_{il}\}$ from the densities of $\{x_i + x_l\}$. Difficulties may arise because of the possible equalities of some of the mixing weights or of some of the other parameters across the different component distributions. However, Stanat shows how to find another *single* linear combination, $\mathbf{v}^{\mathrm{T}}\mathbf{x}$, decomposition of whose density resolves these difficulties.

Example 4.7.5 Mixture of bivariate normals

Tarter and Silvers (1975) base a slightly different method on orthogonal expansion density estimates. They show that the bivariate normal mixture density

$$p(\mathbf{x}) = \sum_{j=1}^{k} \pi_j \phi_2(\mathbf{x}|\boldsymbol{\mu}_j, \Sigma_j)$$

can be represented, within the unit square, as

$$p(\mathbf{x}) = \sum_{\mathbf{r} \in N} B_{\mathbf{r}} \exp(2\pi \mathrm{i} \mathbf{r}^{\mathrm{T}} \mathbf{x}), \tag{4.7.11}$$

where N is the set of all 2-tuples of integers, $\mathrm{i} = \sqrt{(-1)}$ and, approximately,

$$B_{\mathbf{r}} = \sum_{j=1}^{k} \pi_j \exp(-2\pi i \mathbf{r}^{\mathrm{T}} \boldsymbol{\mu}_j - 2\pi^2 \mathbf{r}^{\mathrm{T}} \Sigma_j \mathbf{r}),$$

where, throughout, π without a subscript denotes 3.14159.... (If necessary, transformations are made so that the data do lie in the unit square.) Given data in the form of a random sample, $\mathbf{x}_1, \ldots, \mathbf{x}_n$, a practicable estimate of (4.7.11) can be obtained by

(a) replacing $B_{\mathbf{r}}$ by an estimate,

$$\hat{\mathbf{B}}_{\mathbf{r}} = n^{-1} \sum_{s=1}^{n} \exp(-2\pi i \mathbf{r}^{\mathrm{T}} \mathbf{x}_s);$$

(b) restricting summation to a finite subset, N_0, of N. The resulting bivariate density estimate,

$$\hat{p}(\mathbf{x}) = \sum_{\mathbf{r} \in N_0} \hat{B}_{\mathbf{r}} \exp(2\pi i \mathbf{r}^{\mathrm{T}} \mathbf{x}),$$

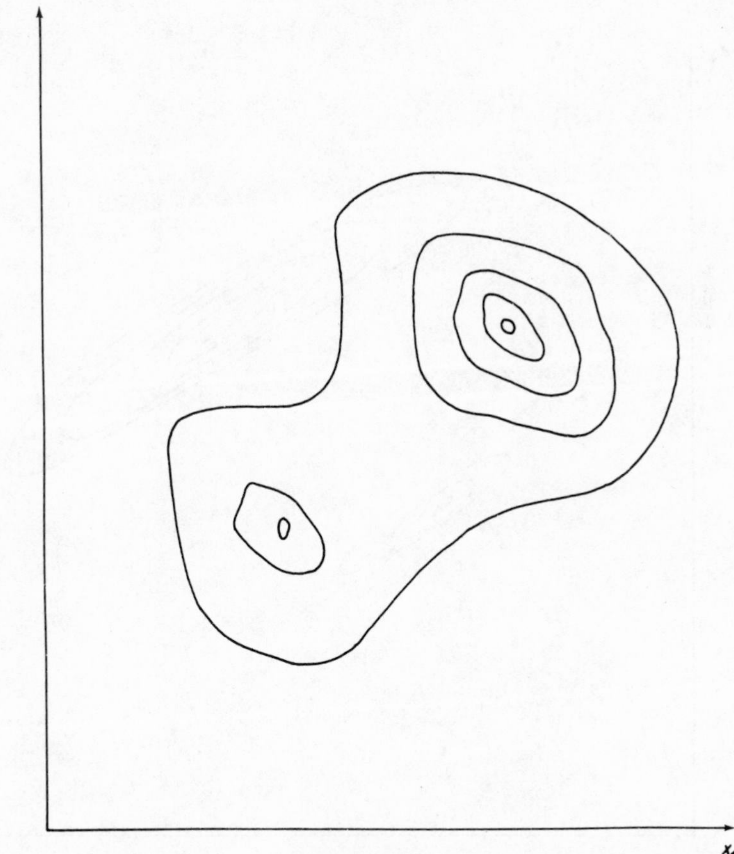

Figure 4.7.4

can then be displayed in the form of equiprobability contours, as shown schematically in Figure 4.7.4. If, instead of $\hat{p}(\mathbf{x})$, we consider $\hat{p}_\lambda(\mathbf{x})$, where

$$\hat{p}_\lambda(\mathbf{x}) = \sum_{\mathbf{r}\in N_0} \hat{B}_\mathbf{r} \exp(2\pi i \mathbf{r}^\mathrm{T}\mathbf{x} + 2\pi^2 \mathbf{r}^\mathrm{T}\Lambda\mathbf{r}),$$

in which $\Lambda = \begin{pmatrix} 0 & \lambda \\ \lambda & 0 \end{pmatrix}$, then this would give an estimate of a density which approximates to the mixture density

$$p_\lambda(\mathbf{x}) = \sum_{j=1}^{k} \pi_j \phi_2(\mathbf{x} \mid \boldsymbol{\mu}_j, \Sigma_j - \Lambda).$$

Note that this requires λ to be such that $\Sigma_j - \Lambda$ is positive definite, for each j, and that the component variances remain unchanged. If λ is chosen to make $\Sigma_j - \Lambda$, say, singular (Tarter and Silvers call this *covariance reduction*), then the equiprobability contours of the jth component density will degenerate to

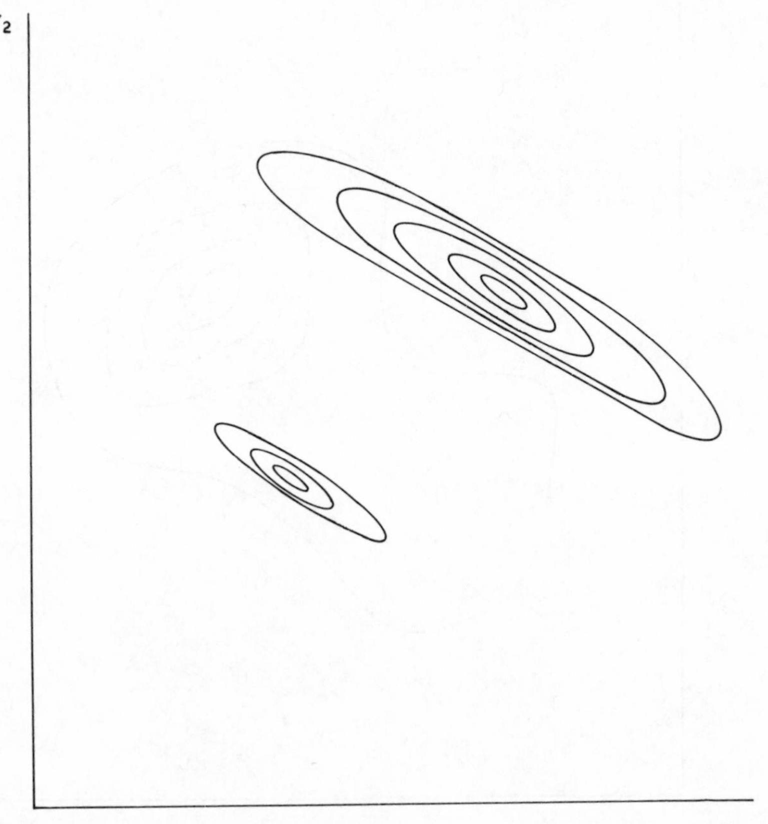

Figure 4.7.5

straight-line segments. If all the Σ_j's are nearly the same, this will lead to good separation of the components. (Figure 4.7.5 gives a schematic illustration in just such a case, corresponding to the original density estimate in Figure 4.7.4.) If a component density has been isolated, with or without covariance reduction, analysis of the position and shape of the equiprobability contours provides estimates of the mixing weight and the parameters of the component densities.

Example 4.7.6 Mixture of lognormals

If $p(x)(x > 0)$ denotes the density of a mixture of lognormals, the transformation

$$q(y) = p(e^y)e^y$$

yields, in $q(y)$, the density of a normal mixture, which can be dealt with as in Example 4.7.3.

In examples of case (b) in which the component densities differ only in scale, test functions of the following type are used:

$$p_\lambda(x) = \sum_{j=1}^{k} \pi_j q(x, \gamma(\sigma_j), \lambda),$$

where

(a) $q(x, \gamma, \lambda)$ is a unimodal density function with mode at γ and 'narrowness' dictated by λ;
(b) $\gamma(\sigma)$ is a strictly monotonic function of σ;
(c) the curve of $p_\lambda(x)$ may be constructed from the curve of $p(x)$.

Clearly, once a suitable $q(x, \gamma, \lambda)$ has been discovered, we are back in the position of identifying the number of components in the mixture by 'formant changing' in $p_\lambda(x)$.

Example 4.7.7 Mixture of exponentials (Medgyessy, 1977, p. 146)

Suppose

$$p(x) = \sum_{j=1}^{k} \pi_j \theta_j^{-1} \exp(-x/\theta_j), \qquad x > 0, 0 < \theta_1 < \cdots < \theta_k < \lambda.$$

Define

$$p_\lambda(x) = \sum_{j=1}^{k} \pi_j \frac{\sigma - 1}{B(\lambda^{-1} + 1, \lambda^{-1}(\sigma - 1)^{-1})} \theta_j^{-1} [\exp(-x/\theta_j) - \exp(-\sigma x/\theta_j)]^{1/\lambda}, \quad x > 0,$$

with λ^{-1} an integer, $B(a, b)$ the beta function, and $\sigma > 1$, all given. The associated q can be shown to be a unimodal density with mode

$$\gamma = \theta(\log \sigma)/(\sigma - 1),$$

and whose 'narrowness' increases as λ^{-1} increases.

Furthermore, using binomial expansions we may write

$$p_\lambda(x) = \frac{\sigma - 1}{B(\lambda^{-1} + 1, \lambda^{-1}(\sigma - 1)^{-1})} \sum_{r=0}^{\lambda^{-1}} \binom{\lambda^{-1}}{r} (-1)^r p\{[r(\sigma - 1) + \lambda^{-1}]x\},$$

which is devoid of any explicit mention of the parameters.

In some examples, such a direct method is not available but it is possible to manoeuvre so as to use the original approach. Illustrations are given by Medgyessy (1977, pp. 151–154) for mixtures of normal densities with zero means and for mixtures of exponentials.

The analysis of discrete data proceeds on very similar lines. For the case of distinct modes a basic 'narrowing' is made to give a test function on the discrete sample space which can be calculated directly from the original probabilities alone. When the modes are all the same, the procedure described above for a continuous sample space is adapted suitably.

Example 4.7.8 Mixture of binomials (Medgyessy, 1977, p. 195)

Suppose

$$p(x) = \sum_{j=1}^{k} \pi_j \binom{N}{x} \theta_j^x (1 - \theta_j)^{N-x}, \qquad x = 0, 1, \ldots, N,$$

with N known, $\frac{1}{2} < \theta_j \leqslant 1$, $j = 1, \ldots, k$, and such that all component modes are distinct.

Reparameterizing by $e^{-\phi_j} = \theta_j$, $j = 1, \ldots, k$, we take, as the test function,

$$p_\lambda(x) = \sum_{j=1}^{k} \pi_j \binom{N}{x} e^{-(\phi_j - \lambda)x} [1 - e^{-(\phi_j - \lambda)}]^{N-x},$$

and $0 < \lambda < \phi_1 < \cdots < \phi_k \leqslant \log 2$, say.

By an argument based this time on probability generating functions, it can be shown that there is a direct relationship between the $p(\cdot)$ and the $p_\lambda(\cdot)$, defined by

$$p_\lambda(x) = \sum_{y=x}^{N} e^{\lambda x} \binom{y}{x} (1 - e^{\lambda})^{y-x} p(x).$$

This allows calculation of the test function for any λ, and variation of λ reveals the number of components.

As has been emphasized, these procedures are based on knowledge of the density curve itself and they are aimed principally at finding out how many components there are. So far as the analysis of statistical data is concerned, application is encouraged by the statement of Medgyessy (1977) that the method will work if $p(x)$ is subject to a small amount of error. The technique of Example 4.7.3 for fitting a normal mixture could be applied if, say, a smooth kernel-based density estimate were generated from the data. As yet this suggestion has not been developed, although the method of Gregor (1969), reported earlier, incorporates a three-point moving-average smoothing of histogram frequencies before attempting the decomposition.

The principle of smoothing a set of mixture data before trying to decompose the underlying density has been investigated from a slightly different angle by Taylor (1965). Suppose it is felt that data, in histogram form, come from a normal mixture. The suggestion is to apply a moving average transformation to the histogram data and to relate the resulting picture to the corresponding transformation of the associated probabilities generated by a normal mixture. The practice goes back to some of the earlier work on electrophoresis in that, from the smoothed picture, the extreme left-hand 'hump', say, is picked off and the parameters of the corresponding normal component, including the mixing weight, are estimated using somewhat empirical procedures. The smoothed version of this fitted component is then 'subtracted' from the smoothed version of the histogram and the procedure is repeated as often as necessary. Taylor (1965) uses the method to decompose a set of data of 162 lamphrey lengths into seven components. Not only should the moving average operation filter off some of the small-sample variability, but it can be shown that subjective bias in the procedure by which the lamphrey are measured may well be corrected.

CHAPTER 5

Learning about the components of a mixture

In this chapter we look at two types of ignorance about the components of a mixture:

(a) ignorance of the number of components;
(b) ignorance of the indicator vector of a given observation from the mixture.

In the context of (b) we shall assess, in Section 5.7, the information that mixture data might contribute to the effectiveness of a discriminant rule.

The remainder of the chapter is devoted to problem (a).

5.1 INTRODUCTION

The most fundamental 'parameter' in the definition of a finite mixture distribution is k, the number of component subpopulations. Its importance results partly from technical considerations, in that without knowing k we cannot go further in the process of estimating the mixture density, but it is also often the crucial parameter so far as applications are concerned, as illustrated by the following hypothetical questions.

(a) Suppose data are accumulated from clinical tests on a group of patients. Do they suggest that there are two underlying subgroups, possibly corresponding to two different disease classes, or is there only one? Murphy (1964) asks a version of this question in relation to hypertension: is it valid to suggest the existence of a distinct condition called 'essential hypertension'?
(b) Do data on the length of a large number of fish suggest the presence of more than one year's spawning in the sample?
(c) Given a sediment sample that has undergone grain size analysis, how many minerals are present?

(d) In latent structure analysis, how many latent classes are required to provide a reasonable model?

Questions like these are clearly of great practical importance. They point to a close link with cluster analysis and some of them, such as (a) and (b), seem tailormade for the application of statistical testing procedures.

As we have already mentioned, many papers in applied fields talk not in terms of mixtures but of multimodal distributions. In direct applications of mixtures, the component densities involved usually differ at least in location so that, if the component densities are appreciably different, the mixture is multimodal. We have emphasized in Section 3.3.1 that even location mixtures need not be multimodal, the latter being a stronger concept, but we shall include in the present chapter inferential procedures for assessing modality. Rejection of unimodality can be regarded as conservative rejection of the pure-component hypothesis.

The techniques to be discussed vary greatly in their degree of formality. We shall see that in this context informal methods certainly play an important role, if only for the fact that the establishment of a general formal procedure has thus far proved impossible.

5.2 INFORMAL TECHNIQUES

In this section we merely recall the methodology of Sections 4.1 and 4.7. In both, the assessment of k was considered as an inevitable part of the process, whether by the successive stripping off of components by Medgyessy's method, or from the plots obtained in Section 4.1 by the methods of Tanner, Bhattacharya, Harding, or even using the familiar histogram. Some lack of identifiability is inevitable. Is a unimodal but squat histogram suggestive of a pure platykurtic component or of a mixture of two poorly separated normal densities?

5.3 FORMAL TECHNIQUES FOR SPECIAL CASES

The formal procedures envisaged here and in Section 5.4 will all involve statistical tests. If it is simply a question of whether or not the data come from a single component then we may always fall back on 'omnibus' tests of that null hypothesis. By the nature of omnibus tests, we stand to lose power by not taking into account the intended alternative mixture hypothesis but at least, for what it is worth, the critical region of the test will be well defined.

Example 5.3.1 Mixture of normal densities

Many tests of normality have been developed, quite apart from the distribution-free goodness-of-fit tests such as the chi-squared and Kolmogorov–Smirnov tests. Depending on the type of mixture envisaged, tests based on the standard coefficients of skewness and kurtosis can be used as well as other normal specific procedures; see, for instance, Oja (1981, 1983), Hall and Welsh (1983), and Spiegelhalter (1983).

Example 5.3.2 Mixture of two known densities

Let

$$p(x) = \pi_1 f_1(x) + (1 - \pi_1) f_2(x)$$

and suppose that $t(X)$ is some statistic whose means and variances are (μ_{t1}, μ_{t2}) and $(\sigma_{t1}^2, \sigma_{t2}^2)$, corresponding to the two components. Then, for the mixture,

$$\mathbb{E}t(X) = \pi_1 \mu_{t1} + (1 - \pi_1)\mu_{t2} = \mu_t(\pi_1)$$
$$\operatorname{var} t(X) = \pi_1 \sigma_{t1}^2 + (1 - \pi_1)\sigma_{t2}^2 + 2\pi_1(1 - \pi_1)(\mu_{t1} - \mu_{t2})^2 = \sigma_t^2(\pi_1)$$

If m_t denotes the sample moment corresponding to t, then, asymptotically, m_t is normally distributed, with means and variances as follows:

(a) if $\pi_1 = 1, m_t \sim N(\mu_{t1}, n^{-1}\sigma_{t1}^2)$;
(b) if $\pi_1 = 0, m_t \sim N(\mu_{t2}, n^{-1}\sigma_{t2}^2)$;
(c) if $0 < \pi_1 < 1, m_t \sim N[\mu_t(\pi_1), n^{-1}\sigma_t^2(\pi_1)]$.

Using these asymptotic results, we may easily construct approximate tests of $\mathrm{H}_0{:}\pi_1 = 0$ against $\mathrm{H}_1{:}\pi_1 > 0$ or $\mathrm{H}_0{:}\pi_1 = 1$ against $\mathrm{H}_1{:}\pi < 1$ and evaluate the power thereof (see Tiago de Oliveira, 1965).

Example 5.3.3 Mixture of two known densities which are symmetric and have equal variances

In this case, Johnson (1973) unconventionally takes, as the null hypothesis, H_0:mixture of two known densities which are symmetrical and have equal variances.

Suppose the known means are μ_1, μ_2 and that the sample mean is $\bar{x}$. Let

$$\hat{\pi} = (\bar{x} - \mu_2)/(\mu_1 - \mu_2),$$

the moment estimator of the first mixing weight π. Suppose, for chosen c,

$$P_j = \int_{-\infty}^{c} f_j(x)\,\mathrm{d}x, \qquad j = 1, 2,$$

and that

$$y_i = \begin{cases} 1, & \text{if } x_i < c, \\ 0, & \text{otherwise}, \ i = 1, \ldots, n. \end{cases}$$

Let $\pi^* = (\bar{y} - P_2)/(P_1 - P_2)$. Both $\hat{\pi}$ and π^* are unbiased estimators of π and, if H_0 is true, $\hat{\pi}$ and π^* should be similar. A possible test statistic is thus the standardized version of

$$\hat{\pi} - \pi^*.$$

A helpful feature is that $\text{var}(\hat{\pi}-\pi^*)$ is independent of π, if $c=\frac{1}{2}(\mu_1+\mu_2)$. Johnson (1973) gives values for the power of 5 per cent size tests of this type for the normal components case against the alternative

$$\mathrm{H}_1: \qquad \text{distribution is } N(\mu, \sigma^2).$$

A special circumstance arises if $\mu=\frac{1}{2}(\mu_1+\mu_2)$ because then the power remains constant as the sample size increases. Some further tests are briefly described by Johnson (1973).

Two-sample versions of the test statistics $\hat{\pi}$ and π^* are used by Choi (1979) to test whether or not two mixtures of the same two known component densities are identical. Comparison is made with the Wilcoxon–Mann–Whitney test and the two-sample Kolmogorov–Smirnov test. If the components are normal, the Kolmogorov–Smirnov test is less powerful than the three others, which are all comparable. With exponential components, the Kolomogorov–Smirnov test again does worst, but the test based on $\hat{\pi}$ is also comparatively poor unless the exponential parameters differ appreciably. In view of its simplicity, the test based on π^* seems the most attractive. (The recommendation is that c should be taken to be the sample median of the combined sample—a data-based choice!)

Example 5.3.4 Mixture of two univariate normals

H_0: $x_1,\ldots,x_n \sim N(\mu,\sigma_0^2)$, independently.
H_1: for some n_1 and some permutation τ of $\{1,\ldots,n\}$,

$$x_{\tau(1)},\ldots,x_{\tau(n_1)} \sim N(\mu_1,\sigma_1^2),\ x_{\tau(n_1+1)},\ldots,x_{\tau(n)} \sim N(\mu_2,\sigma_1^2).$$

The test statistic used is the maximum ratio of between sum of squares to within sum of squares,

$$F_{\max} = \max \frac{n_1 n_2(\bar{x}_1-\bar{x}_2)^2}{[(n_1-1)s_1^2+(n_2-1)s_2^2](n_1+n_2)},$$

where the maximum is over all partitionings of the data set into two groupings and the other notation is as usual. Calculation of $F_{\max}$ in practice is eased by the fact that the optimum partitioning is given by a single cut-off in the order statistic; see Engelman and Hartigan (1969), who give critical values $F_m(\alpha)$ for $F_{\max}$, for tests of size $(1-\alpha)$. Approximately, for large n,

$$\log_e[F_m(\alpha)+1] = \log_e(1-2/\pi) + z_\alpha(n-2)^{-1/2} + 2.4(n-2)^{-1},$$

where z_α is the 100α-percentile of the standard normal distribution.

Extension of this technique to higher dimensions and more than two components is discussed by Hartigan (1977). The $F_{\max}$ in Example 5.3.4 is asymptotically normally distributed, but this may not hold in more than one dimension. Hartigan (1977) makes some conjectures about the asymptotics.

This example brings us close to cluster analysis. In Section 4.3 we encountered a clustering-based method for classification of observations and parameter estimation originally proposed by Scott and Symons (1971). This relied on fixing k, but the whole procedure can be imbedded in one of the admittedly *ad hoc* routines for choosing the number of clusters (see Everitt, 1980, Section 3.4).

Example 5.3.5 Mixture of two Poissons

Let

$$p(x) = \pi_1 \mathrm{Po}(x, \theta_1) + (1 - \pi_1)\mathrm{Po}(x, \theta_2), \qquad x = 0, 1, \ldots,$$

where $\mathrm{Po}(x, \theta)$ denotes a Poisson probability mass function with mean θ. Then

$$\mathbb{E}(X) = \pi_1\theta_1 + (1 - \pi_1)\theta_2$$

and

$$\mathrm{var}(X) = \pi_1\theta_1 + (1 - \pi_1)\theta_2 + 2\pi_1(1 - \pi_1)(\theta_1 - \theta_2)^2.$$

Thus, if $0 < \pi_1 < 1$, $\mathrm{var}(X) - \mathbb{E}(X) > 0$, and a one-sided test based on $s^2 - \bar{X}$ suggests itself, where $\bar{X}$ is the sample mean and s^2 the sample variance. If there is a pure component, with mean θ,

$$\sqrt{n}(s^2 - \bar{X})/\sqrt{(1 - 2\sqrt{\bar{X}} + 3\bar{X})} \sim N(0, 1),$$

approximately (see Tiago de Oliveira, 1965).

5.4 GENERAL FORMAL TECHNIQUES

When we introduced the topic of this chapter in Section 1.2.2, we noted that certain obvious questions of interest could be formulated in hypothesis testing terms. In many applications, the problem reduces to one of testing between

$$\mathrm{H_o}: \quad \text{single normal component}$$

and

$$\mathrm{H_a}: \quad \text{mixture of two normals.}$$

Although we cannot hope for, say, uniformly most powerful tests, the fact that $\mathrm{H_o}$ can be regarded as a special case of $\mathrm{H_a}$ might encourage us to try the generalized likelihood ratio test, with its convenient asymptotic theory based on χ^2 distributions. However, even the introductory discussion in Section 1.2.2 suffices to demonstrate that the problem is not so straightforward.

Example 5.4.1 Mixture of two known densities

$$p(x \mid \pi) = \pi f_1(x) + (1 - \pi)f_2(x), \qquad 0 \leqq \pi \leqq 1.$$

Suppose $f_1(\cdot)$ is thought to be the main underlying density, with $f_2(\cdot)$ a possible contaminant. A natural null hypothesis is, therefore,

$$\mathrm{H_o}: \quad \pi = 1.$$

Given H_o and data $x = (x_1, \ldots, x_n)$, the likelihood is

$$L_o(x) = \prod_{i=1}^{n} f_1(x_i).$$

Let

$$L_a(x) = \prod_{i=1}^{n} [\hat{\pi} f_1(x_i) + (1 - \hat{\pi}) f_2(x_i)],$$

where $\hat{\pi}$ is the maximum likelihood estimator of π under the more general mixture model and let

$$2 \log \lambda = 2 \log [L_a(x)/L_o(x)].$$

If the usual asymptotic theory is valid we should expect that for large n, under H_o,

$$2 \log \lambda \sim \chi^2(1), \tag{5.4.1}$$

approximately. Note that, in this example, the number of degrees of freedom does not seem to be in doubt.

Consider, however, the value of $\hat{\pi}$. Let

$$\frac{\partial \mathscr{L}(\pi)}{\partial \pi} = \sum_{i=1}^{n} \frac{f_1(x_i) - f_2(x_i)}{\pi[f_1(x_i) - f_2(x_i)] + f_2(x_i)}$$

Then (cf. Section 4.3) if

$$\left.\frac{\partial \mathscr{L}}{\partial \pi}\right|_{\pi=0} < 0, \hat{\pi} = 0; \qquad \text{if } \left.\frac{\partial \mathscr{L}}{\partial \pi}\right|_{\pi=1} > 0, \hat{\pi} = 1;$$

otherwise $\hat{\pi}$ satisfies $\dfrac{\partial \mathscr{L}}{\partial \pi} = 0$. Now

$$\left.\frac{\partial \mathscr{L}}{\partial \pi}\right|_{\pi=1} = n - \sum_{i=1}^{n} r(x_i),$$

where $\quad r(x_i) = f_2(x_i)/f_1(x_i).$
Provided $f_1(\cdot)$ and $f_2(\cdot)$ are not identical,

$$\mathbb{E}[r(X) | H_o] = 1,$$

so that

$$\mathbb{E}[n - \sum_{i=1}^{n} r(X_i) | H_o] = 0$$

and, with positive probability, for all n,

$$\left.\frac{\partial \mathscr{L}}{\partial \pi}\right|_{\pi=1} > 0,$$

and therefore $2 \log \lambda = 0$.

In fact, asymptotically, under H_o, this occurs with probability $\frac{1}{2}$, so that the distributional result (5.4.1) will not hold.

The crux of the problem here is the need to impose consistency on $\hat{\pi}$, combined with the fact that the null value, 1, of π is on the boundary of the parameter space. Without the constraints, the problems would disappear.

Asymptotic theory does, however, suggest the following approximate test of H_0 for Example 5.4.1. Evaluate $\hat{\pi}$ and reject H_0 only if

$$\hat{\pi} < 1 \qquad \text{and} \qquad 2\log\lambda > \chi^2(2\alpha \mid 1),$$

where $\chi^2(\beta \mid 1)$ is the $100(1-\beta)$ percentile of the $\chi^2(1)$ distribution. This will give a significance test of approximate size α.

In more complicated problems, use of the GLR test statistic is much more difficult. In the early applications, the associated problems either were not recognized or were investigated using very restricted numerical studies. Wolfe (1970) used the chi-squared approximation to test whether a mixture of k d-variate normals gave a satisfactory fit to a set of data as opposed to a $(k+1)$—component mixture. He suggested allocating $(d+1)$ degrees of freedom to $2\log\lambda$ if the covariance matrices are assumed equal and $(d+1)(\frac{1}{2}d+1)$ otherwise, but subsequently withdrew this suggestion and, in a later report (Wolfe, 1971), proposes an approximation for a test between

$$\begin{aligned} &H_0: && k'\text{-component mixture of } d\text{-variate normals,} \\ &H_a: && k\text{-component mixture } (k > k'). \end{aligned}$$

Given that the covariance matrices are equal, he suggests, on the basis of a small simulation study, that, approximately,

$$2n^{-1}(n-1-d-\tfrac{1}{2}k)\log\lambda \sim \chi^2[2d(k-k')].$$

Everitt (1981a) carries out a more extensive simulation for the simple case $k'=1$, $k=2$, for d up to 10 and for n (sample size) between 25 and 500. He finds that Wolfe's suggestion is unsatisfactory unless n is at least about $10d$. Everitt (1981a) also shows that the power of the test is poor unless the component densities are well separated—as we have now come to expect!

If, under H_a,

$$p(x) = \pi f(x \mid \boldsymbol{\theta}_1) + (1-\pi) f(x \mid \boldsymbol{\theta}_2), \qquad 0 < \pi < 1,\ \boldsymbol{\theta}_1 \neq \boldsymbol{\theta}_2,$$

where $\boldsymbol{\theta}$ is q-dimensional, then Hartigan (1977) conjectures that the asymptotic null distribution of $2\log\lambda$ might be 'somewhere between' those of $\chi^2(q)$ and $\chi^2(q+1)$. The latter is justified by the difference in numbers of parameters and the former by the fact that H_0 is defined by the q constraints $\boldsymbol{\theta}_1 = \boldsymbol{\theta}_2$. He neglects our naive suggestion of Chapter 1 that, since H_0 can be defined by $\pi = 1$, $\chi^2(1)$ is a candidate!

The straight likelihood ratio test is also applied by Hasselblad (1969) and Aitkin and Tunnicliffe Wilson (1980), although the latter paper recognizes the theoretical incorrectness of the approximation. Binder (1978a) mentions a simulation study which refutes the standard '$2\log\lambda$' result and he also (1978b)

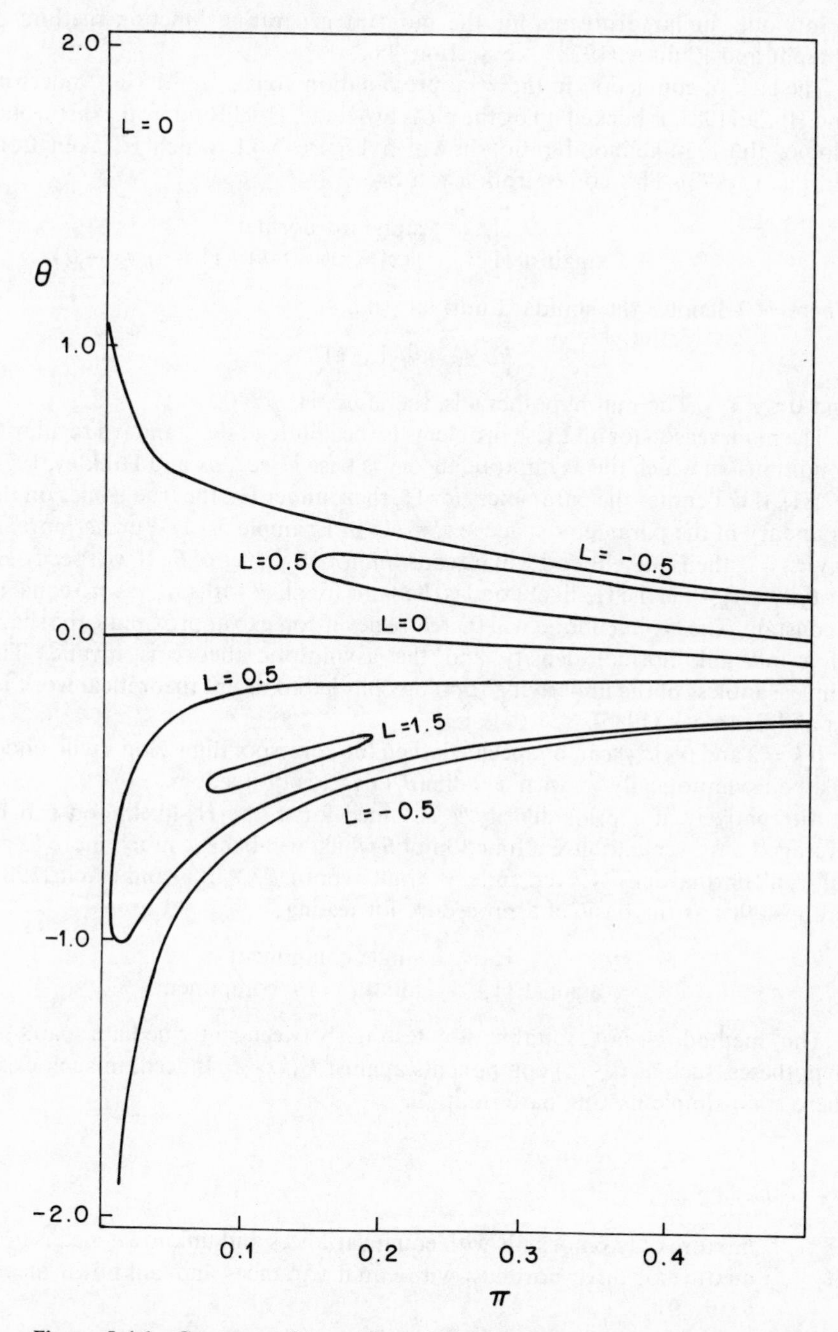

Figure 5.4.1 Contours of logarithm of likelihood ratio (L). Reproduced by permission of Academic Press from Hartigan (1977)

points out similar problems for the moment generating function method of Quandt and Ramsey (1978); see Section 4.6.

The lack of conviction in the χ^2 approximation voiced by Aitkin, Anderson, and Hinde (1981) is backed up by the plots of Minder (1980) and by the disturbing plot of the log-likelihood ratio shown in Figure 5.4.1, which is taken from Hartigan (1977). This comes from a test of

$$\begin{aligned} \mathrm{H_o}: &\quad \text{standard normal} \\ \text{against } \mathrm{H_a}: &\quad p(x) = \pi\phi(x-\theta) + (1-\pi)\phi(x-\theta'), \end{aligned}$$

where $\phi(\cdot)$ denotes the standard normal p.d.f.,

$$\theta' = -\pi\theta/(1-\pi)$$

and $0 \leqslant \pi \leqslant \frac{1}{2}$. The null hypothesis is, therefore, $\mathrm{H_o}:\theta = 0$.

The main reason for all these problems is the failure of the standard regularity conditions on which the asymptotic theory is based (see Cox and Hinkley, 1974, p. 281). If $\boldsymbol{\psi}$ denotes the parameters for $\mathrm{H_a}$ then, under $\mathrm{H_o}$, the true $\boldsymbol{\psi}_o$ lies on the boundary of the parameter space (e.g. $\pi = 1$ in Example 5.4.1). Furthermore, if, say, $\pi_j = 0$, the $\mathrm{H_a}$-likelihood will be constant for all values of $\boldsymbol{\theta}_j$. If we specify $\mathrm{H_0}$ by $\boldsymbol{\theta}_{k-1} = \boldsymbol{\theta}_k$, then the $\mathrm{H_a}$-likelihood is the same for all $\boldsymbol{\psi}$ with $(\pi_{k-1} + \pi_k)$ equal to a constant. The $\mathrm{H_a}$-likelihood will therefore never come to approximate the shape of a full-rank normal density and the asymptotic theory is invalid. The unpleasantness of the underlying topology has led to recent theoretical work by Li and Sedransk (1985).

If $k = 2$ and (π_1, π_2) can be specified, then the corresponding $2\log\lambda$ will, under $\mathrm{H_0}$, be asymptotically χ^2 with $d = \dim(\boldsymbol{\theta}_1)$ degrees of freedom.

Alternatively, if a prior density is assumed for $\boldsymbol{\pi}$, the $\mathrm{H_a}$-likelihood can be averaged out over $\boldsymbol{\pi}$ to give a function of $\boldsymbol{\theta}$ which will behave much more like a full-rank normal density, even under the null hypothesis. Aitkin and Rubin (1985) propose this as the basis of a procedure for testing

$$\begin{aligned} \mathrm{H_0}: &\quad \text{single component} \\ \text{against } \mathrm{H_a}: &\quad \text{mixture of } k \text{ components.} \end{aligned}$$

The method is not suitable for testing between intermediate pairs of hypotheses, such as $(k-1)$ components against k $(k > 2)$. Indeed, in such cases there is no simple nesting pattern at all.

Example 5.4.2

$\mathrm{H_0}$: mixture of two normals with equal variances and unknown means θ, θ'.

$\mathrm{H_a}$: mixture of three normals, with equal variances and unknown means $\theta_1, \theta_2, \theta_3$.

$\mathrm{H_0}$ is clearly a special version of $\mathrm{H_a}$, but it could be obtained from $\mathrm{H_a}$ by equating any pair from θ_1, θ_2, and θ_3.

Even the 'fully categorized data' version of this example is hard to deal with. In this case there are three independent samples, each from a normal population; the variances are assumed equal, but not the means.

The number of distinct means is either three, two, or one. The difficulty is that if the number is 'two', there are three choices for the odd one out. The test of 'one' against 'three' is just the one-way analysis of variance F-test, but the test of 'two' against 'three' is not at all simple. Certainly no recognized statistical tables can be used. Close parallels to this problem exist in the theory of change-point inference (Feder, 1975). The fact that we have trouble with this 'testing' aspect of the multiple comparisons problem does not leave much hope for the incomplete-data version, which is our mixture example.

In the absence of a general procedure, Aitkin, Anderson, and Hinde (1981) carry out a significance test by Monte Carlo methods for an example involving latent class models. Having estimated the parameters of the simpler model, they simulate nineteen independent sets of data and compare the resulting values of $2\log\lambda$ with that from the original data. Unfortunately, such simulation can be very expensive because of the time taken to compute maximum likelihood estimates.

An alternative general approach for testing a null hypothesis is that based on *locally optimal scores* (Neyman, 1959).

Example 5.4.3 (revisited) *Mixture of two known densities*

Define

$$u_n = (\sqrt{n}\gamma)^{-1}\left[\sum_{i=1}^{n} \frac{\partial}{\partial \pi} \log p(x_i|\pi)\bigg|_{\pi=1}\right],$$

where $\gamma^2 = \text{var}\,[(\partial/\partial\pi)\log p(X|\pi)|_{\pi=1}]$ and the variance is calculated under the null hypothesis, H_0: $\pi = 1$. Thus

$$u_n = (\sqrt{n}\gamma)^{-1} \sum_{i=1}^{n} [1 - f_2(x_i)/f_1(x_i)]$$

and

$$\gamma^2 = \text{var}\left[\frac{f_2(X)}{p(X|\pi)}\bigg|_{\pi=1}\right] = \int \frac{[f_2(x)]^2}{f_1(x)}\,dx - 1.$$

Under H_0, asymptotically, $u_n \sim N(0, 1)$ and H_0 is rejected for large negative values of u_n (see Durairajan and Kale, 1979).

Example 5.4.4 A two-parameter mixture

Suppose

$$p(x|\pi,\theta) = \pi f(x|1) + (1-\pi)f(x|\theta),$$

where $0 < \pi < 1, \theta \geq 1$, and $f(\cdot|\cdot)$ is of known functional form. Note that, although there are two parameters, one of them can be regarded as a nuisance parameter which 'disappears' under the null hypothesis expressed in terms of the other.

Suppose, for example, we express H_0 by $\theta = 1$, and consider, for given π, the locally optimal score statistic, $u_n(\pi)$. Since $p(x|\pi, \theta)$ is linear in π, $u_n(\pi)$ turns out to be independent of π:

$$u_n(\pi) = u_n = (\sqrt{n\gamma})^{-1} \sum_{i=1}^{n} \frac{\partial}{\partial\theta} \log f(x_i|\theta)\bigg|_{\theta=1},$$

with
$$\gamma^2 = \operatorname{var}\left[\frac{\partial}{\partial\theta} \log f(X|\theta)\bigg|_{\theta=1}\right].$$

We are left with the locally optimal scores test related to the 'pure' parametric density $f(x|\theta)$ at $\theta = 1$.

Example 5.4.5 A two-parameter exponential mixture

Suppose
$$p(x|\pi,\theta) = \pi e^{-x} + (1-\pi)\theta e^{-\theta x}, \qquad \theta \geqslant 1,$$

with H_0: $\theta = 1$. Then
$$u_n = (\sqrt{n})^{-1} \sum_{i=1}^{n} (1 - x_i) = \sqrt{n}(1 - \bar{x}).$$

We thus obtain, asymptotically, a similar test of H_0, rejecting H_0 if u_n is large and positive. The critical region and approximate power are easy to calculate, using the central limit theorem.

Of course, in Examples 5.4.3 and 5.4.4 the roles of θ and π can be reversed, as was suggested by Davies (1977) for Example 5.4.4. Then, however, the corresponding score function $u_n(\theta)$ is *not* independent of θ and a straightforward critical region cannot be constructed. Davies (1977) proposes a conservative procedure with a critical region of the form

$$\{\sup_{\theta} u_n(\theta) > c\}.$$

Davies (1977) gives guidelines on how to approximate to c, so as to obtain a test of given size. In principle, the method can be extended to deal with problems involving more than one nuisance parameter, such as more complicated mixture models. The practicalities are not well investigated as yet.

Further general techniques for model choice are those based on Bayes factors (i.e. posterior to prior odds ratios for the null against the alternative) or on penalized likelihoods. The general form of the latter is as follows. Suppose $\boldsymbol{\psi}_i$ is a vector of s_i parameters in model i, $i = 1, \ldots, M$, and that $\hat{\mathscr{L}}_i$ is the maximized log-likelihood in model i. Let

$$C_i = \hat{\mathscr{L}}_i - s_i d,$$

where d is some constant. Then model j is chosen, where $C_j = \max_i\{C_i\}$. The penalty $-s_i d$ is imposed to compensate for the fact that, the larger the number of parameters, the larger $\hat{\mathscr{L}}_i$ will probably be, irrespective of the real value of the extra complication.

This approach includes the specific procedures of Akaike (1974) and Schwartz (1978). Akaike's AIC criterion has been illustrated in a mixture/clustering context by Sclove (1983). A major problem is that the theoretical justifications for these criteria, Akaike's in particular, rely on the same conditions as the usual asymptotic theory of the GLR test.

Standard goodness-of-fit tests tend to be used in latent class analysis. Models are fitted with increasing k until the fit is acceptable. The problems associated with multiple 'correlated' significance tests tend to be ignored.

Not all problems involving tests about mixtures create such difficulties. Choi (1979) and McLachlan, Zawoko, and Ganesalingam (1982) develop a variety of tests, including the GLR test, for assessing whether two samples come from the same, or different, mixtures of two components.

5.5 THE STRUCTURE OF MODALITY

Although we have stressed that multimodality is not equivalent to the existence of a mixture, rejection of a null hypothesis of unimodality would suggest that the pure-component hypothesis should also be rejected, given that the underlying pure components are believed to be unimodal. Furthermore, it can be easier to establish formal tests for bimodality, or even bitangentiality, than was the case in Section 5.4. Before discussing this, we look in more detail at the structure of unimodality and multimodality.

The only case where there exists a large amount of general theory is for $k = 1$, unimodal densities, for which the following variety of necessary and sufficient conditions exist:

(a) The derivative of the density function, assuming its existence, has at most one distinct change of sign and, if there is one, the change is from positive to negative.
(b) The logarithm of the density function is concave (Ibragimov, 1956).
(c) If the mode is at the origin then the random variable of interest, X, can be written $X = UY$, where U and Y are independent random variables and $U \sim \text{Un}(0, 1)$; see Khinchin (1938) and Isii (1958).
(d) The cumulative distribution is either convex everywhere on the sample space, or concave everywhere, or changes once from convex to concave (Olshen and Savage, 1970). Note that Ghosh (1978) characterizes *bimodality* by the following pattern for the cumulative distribution function: convex, concave, convex, and finally concave. If one mode is at a boundary of the sample space then one or other of the four parts of the pattern disappears. These characteristics extend to multimodality.

The concept of unimodality is generalized by Olshen and Savage (1970) and Medgyessy (1977, Chapter II). Keilson and Gerber (1971) derive a result, parallel to (b), for the discrete case. Given probabilities $\{p_n\}$, unimodality is characterized by the existence of an M such that

$$p_n \geqslant p_{n-1}, \qquad n \leqslant M,$$
$$p_{n+1} \leqslant p_n, \qquad n \geqslant M.$$

This is equivalent to log-concavity of the $\{p_n\}$; that is,

$$p_n^2 \geqslant p_{n-1} p_{n+1},$$

for all n apart from at the boundaries.

We have already commented that not all mixture densities are multimodal. Clearly, all scale mixtures of normals with the same mean, and all exponential mixtures, are unimodal. Multimodality, and the confusing non-equivalence with mixtures, arises in the consideration of mixtures of densities with different location parameters. These may or may not be unimodal, depending on the degree of separation. By far the most work has been done on the following familiar example.

Example 5.5.1 Mixture of two univariate normals (Helguero, 1904; Harris and Smith, 1949; Eisenberger, 1964; Wessels, 1964; Robertson and Fryer, 1969; Behboodian, 1970b).

Let

$$p(x|\psi) = \pi\phi(x|\mu_1, \sigma_1) + (1-\pi)\phi(x|\mu_2, \sigma_2), \qquad -\infty < x < \infty.$$

Since the modality of p is unaffected by location and scale changes the important parameters are $\pi, \Delta = (\mu_2 - \mu_1)/\sigma_1$, and $\sigma = \sigma_2/\sigma_1$. Suppose, with no loss of generality, we take $\Delta \geqslant 0$. Then the following represents a complete description of the parameter space in terms of bimodality (Robertson and Fryer, 1969).

(a) p is unimodal if $0 \leqslant \Delta \leqslant \Delta_0$, where

$$\Delta_0 = [2(\sigma^4 - \sigma^2 + 1)^{3/2} - (2\sigma^6 - 3\sigma^4 - 3\sigma^2 + 2)]^{1/2}/\sigma,$$

(b) For $\Delta > \Delta_0$, p is bimodal if and only if π lies in the open interval (π_1, π_2), where

$$\pi_i^{-1} = 1 + \frac{\sigma^3 y_i}{\Delta - y_i} \exp\{-\tfrac{1}{2}y_i^2 + \tfrac{1}{2}[(y_i - \Delta)/\sigma]^2\}, \qquad i = 1, 2,$$

and (y_1, y_2) are the roots of

$$(\sigma^2 - 1)y^3 - \Delta(\sigma^2 - 2)y^2 - \Delta^2 y + \Delta\sigma^2 = 0,$$

with $0 < y_1 < y_2 < \Delta$.

(c) Otherwise, p is unimodal.

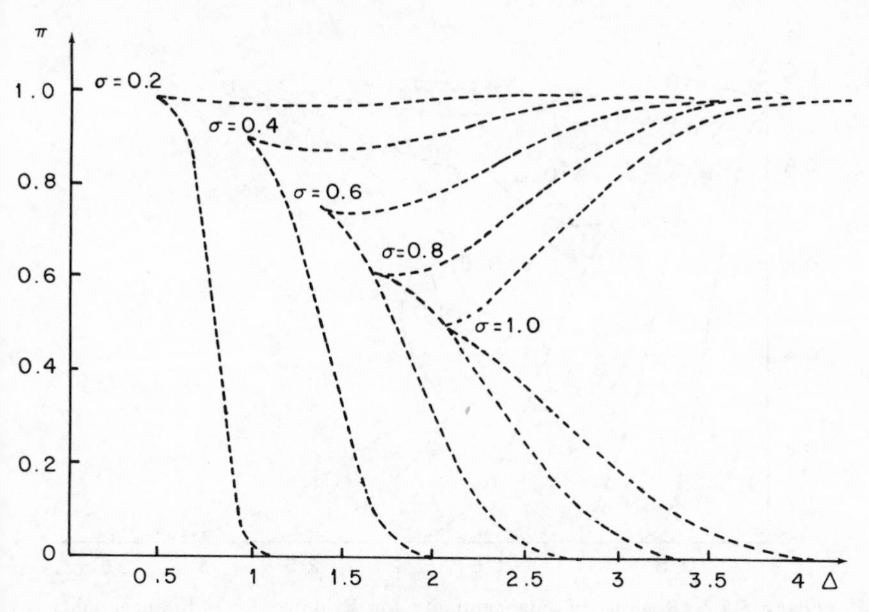

Figure 5.5.1 Regions of bimodality for Example 5.5.1. Bimodality obtains to the right of the dotted curves. Reproduced by permission of the Chief Editor, Scandinavian Actuarial Journal, from Robertson and Fryer (1969)

Figure 5.5.1, from Robertson and Fryer (1969), shows the bimodal range, in terms of π and Δ, for several values of σ. Given σ, the region to the right of the dotted curve corresponds to bimodality.

Interesting special results are as follows:

(d) If $\Delta \leqslant 2\min(1, \sigma)$, p is unimodal.
(e) If $\Delta \geqslant \frac{3}{2}\sqrt{3}\min(1, \sigma)$, then (π_1, π_2) exist such that p is bimodal for $\pi_1 < \pi < \pi_2$.
(f) If $\sigma = 1$ (equal variances) there is a critical Δ_π such that p is bimodal for $\Delta > \Delta_\pi$: for example, $\Delta_{0.5} = 2$ and $\Delta_{0.9} = 3.3$
(g) The separation of the modes is less than $|\mu_1 - \mu_2|$.

Clearly, in the current example, less separation is required for bitangentiality than for bimodality. Figure 5.5.2 displays bitangentiality regions obtained from numerical calculations by Robertson and Fryer (1969). They work from the characterization that bitangentiality is equivalent to the existence of four points of inflexion on $p(x \mid \psi)$, as opposed to just two, and provide a fairly full analysis of the phenomenon. Note that the bimodality regions are, of course, subsets of those for bitangentiality. Note also the following special remarks:

(a) For $\sigma^2 < 5 - \sqrt{24}$, there is a range of π for which bitangentiality occurs even at $\Delta = 0$.
(b) For $\sigma = 1$, the critical Δ_π for bitangentiality is quite stable, as π varies.

Wessels (1964) also gives quite a detailed study of this example.

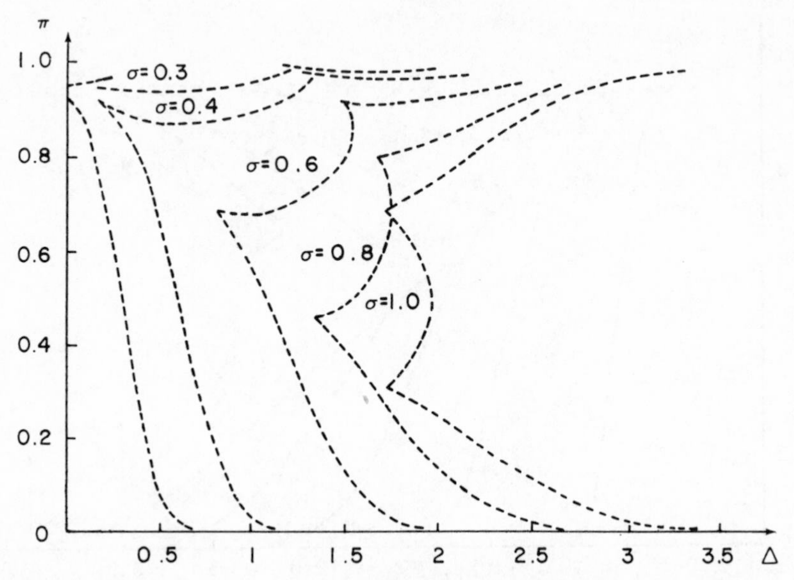

Figure 5.5.2 Regions of bitangentiality for Example 5.5.1. Bitangentiality obtains to the right of the dotted curves. Reproduced by permission of the Chief Editor, Scandinavian Actuarial Journal, from Robertson and Fryer (1969)

Example 5.5.2 Mixture of two multivariate normals

Let

$$p(x) = \pi\phi_d(\mathbf{x}|\boldsymbol{\mu}_1, \Sigma_1) + (1-\pi)\phi_d(\mathbf{x}|\boldsymbol{\mu}_2, \Sigma_2),$$

where $\phi_d(\mathbf{x}|\boldsymbol{\mu}, \Sigma)$ denotes the density function for a d-dimensional normal random vector with mean $\boldsymbol{\mu}$ and covariance matrix Σ. The modality of the density surface depends on the univariate sections. These can be represented by mixture densities of the form

$$q(y) = \pi\phi(y|\mu_1, \sigma_1) + (1-\pi)\phi(y|\mu_2, \sigma_2), \qquad -\infty < x < \infty,$$

where $\mu_j = \mathbf{c}^{\mathrm{T}}\boldsymbol{\mu}_j, \quad j = 1, 2$

$\sigma_j^2 = \mathbf{c}^{\mathrm{T}}\Sigma_j\mathbf{c}, \quad j = 1, 2,$

and $\mathbf{c}$ is an arbitrary unit vector.

Define

$$\delta^2(\mathbf{c}) = \frac{\mathbf{c}^{\mathrm{T}}(\boldsymbol{\mu}_2 - \boldsymbol{\mu}_1)(\boldsymbol{\mu}_2 - \boldsymbol{\mu}_1)^{\mathrm{T}}\mathbf{c}(\mathbf{c}^{\mathrm{T}}\Sigma_1\mathbf{c} + \mathbf{c}^{\mathrm{T}}\Sigma_2\mathbf{c})}{2(\mathbf{c}^{\mathrm{T}}\Sigma_1\mathbf{c})(\mathbf{c}^{\mathrm{T}}\Sigma_2\mathbf{c})}.$$

Then, from result (e) in Example 5.5.1, there is no bimodality, for any π, if

$$\sup_{\mathbf{c}} \delta^2(\mathbf{c}) < 27/8.$$

Calculation of the extreme $\mathbf{c}$ is not trivial. For the special case $\Sigma_1 = \Sigma_2 = \Sigma$, the modes will fall on the section through $\boldsymbol{\mu}_1$ and $\boldsymbol{\mu}_2$. Furthermore,

$$\delta^2(\mathbf{c}) = \frac{\mathbf{c}^{\mathrm{T}}(\boldsymbol{\mu}_2 - \boldsymbol{\mu}_1)(\boldsymbol{\mu}_2 - \boldsymbol{\mu}_1)^{\mathrm{T}}\mathbf{c}}{\mathbf{c}^{\mathrm{T}}\Sigma\mathbf{c}},$$

$$\mathbf{c} \propto \Sigma^{-1}(\boldsymbol{\mu}_2 - \boldsymbol{\mu}_1),$$

corresponding to the familiar linear discriminant function. Konstantellos (1980) considers Example 5.5.2 in more detail.

Example 5.5.3 Mixture of two von Mises densities

Let

$$p(x) = \pi_1 M(x; 0, K_1) + (1 - \pi_1)M(x; \theta_0, K_2), \qquad 0 < x < 2\pi,$$
$$0 \leqslant \theta_0 \leqslant \pi, K_1, K_2 > 0,$$

where $\quad M(x; \theta, K) = [2\pi I_0(K)]^{-1} \exp[K \cos(x - \theta)]$

and $I_0(K)$ is the zero-order modified Bessel function of the first kind (Mardia, 1972, p. 57). Note that π, unsubscripted, is used to represent 3.14159.... Detailed conditions for determining whether there is unimodality or bimodality are given by Mardia and Sutton (1975). When $\theta_0 = \pi$, bimodality obtains for an intermediate range of π_1, for any K_1, K_2. For other θ_0, the situation is more complex.

Mardia (1972, Sections 3.6, 5.1, and 5.4.3) uses a concatenation of k von Mises densities on $(0, 2\pi/k)$ as a k-modal density on the circle; bimodal distributions on the sphere are discussed in Mardia (1972, Section 8.5).

In Section 2.2, we mentioned, as alternatives to mixture densities, other parametric densities which may manifest multimodality.

Example 5.5.4 The quartic exponential density

Let

$$p(x) = C(\boldsymbol{\alpha}) \exp[-(\alpha_1 x + \alpha_2 x^2 + \alpha_3 x^3 + \alpha_4 x^4)], \qquad \alpha_4 > 0, -\infty < x < \infty,$$

where $C(\boldsymbol{\alpha})$ is the normalizing constant. As mentioned in Section 2.2, the stationarity equation for $p(x)$ is a cubic which has one or three real roots, depending on whether $p(x)$ is unimodal or bimodal. Transformation to $y = x + \alpha_3/4\alpha_4$ gives

$$p(y) \propto \exp[-(\beta_1 y + \beta_2 y^2 + \beta_4 y^4)],$$

where
$$\beta_1 = (8\alpha_1\alpha_4^2 - 4\alpha_2\alpha_3\alpha_4 + \alpha_3^3)/8\alpha_4^2,$$
$$\beta_2 = (8\alpha_2\alpha_4 - 3\alpha_3^2)/8\alpha_4,$$
$$\beta_3 = \alpha_4.$$

From the theory of cubic equations (Korn and Korn, 1968, p. 23), bimodality follows if and only if $\delta < 0$, where $\delta = 27\beta_1^2\beta_4 + 8\beta_2^3$. Note that this implies $\beta_2 < 0$.

The special form

$$p(x) \propto \exp\{-[\beta(x-\mu)^2 + \gamma(x-\mu)^4]\}, \qquad \gamma > 0,$$

provides a three-parameter density which is symmetric about μ. Bimodality, with modes at $\pm\sqrt{(-\beta/2\gamma)}$, occurs if and only if $\beta < 0$ (see Matz, 1978).

As remarked in Section 2.2, more general exponential-family densities exist which allow more than two modes. Densities based on Gram–Charlier and Edgeworth expansions can also exhibit multimodality and we close this

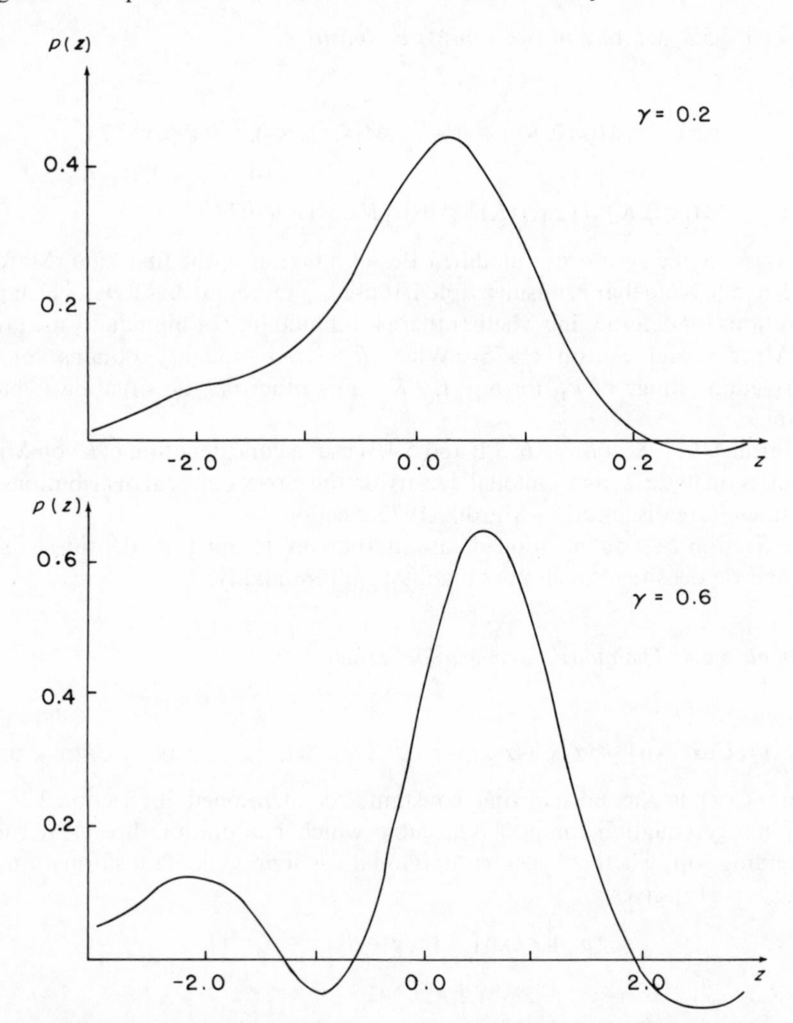

Figure 5.5.3 Curves for the truncated expansion from equation (5.5.1), for $\gamma = 0.2$ and $\gamma = 0.6$

subsection with a simple example based on the first two terms of the Gram–Charlier expansion.

Example 5.5.5 A simple truncated Gram–Charlier expansion

$$p(z) = [1 + \gamma H_3(z)]\phi(z), \tag{5.5.1}$$

where $H_3(z) \equiv -(z^3 - 3z)$ is the third-order Hermite polynomial and $z = (x - \mu)/\sigma$ is a standardized variable.

For any γ, $p(z)$ integrates to one. As γ increases from zero, a secondary mode gradually appears but simultaneously it becomes obvious that, for some z, $p(z)$ goes negative; see Figure 5.5.3. A symmetric 'true' density in the spirit of (5.5.1) is given by

$$p(z) = (1 + \gamma z^2)\phi(z)/(1 + \gamma), \qquad \gamma > 0, -\infty < z < \infty. \tag{5.5.2}$$

This is bimodal, with modes at $z = \pm\sqrt{(2 - \gamma^{-1})}$, if and only if $\gamma > \frac{1}{2}$.

As in the case of Example 5.5.4, the inclusion of more terms in the expansions allows the possibility of more than two modes.

5.6 ASSESSMENT OF MODALITY

An early non-parametric approach for grouped data is described by Haldane (1952), based on the assessment of local curvature of the histogram.

A general approach with parametric models depends on asymptotic theory. Suppose the parameter vector is ψ, of length r, that $\hat{\psi}$ denotes the maximum likelihood estimator, and that $V(\psi)$ is the asymptotic covariance matrix of $\hat{\psi}$. Then, as a large-sample approximation, given appropriate regularity conditions,

$$(\psi - \hat{\psi})^{\mathrm{T}} V(\psi)^{-1}(\psi - \hat{\psi}) \sim \chi^2(r),$$

or, more usefully,

$$(\psi - \hat{\psi})^{\mathrm{T}} K(\hat{\psi})(\psi - \hat{\psi}) \sim \chi^2(r), \tag{5.6.1}$$

where $K(\hat{\psi})$ is the sample information matrix, evaluated at $\hat{\psi}$. This can be used to obtain a region for ψ with an approximate confidence cover of any prescribed level. If this region can be shown to lie completely within the part of the parameter space which represents bimodality or bitangentiality, then there is sufficiently strong evidence for rejecting the complementary null hypothesis (recall Figure 3.3.1 for normal mixtures). The problem about applying this is the irregularity of the maximum likelihood theory for mixtures near the subspace of the parameter space corresponding to the single-population hypothesis (see Section 5.4). In some examples, $\hat{\psi}$ may lie far enough away from this subspace to permit the use of (5.6.1).

For the non-mixture densities, such as those in Examples 5.5.4 and 5.5.5, for example, there is more regularity. For the quartic exponential, all is well because of the properties of exponential family densities. For the symmetric case, for which

$$p(x) \propto \exp\{-[\beta(x-\mu)^2 + \gamma(x-\mu)^4]\}, \qquad \gamma > 0, -\infty < x < \infty,$$

unimodality is equivalent to $\beta > 0$. Approximately, $(\beta - \hat{\beta})/\sqrt{V_\beta(\hat{\psi})} \sim N(0, 1)$, where $V_\beta(\psi)$ is the asymptotic variance of β, calculated in the usual way; this can form the basis of an approximate test of unimodality.

Similarly, for the density given by (5.5.2), with non-standardized form

$$p(x) = \left[1 + \gamma\left(\frac{x-\mu}{\sigma}\right)^2\right]\phi(x|\mu, \sigma)/(1+\gamma),$$

we may test for unimodality between

$$H_0: \quad \gamma \leqslant \tfrac{1}{2},$$

and

$$H_1: \quad \gamma > \tfrac{1}{2}.$$

A similar procedure could be applied in the other examples, such as the non-symmetric quartic exponential, although it would be more difficult to assess whether the confidence region lay within, say, the bimodal part of the parameter space (see Cobb, Koppstein, and Chen, 1983). For Example 2.2.9(c), an approximate confidence region for (β_1, β_2) may be drawn as an ellipse on Figure 2.2.3. If there is a great deal of data, the 'area' covered by the confidence region will be very small, so that we may base our conclusion on the value of $\hat{\psi}$; see Mardia and Sutton (1975) for an illustration of this for von Mises mixtures.

General approaches to modality are possible through non-parametric density estimation. One method is to carry out maximum likelihood estimation without imposing parametric assumptions, but with various restrictions on the underlying density:

(a) with the restriction of unimodality;
(b) with the restriction (less rigid) of bimodality.

This will involve the use of isotonic fitting techniques (Barlow *et al.*, 1972). The corresponding likelihood ratio suggests itself as a test statistic, but as yet the distribution theory on which a significance test might be based is only conjectured; see Hartigan (1977) who also suggests, without providing detailed results, some other approaches to testing bimodality.

Another non-parameteric density estimation procedure, the kernel method, also permits assessment of modality under certain circumstances (Silverman, 1981).

Given a set of independent and identically distributed univariate observations $D = (x_1, \ldots, x_n)$, the kernel-based density estimate is given (see Section 2.2) by

$$p(x|D, \lambda) = (n\lambda)^{-1} \sum_{i=1}^{n} K[(x - x_i)/\lambda].$$

Suppose that the smoothing parameter, λ, is to be chosen from the range $(0, \infty)$, that the kernel function, $K(\cdot)$, has a unique mode, at zero, and that, for ease of discussion only, no two of $x_1, \ldots, x_n$ are equal. For very small λ, the density estimate will have n modes, one corresponding to each data point. As λ increases, it seems plausible that the number of modes should decrease.

Define the *k-critical smoothing parameter*, $\lambda_c(k)$, by

$$\lambda_c(k) = \inf\{\lambda; p(\cdot|D, \lambda) \text{ has at most } k \text{ modes}\}.$$

If $\lambda_c(k)$ is 'large', it is natural to reject the hypothesis that the underlying density has k modes or fewer: we are having to smooth inordinately much in order to obtain a density estimate with so few modes. The assessment of significance and, indeed, calculation of $\lambda_c(k)$ are eased if we can assert that

$$p(\cdot|\lambda, D) \textit{ has more than } k \textit{ modes if and only if } \lambda < \lambda_c(k). \qquad (*)$$

If this is true, the modality of $p(\cdot|\lambda, D)$ determines whether $\lambda < \lambda_c(k)$. With a view to assessing modality, suppose we adopt a null hypothesis of k modes. Then, if (*) is true, the significance level of this against the alternative of more than k modes is

$$\Pr(\lambda_c(k) > \lambda_0 | \text{a random } n\text{-sample from } p(\cdot)), \qquad (5.6.2)$$

where λ_0 is the value of $\lambda_c(k)$ obtained from the *given* n-sample, D, and $p(\cdot)$ is the underlying density function. Silverman (1981) shows how to apply a *bootstrap* procedure with which to use D to simulate density estimates corresponding to n-samples from $p(\cdot)$ and thereby to obtain a Monte Carlo estimate of (5.6.2). Again, (*) is useful here because whether $\lambda_c(k) > \lambda_0$ is true for a particular simulation can be seen by counting the number of modes in the simulated density estimate with smoothing parameter λ_0. In practice, the k-modes hypotheses are tried successively with $k = 1, 2, \ldots$ until a satisfactory fit is obtained.

Whether (*) can be shown to hold is vital. It is a plausible result in general, but so far it has only been proved for the case of a Gaussian kernel function

$$K(z) = (2\pi)^{-1/2} \exp(-\tfrac{1}{2}z^2), \qquad -\infty < z < \infty$$

(see Silverman, 1981). In this context, therefore, the choice of kernel function itself is important.

Silverman (1983) consolidates the method with detailed theoretical analysis in which, subject to certain regularity conditions, the following are established:

(a) When the true density, p, does indeed have at most k modes, $\lambda_c(k)$ tends stochastically to zero.
(b) When p has more than k modes, $\lambda_c(k)$ does not tend to zero.
(c) When p has at most k modes, the rate of convergence of $\lambda_c(k)$ to zero is the same as that which is optimal for uniform estimation of the density using the kernel method.

For a recent test for unimodality based on the empirical distribution function, see Hartigan and Hartigan (1985).

5.7 DISCRIMINANT ANALYSIS

Suppose an individual is known to belong to one or other of k classes, but it is not known to which. The objective of discriminant analysis is to make inferences about the unknown class membership, c. With a view to aiding this inference, a so-called feature vector of measurements, x, is obtained from the individual. Crucial quantities, so far as discriminant analysis is concerned, are the probabilities

$$\{p(c|x), c = 1, \ldots, k\}.$$

Statistical discriminant analysis involves the construction of models for these probabilities and the development of related inference procedures. In practice, the $\{p(c|x)\}$ will not be known, but they may be estimated using data from other individuals in the same set of classes. These individuals form the so-called *training data*.

Sometimes, the relative values of $\{p(c|x)\}$, given x, are used in conjunction with a loss function to assign the individual to one of the classes. In the medical context, this constitutes a diagnosis and may dictate the subsequent treatment. For the simplest zero–one loss functions, the decision is based on the relative sizes of the $\{p(c|x)\}$. They, or their ratios, called *odds ratios*, are therefore of great importance. The assignment rules are *likelihood ratio* rules.

Usually, particularly in textbook discussions, the training data consist of only *fully categorized* cases. The books by Lachenbruch (1975) and Hand (1981) cover the general topic. However, in some applications it may well be that there are also a number of uncategorized cases. In this section, we consider the special question of whether these data contribute to the discriminatory power of the total training set. This question is particularly important in the following circumstances:

(a) There is little or no fully categorized data.
(b) Diagnosis of the new individual cannot wait until the other data have been fully categorized.
(c) It is very costly to confirm the categories of the training data (Brown, 1976). In some circumstances, it may be possible to weigh the cost of categorization against the extra discriminatory power that may be gained thereby.

A cornerstone of the discussion is the joint probability function

$$\{p(x, c); x \in \mathscr{X}, c = 1, \ldots, k\}$$

and the two factorizations that arise from it:

$$p(x, c) = p(x|c)\,p(c) \tag{5.7.1}$$

and

$$p(x, c) = p(c|x)\,p(x). \tag{5.7.2}$$

Although these are equivalent formulations, two different approaches to parametric statistical modelling have been developed, one associated with each factorization. The following terminology is due to Dawid (1976).

(a) Sampling Paradigm

Introduce parametric models for the two factors of (5.7.1), so that

$$p(x,c) = p(x|c,\boldsymbol{\theta})\,p(c|\boldsymbol{\pi}). \tag{5.7.3}$$

Thus the prior probabilities of class membership and the conditional sampling densities of x are modelled. The marginal density for x is of the mixture form and the probabilities of interest, from the point of view of discriminant analysis, are obtained, by Bayes' theorem, as

$$p(c|x) = p(x|c,\boldsymbol{\theta})\,p(c|\boldsymbol{\pi})/\sum_{c'} p(x|c',\boldsymbol{\theta})\,p(c'|\boldsymbol{\pi}), \qquad c = 1,\ldots,k.$$

(b) Diagnostic Paradigm

In contrast to (a), we model the factors of (5.7.2):

$$p(x,c) = p(c|x,\boldsymbol{\beta})\,p(x|\boldsymbol{\gamma}). \tag{5.7.4}$$

A popular example is the linear logistic model, in which

$$p(c|x,\boldsymbol{\beta})/p(k|x,\boldsymbol{\beta}) = \exp(\boldsymbol{\beta}_c^{\mathrm{T}} x), \qquad c = 1,\ldots,k-1.$$

Of course, given a parametric model for (5.7.3), there is an equivalent one for (5.7.4) with $\boldsymbol{\beta}$ and $\boldsymbol{\gamma}$ as functions of $(\boldsymbol{\theta}, \boldsymbol{\pi})$. Generally, however, the two approaches produce different models for $p(x, c)$. Typically, in the sampling paradigm, $\boldsymbol{\theta}$ and $\boldsymbol{\pi}$ are chosen to be distinct parameters and, in the diagnostic paradigm, $\boldsymbol{\beta}$ and $\boldsymbol{\gamma}$ are distinct. Sometimes $\boldsymbol{\gamma}$ is not considered at all. It is undoubtedly true that, so far, the sampling paradigm has been extremely popular. However, Dawid (1976) argues that, in some contexts, it may be more realistic to model the factors in the diagnostic paradigm. Often the results achieved by corresponding versions of the two paradigms are quite similar (Efron, 1975), but there is a disturbing qualitative difference between the two paradigms so far as the usefulness of uncategorized cases in discrimination is concerned. We shall examine this problem using a likelihood-based approach.

If the diagnostic paradigm is used, with $\boldsymbol{\beta}$ and $\boldsymbol{\gamma}$ distinct, then the uncategorized data $\{x_i\}$ contribute factors $\{p(x_i|\boldsymbol{\gamma})\}$. They do not give us any information at all about $\boldsymbol{\beta}$, and therefore about $\{p(c|x,\boldsymbol{\beta})\}$. On the other hand, if the sampling paradigm is used, and is the correct model, the uncategorized data provide information about the mixture model for $p(x)$. If the mixture is identifiable, and we have a very large amount of data available, then we are close to knowing the true mixture density and, therefore, the true 'optimal' likelihood ratio discriminant rules! This happy eventuality occurs even in the extreme case of M0 data, in which there is no fully categorized observation!

We are left in practice with a real dilemma. If we adopt the diagnostic paradigm when it is false, we shall lose valuable information: if we wrongly use the sampling paradigm, we are unnecessarily incorporating useless data.

Usually, the two sets of 'distinct' parameterizations are not equivalent. An exception is the case where the sample space for the feature variable is multinomial. In this case, the complete-data sample space is that of a $k \times l$ contingency table, where k is the number of classes and l is the number of cells in the feature variable's multinomial distribution. The cell probabilities $\{\theta_{jh}\}$ have the two equivalent parameterizations defined as in Example 4.3.8, one representing each paradigm. Whichever paradigm is used for maximum likelihood estimation (the diagnostic paradigm is by far the easier, as indicated earlier), the discriminatory information is concentrated in the fully categorized data. Although we have talked in terms of 'parameterizations' for this example, it is worth bearing in mind the fact that this model can be regarded as *the non-parametric* model for discrete data and this is the real reason for the equivalence.

If the sampling paradigm is appropriate, so that the underlying model for the uncategorized cases is a parametric mixture model, then, given identifiability of the mixture, it does appear to be worth while to use a discriminant rule which includes them. We shall consider in detail the familiar problem in which there are two classes and, within class j,

$$x \sim N_d(\boldsymbol{\mu}_j, \Sigma), \qquad j = 1, 2,$$

where d denotes the dimensionality of the feature vector, x. The marginal distribution of x is therefore that of a mixture of the two multivariate normal distributions.

With the assumption of equal covariance matrices and given mixing weights, or incidence rates, π_1, π_2, the rule for assigning an unconfirmed feature vector $\mathbf{y}$ is based on the linear discriminant function (LDF)

$$L(\mathbf{y}) = \boldsymbol{\delta}^{\mathrm{T}}\mathbf{y} + \beta, \tag{5.7.5}$$

where $\boldsymbol{\delta} = \Sigma^{-1}(\boldsymbol{\mu}_2 - \boldsymbol{\mu}_1)$,

$$\beta = \log(\pi_2/\pi_1) - \tfrac{1}{2}\boldsymbol{\delta}^{\mathrm{T}}(\boldsymbol{\mu}_2 + \boldsymbol{\mu}_1)$$

(cf. Example 4.3.4).

If $L(\mathbf{y}) > 0$, $\mathbf{y}$ is assigned to the second category and, otherwise, to the first.

In practice, of course, the parameters are unknown. Usually, estimates are substituted for π_1, π_2, $\boldsymbol{\mu}_1$, $\boldsymbol{\mu}_2$ and Σ and the resulting discriminant function

$$\hat{L}(\mathbf{y}) = \hat{\boldsymbol{\delta}}^{\mathrm{T}}\mathbf{y} + \hat{\beta} \tag{5.7.6}$$

is used in the same manner as $L(\mathbf{y})$.

For any assignment rule, its efficiency may be judged on the basis of the consequent *expected error rate*—the expected value of the probability of misclassifying a 'random' categorized case. If all parameters are known and the LDF (5.7.5) is used, this probability is easy to evaluate and is given by

$$\pi_1\Phi(-\tfrac{1}{2}\Delta + \lambda) + \pi_2\Phi(-\tfrac{1}{2}\Delta - \lambda), \tag{5.7.7}$$

where $\lambda = \log(\pi_2/\pi_1)$ and $\Delta^2 = (\boldsymbol{\mu}_1 - \boldsymbol{\mu}_2)^{\mathrm{T}}\Sigma^{-1}(\boldsymbol{\mu}_1 - \boldsymbol{\mu}_2)$, the squared Mahalanobis distance. If, however, an estimated rule such as (5.7.6) has to be used, then the expected value of the misclassification probability has to be worked out over the variation underlying **y**, $\hat{\beta}$ and $\hat{\boldsymbol{\delta}}$. Explicit calculations are feasible only asymptotically, when the size of the data set yielding $\hat{\beta}$ and $\hat{\boldsymbol{\delta}}$ tends to infinity. To

Table 5.7.1 Asymptotic relative efficiency of normal discrimination based on a sample with fraction γ unclassified. Reproduced by permission of the American Statistical Association from O'Neill (1978)

π_1	Δ	$\text{ARE}_1(\lambda, \Delta, \gamma)$	$\text{ARE}_\infty(\lambda, \Delta, \gamma)$	$q(\lambda, \Delta, \gamma)$
		(a) $\gamma = 1$		
0.5	2	.100	.100	1
	2.5	.213	.213	1
	3	.360	.360	1
	3.5	.513	.513	1
	4	.655	.655	1
0.667	2	.085	.122	1.61
	2.5	.194	.238	1.35
	3	.338	.382	1.24
	3.5	.496	.535	1.18
	4	.642	.672	1.15
0.9	2	.058	.199	5.82
	2.5	.134	.339	3.95
	3	.254	.490	3.08
	3.5	.403	.630	2.65
	4	.558	.749	2.42
		(b) $\gamma = 0.5$		
0.5	2	.550	.550	1
	2.5	.607	.607	1
	3	.680	.680	1
	3.5	.756	.756	1
	4	.828	.828	1
0.667	2	.543	.561	1.15
	2.5	.597	.619	1.14
	3	.669	.691	1.13
	3.5	.748	.767	1.13
	4	.821	.836	1.12
0.9	2	.573	.600	1.78
	2.5	.590	.670	1.78
	3	.636	.745	1.87
	3.5	.704	.815	1.96
	4	.780	.874	2.03

ascertain the value of uncategorized observations, such asymptotic error rates have to be evaluated for two situations: firstly, when all observations are fully categorized and, secondly, under the assumption that, on average, a proportion, γ, say, of the observations are uncategorized. The ratio of the asymptotic error rates for the two cases mentioned above gives a measure of asymptotic relative efficiency (ARE) of the incomplete data against the complete. If maximum likelihood estimates are used in (5.7.6), possibly using the relevant version of the EM algorithm, it turns out that this ARE depends only on π_1 (or, equivalently, λ), Δ, d, and γ. A selection of values of $\text{ARE}_d(\lambda, \Delta, \gamma)$ are given in Table 5.7.1, which has been extracted from O'Neill (1978).

Although Table 5.7.1 gives values only for $d = 1, \infty$, all remaining values may be calculated from the formula

$$\text{ARE}_d(\lambda, \Delta, \gamma) = \frac{q(\lambda, \Delta, \gamma)\,\text{ARE}_1(\lambda, \Delta, \gamma) + (d-1)\,\text{ARE}_\infty(\lambda, \Delta, \gamma)}{q(\lambda, \Delta, \gamma) + (d-1)}$$

The appropriate $\{q(\lambda, \Delta, \gamma)\}$ also appear in Table 5.7.1; see O'Neill (1978) for further details.

Various interesting features can be seen in Table 5.7.1. For the symmetric mixture with $\lambda = 0\,(\pi_1 = \pi_2 = 0.5)$, $\text{ARE}_d(0, \Delta, \lambda)$ is independent of the dimensionality, d. Furthermore, unless π_1 is near 1 (or 0, since the results are symmetric in π_1 about 0.5) the ARE does not vary much with d. The separation between the component distributions, measured by Δ, is more influential, as usual, and the usefulness of uncategorized cases is quite surprising when Δ is large. For instance, $\text{ARE}_1(0, 4, 1) = 0.655$, where '1' as the third argument means that the data set contained no confirmed case at all. This further emphasizes the importance of choosing correctly between the sampling and diagnostic paradigms and, to some extent, counters the negative opinion expressed by Little (1978), to the effect that it is not worth including uncategorized cases in the discriminant function.

The above results, which were published almost simultaneously by O'Neill (1978) and Ganesalingam and McLachlan (1978), are asymptotic. There is, however, empirical evidence from simulation studies and application to real data that the small-sample behaviour is also encouraging. Ganesalingam and McLachlan (1979a) report on experiments with $d = 1, 2, 3$ and sample sizes as low as 20. They compare average error rates for the cases $\gamma = 0$ (all data fully categorized) and $\gamma = 1$ (all data uncategorized) with those from the optimal rule based on the true parameter values. The results in Table 5.7.2 (Table IV of Ganesalingam and McLachlan, 1979a) show that the simulated efficiencies, relative to the optimal rule, are generally higher than the asymptotic values. It seems that, for instance, the asymptotic relative efficiencies are a reliable guide, for the univariate case, only if $n \geq 100$ and $\Delta \geq 2$. Furthermore, estimation of the parameters β, $\boldsymbol{\delta}$ of the LDF is inefficient if the data are uncategorized. In spite of this, however, the error rate incurred by the estimated LDF is not much worse than that for fully categorized data.

It is clear that, provided the normal mixture model is appropriate and that the component densities are well separated according to Δ, uncategorized cases are

Table 5.7.2 Asymptotic and, in parentheses, simulated efficiencies (%) of the linear discriminant rule based on uncategorized data relative to that based on fully categorized data; $n =$ total sample size. Reproduced by permission of Gordon and Breach, Science Publishers, Inc., from Ganesalingam and McLachlan (1979a).

	$d = 1. n = 20$		$d = 2. n = 20$		$d = 3. n = 40$	
Δ	$\pi_1 = 0.25$	$\pi_1 = 0.5$	$\pi_1 = 0.25$	$\pi_1 = 0.5$	$\pi_1 = 0.25$	$\pi_1 = 0.5$
1	0.25 (33.01)	0.51 (25.12)	0.34 (46.71)	0.51 (63.11)	0.42 (25.00)	0.51 (43.39)
2	7.29 (22.05)	10.08 (17.74)	9.36 (25.73)	10.08 (16.26)	10.51 (16.28)	10.08 (14.51)
3	31.41 (19.57)	35.92 (23.54)	35.13 (43.91)	35.92 (29.63)	36.78 (29.01)	35.92 (23.46)

quite informative, despite the fact that an iterative technique will be required to obtain estimates of parameters. Figure 5.7.1 shows the discriminant boundaries obtained with the two-dimensional sepal data from the *Iris versicolour* and *Iris setosa* samples of Fisher (1936). In the diagram, taken from O'Neill (1978), the boundaries that result from taking the data as 'fully categorized' and 'uncategorized' are remarkably close together. The two samples are, however, very well separated.

Further empirical results are given by Titterington (1976) and Makov (1980a) using the sequential updating methods described in Chapter 6. In spite of the comparative ease of implementation of the sequential procedures, the use of maximum likelihood estimates, when practicable, is clearly desirable. Han (1978) considers the situation where there are some fully categorized cases and a further sample, *all* of which come from the same, but unknown, component population. (The underlying model is assumed to be a mixture of two multivariate normals with equal covariance matrices.) The problem of updating a logistic discriminant function is mentioned in Anderson (1979), who suggests the use of maximum likelihood estimates from Ml data. Of course, the logistic approach is based on the diagnostic paradigm and it would appear that, on the basis of previous remarks, the uncategorized cases have nothing to offer.

However, Anderson (1979) mixes the two parameterizations, using the $\boldsymbol{\pi}$ in (5.7.3) and $\boldsymbol{\beta}$ in (5.7.4) as basic parameters. Thus γ in (5.7.4) is expressed as a function of $\boldsymbol{\beta}$ and $\boldsymbol{\pi}$ and the uncategorized data are therefore worth incorporating into the discriminant rule.

Ganesalingam and McLachlan (1979b) compare, empirically, the discriminant rule from maximum likelihood with one estimated using the 'cluster analysis' approach of Section 4.3.4.

McLachlan (1975, 1977) discusses a method which lies somewhere between

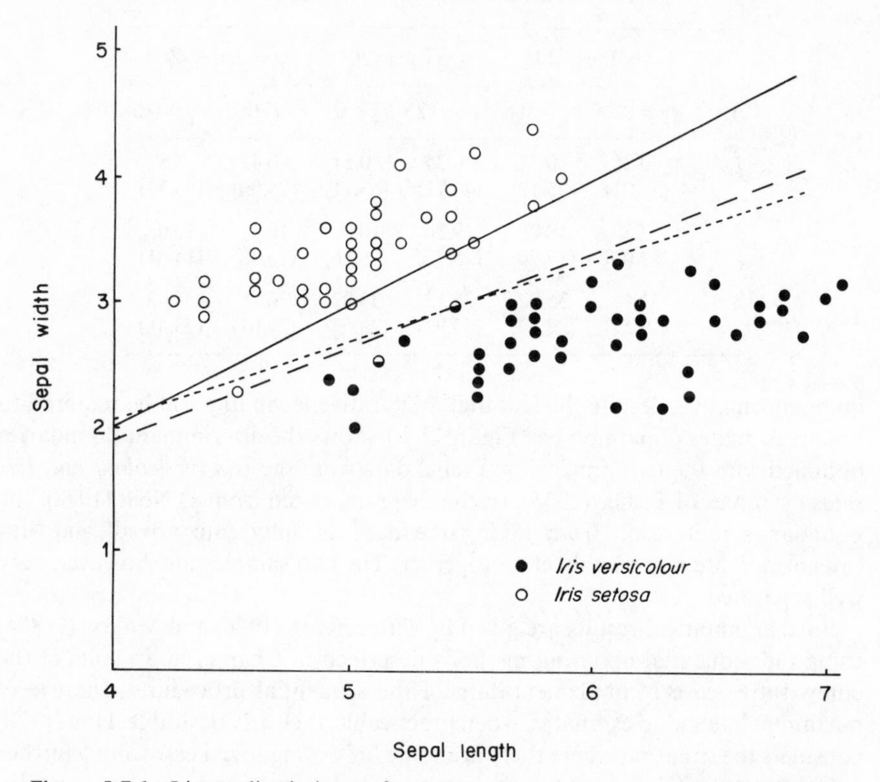

Figure 5.7.1 Linear discriminants for *Iris setosa* and *Iris versicolour* data. Reproduced by permission of the American Statistical Association from O'Neill (1978)

these two approaches and which is used for M2 data, containing both fully categorized and uncategorized cases. The method is based on an iterative procedure somewhat reminiscent of, but not equivalent to, the EM algorithm.

In general, these parametric procedures rely either on maximum likelihood estimation or on some approximation thereto, possibly involving sequential incorporation of the uncategorized cases. Given consistency, it should pay, on average, to incorporate the uncategorized cases if an 'imitator' of the optimal discriminant rule is then used. As is clear from the above, most of the detailed work has been concentrated on mixtures of two multivariate normals with equal covariance matrices.

More general procedures are possible if the assumption of a parametric model is dropped. Several *ad hoc* suggestions are made by Murray and Titterington (1978) based on methods described in Example 4.3.9. Also mentioned in Example 4.3.9, in the context of density estimation, was the penalized likelihood method. It can be used to obtain non-parametric estimates of the density ratio itself, thus providing a direct non-parametric version of the linear logistic approach. Suppose $k = 2, d = 1$, and we have a set of mixture data along with a set of observations from the first component. The two sample sizes are n and n_1, respectively, and the densities are $p(x)$ and $f_1(x)$. The objective is to estimate

$$v(x) = p(x)/f_1(x) = \pi_1 + (1 - \pi_1)\frac{f_2(x)}{f_1(x)}.$$

A plot of $v(x)$ against x is approximately constant, at level π_1, in regions where $f_2(x)/f_1(x)$ is small. If n and n_1 are reasonably large, we may regard the data as realizations of two independent inhomogeneous Poisson processes and, if an estimate $\hat{\mu}(x)$ can be made of $\mu(x)$, the ratio of the intensity of the mixture process to that of the pure process, an estimate of $v(x)$ is given by

$$\hat{v}(x) = (n_1/n)\,\hat{\mu}(x).$$

Suppose the combined order statistic is $z_1, \ldots, z_N$, where $N = n + n_1$. Then, so far as the intensity ratio is concerned, we may restrict our attention (Silverman, 1978) to the conditional log-likelihood

$$\mathscr{L}(\alpha) = \sum_{i=1}^{N} (\varepsilon_i \alpha(z_i) - \log\{1 + \exp[\alpha(z_i)]\}),$$

where $\alpha(z) = \log[\mu(z)]$ and $\varepsilon_i = 1$ if z_i comes from the mixture data, $= 0$ otherwise.

Since we may maximize $\mathscr{L}$ without any parametric restriction, we obtain the degenerate solution $\alpha(z_i) = \infty$ if $\varepsilon_i = 1$ and $\alpha(z_i) = -\infty$ otherwise. To avoid this, a roughness penalty is imposed on the form of $\alpha(\cdot)$. Specifically, we maximize, for some constant K,

$$\mathscr{L}(\alpha) - K \int [\alpha''(x)]^2 \mathrm{d}x,$$

where $\alpha''(\cdot)$ denotes the second derivative of $\alpha(\cdot)$.

The solution for $\alpha(\cdot)$ is the smooth concatenation of piecewise (on the intervals (z_i, z_{i+1})) cubic splines. Choice of K may be achieved either by formal cross-validation (Wahba and Wold, 1975) or, subjectively, by trial and error (Silverman, 1978). A useful feature of the density ratio plot is that the presence and nature of a contaminating distribution can be detected in any clear deviation from a constant level. In this manner, Silverman (1978) shows that a sample of cot-deaths seem to come from a mixture. Some of the babies were similar to a group who died in hospital, but there appeared to be a distinct subpopulation which accounted for about one third of the cases.

CHAPTER 6

Sequential problems and procedures

6.1 INTRODUCTION TO UNSUPERVISED LEARNING PROBLEMS

6.1.1 The problem and its Bayesian solution

In this chapter, we turn our attention to a general class of problems which arise in the context of finite mixture distributions when observations are obtained sequentially, rather than as a one-off sample (although, as we shall see in Sections 6.4.2 to 6.4.4, algorithms based on sequential considerations can prove invaluable as approximate computational devices in non-sequential contexts). Such problems arise routinely in the fields of pattern recognition and signal detection, and general background can be found in Fu (1968), Patrick (1972), and Young and Calvert (1974), as well as in the many references cited in these works. The basic problem is as follows. A sequence of (possibly vector-valued) observations $x_1, \dots, x_n, \dots$ are received, one at a time, and each has to be classified as coming from one of a known number k of exclusive classes $H_1, \dots, H_k$ before the next observation is received. Each decision is made on the basis of knowing all the previous observations, but without knowing whether previous classifications were correct or not. For obvious reasons, such problems are referred to as *unsupervised learning* problems. Defining $\boldsymbol{\psi} = (\boldsymbol{\pi}, \boldsymbol{\theta})$, where $\boldsymbol{\pi} = (\pi_1, \dots, \pi_k)$, $\boldsymbol{\theta} = (\boldsymbol{\theta}_1, \dots, \boldsymbol{\theta}_k)$, we assume that, conditional on $\boldsymbol{\psi}$, the x_n are independent with probability density

$$p(x_n \mid \boldsymbol{\psi}) = \sum_{i=1}^{k} \pi_i f_i(x_n \mid \boldsymbol{\theta}_i). \tag{6.1.1}$$

In other words, we assume that $f_i(x_n \mid \boldsymbol{\theta}_i)$ describes the distribution of x_n, were the latter actually to come from source H_i, and that π_i denotes the chance of any given observation emanating from H_i, $i = 1, 2, \dots, k$. Throughout, we shall implicitly assume that the kinds of regularity conditions discussed in Section 3.1 are satisfied, so that the mixture in (6.1.1) is identifiable. For a sequence of observations, $x_1, \ldots, x_n$, it follows from (6.1.1) that

$$p(x_1, \ldots, x_n | \boldsymbol{\psi}) = \prod_{i=1}^{n} \sum_{j=1}^{k} \pi_j f_j(x_i | \boldsymbol{\theta}_j), \tag{6.1.2}$$

which is a sum of k^n products of component densities, each term in the summation having an interpretation as the probability of obtaining a given one of the k^n possible partitions of the observations among the classes.

Based on the sampling distribution given in (6.1.2) there are, as we saw in Chapter 4, a multitude of standard and not-so-standard inferential procedures that have been proposed for learning about $\boldsymbol{\psi}$. Rather than review each of these separately again in the sequential case, we shall adopt a somewhat different approach, which allows us to give a more unified treatment of the main procedures which have been proposed for unsupervised learning problems. In fact, it turns out in the sequential case to be most illuminating to start with the Bayesian solution.

The Bayesian algorithm for learning about $\boldsymbol{\psi}$ (or the components of interest) involves the specification of a prior density for $\boldsymbol{\psi}$, and the subsequent recursive computation of the posterior density $p(\boldsymbol{\psi} | x_1, \ldots, x_n)$ using

$$p(\boldsymbol{\psi} | x_1, \ldots, x_n) \propto p(x_n | \boldsymbol{\psi}) \, p(\boldsymbol{\psi} | x_1, \ldots, x_{n-1}). \tag{6.1.3}$$

Bayesian classification of x_n is based upon a specified loss structure and, for $i = 1, \ldots, k$, the values of $\Pr(x_n \in H_i | x_1, \ldots, x_n)$, the probability that the nth observation belongs to class H_i, given the observations $x_1, \ldots, x_n$. These probabilities are computed using

$$\Pr(x_n \in H_i | x_1, \ldots, x_n) \propto f_i(x_n | x_1, \ldots, x_{n-1}) \Pr(x_n \in H_i | x_1, \ldots, x_{n-1}), \tag{6.1.4}$$

where (assuming both $\boldsymbol{\pi}$ and $\boldsymbol{\theta}$ are unknown)

$$f_i(x_n | x_1, \ldots, x_{n-1}) = \iint f_i(x_n | \boldsymbol{\theta}_i) \, p(\boldsymbol{\psi} | x_1, \ldots, x_{n-1}) \mathrm{d}\boldsymbol{\pi} \, \mathrm{d}\boldsymbol{\theta} \tag{6.1.5}$$

and

$$\Pr(x_n \in H_i | x_1, \ldots, x_{n-1}) = \iint \pi_i p(\boldsymbol{\psi} | x_1, \ldots, x_{n-1}) \mathrm{d}\boldsymbol{\pi} \, \mathrm{d}\boldsymbol{\theta}. \tag{6.1.6}$$

It is obvious that, for the mixture form inherent in (6.1.1) and (6.1.2), there exist no reproducing (natural conjugate) densities for unsupervised Bayes learning. This means that there is an unavoidable requirement for considerable computer processor time and memory in implementing the Bayesian solution, as we already noted in Section 4.4.

In our subsequent development, it will be convenient to distinguish the following special cases (see, also, the general survey given by Ho and Agrawala, 1968).

Case A

The probabilities $\pi_1, \ldots, \pi_k$ that an observation belongs to class H_i, $i = 1, \ldots, k$, are assumed *unknown*; the conditional probability densities $f_i(x|\boldsymbol{\theta}_i) = f(x|\boldsymbol{\theta}_i, H_i)$ of an observation x, assuming it to come from class H_i, are assumed *completely known* (in the sense that both the functional forms $f_i(\cdot|\boldsymbol{\theta}_i)$ and the parameter vectors $\boldsymbol{\theta}_i$ are assumed known). Such assumptions are often appropriate when large training samples can be made available from each individual class, but there is little initial information regarding the 'mix' of the population under study.

Example 6.1.1 Estimation of crop acreages

An important application having both non-sequential and sequential variants is in the estimation of crop acreages from remote sensors on orbiting satellites. (See Example 2.1.7 and the references there cited.) Here, measurements taken on areas of land vary according to the crop being grown, and total acreage of any single crop can be obtained by estimating the proportions of areas to which it is assigned. Mixture models are used with the $\boldsymbol{\theta}_i$'s, which characterize the various crops, assumed known (since training samples are available in abundance), and the π_i's, which correspond to crop acreage, assumed unknown.

Case B

The class probabilities $\pi_1, \ldots, \pi_k$ are assumed *known*; the conditional densities $f_i(x|\boldsymbol{\theta}_i)$ are assumed to have known functional forms, which involve parameter vectors $\boldsymbol{\theta}_i$, some, or all, of whose components are *unknown*.

Example 6.1.2 Detection of haemophilia

Haemophilia is a sex-linked disease affecting males only. Female carriers, while genetically affected, are in general apparently normal. Techniques for discriminating between carriers and non-carriers among women from families with a history of haemophilia are therefore extremely important.

In one approach to this discrimination problem, two attributes known as factor VIII activity and factor VIII-like antigen are measured. Bivariate normal distributions are assumed for both carriers and non-carriers, with estimates of their means and covariance matrices obtained from training samples. For details, see Hermans and Habbema (1975). The implied model for measurements taken from a suspected carrier is thus in the form of a mixture of two bivariate normal densities with known mixing parameters, which vary from individual to individual since they are prior probabilities for 'carrier' or 'non-carrier' evaluated on the basis of available genetic information on the individual's family. For a sequential treatment of this problem, see Makov (1980a).

Case C

The class probabilities $\pi_1, \ldots, \pi_k$ are assumed *unknown*; the conditional densities $f_i(x \mid \boldsymbol{\theta}_i)$ are assumed to have known functional forms, which involve parameter vectors $\boldsymbol{\theta}_i$, some, or all, of whose components are *unknown*.

Example 6.1.3 Updating a diagnostic system using unconfirmed cases (Titterington, 1976)

In the work by McLachlan (1977), O'Neill (1978), and Ganesalingam and McLachlan (1978) it was shown that information from unconfirmed cases can be important. However, while it is straightforward to update a diagnostic data base with data from confirmed cases, it is not at all obvious what to do when the diagnosis is unconfirmed. In such a case, one possibility is to employ 'fractional updating', whereby the parameters associated with each disease (the $\boldsymbol{\theta}_i$'s) are updated by a 'fraction' of the unconfirmed data, where the 'fractions' are related to the posterior probability of the patient suffering from the various diseases. The proportion of sufferers from each disease (the π_i's) can be estimated in a similar way.

Practical examples are discussed in Titterington (1976) with data from two hypertensive diseases, Conn's syndrome and Cushing's syndrome. Here the practical need for updating with unconfirmed cases is a real one, since discrimination is difficult and there is no large data base of confirmed cases.

6.1.2 Computational constraints and the need for approximations

Unfortunately, typical examples of sequential unsupervised learning problems—such as those mentioned in the previous section—tend to involve observations, often in the form of radar or other signals, which arrive at a high rate and require very fast processing against a background of limited computer memory. For this reason, the formal Bayes learning procedure has sometimes been regarded as of little practical use in this context, since the posterior forms occurring in (6.1.3) to (6.1.6) implicitly involve k^n histories of the process as a result of the complex form of the basic likelihood (6.1.2).

However, as we shall show, many of the *ad hoc* inference and decision procedures which have been advocated can be viewed, in a unified manner, as approximations to the Bayesian solution, or as leading to a related recursive estimation scheme. The formal Bayesian analysis therefore provides a convenient starting point for our discussion.

6.2 APPROXIMATE SOLUTIONS: UNKNOWN MIXING PARAMETERS

6.2.1 The two-class problem: Bayesian and related procedures

We shall illustrate the basic computational problem by considering the case A problem with $k = 2$. Observations, or signals, $x_1, x_2, \ldots, x_n, \ldots$ (which may be vector valued, but, for notational convenience, are here written as scalars), arrive sequentially and, after each x_n is obtained, an immediate decision is required as to which of two possible sources, H_1 or H_2, gave rise to the observation. It is known that each observation comes from one or the other of these two sources, but the actual source is unknown in each case, and there is no feedback as to the correctness of previous decisions.

We assume that, if the source of an observation is known, then the probability distribution of that observation is completely specified by the known density, f_1 for H_1 or f_2 for H_2, but we assume that the probability for source H_1 is unknown. Denoting this prior probability by π (so that $1-\pi$ is the prior probability for H_2), we suppose the observations x_n to be independent, given π, with common density

$$p(x_n|\pi) = \pi f_1(x_n) + (1-\pi)f_2(x_n). \tag{6.2.1}$$

We denote by $p_0(\pi)$ the density for π prior to obtaining any observations, and we write $p(\pi|x_1,\ldots,x_n) = p(\pi|x^n), n \geq 1$, for the posterior density conditional on having observed $x_1,\ldots,x_n$, but without knowing their correct sources. By Bayes theorem, we have

$$p(\pi|x^n) = \frac{p_0(\pi)\prod_{i=1}^{n} p(x_i|\pi)}{\int_0^1 p_0(\pi)\prod_{i=1}^{n} p(x_i|\pi)\mathrm{d}\pi} \tag{6.2.2}$$

$$= \frac{p(x_n|\pi)\,p(\pi|x^{n-1})}{\int_0^1 p(x_n|\pi)\,p(\pi|x^{n-1})\mathrm{d}\pi}. \tag{6.2.3}$$

The form (6.2.2) reveals how the order of complexity increases with n as a result of the mixture form (6.2.1).

The decision regarding the source of an observation x_n depends on the quantity $w_n = \Pr(x_n \in H_1|x_1,\ldots,x_n)$, which is given by

$$w_n = \int_0^1 \Pr(x_n \in H_1|\pi, x_n)\,p(\pi|x^n)\mathrm{d}\pi \tag{6.2.4}$$

$$= \int_0^1 \frac{f_1(x_n)\pi}{p(x_n|\pi)}\,p(\pi|x^n)\mathrm{d}\pi. \tag{6.2.5}$$

Writing $\hat{\pi}^{(n-1)} = \int_0^1 \pi p(\pi|x^{n-1})\mathrm{d}\pi$ and using (6.2.3), we obtain

$$w_n = \frac{f_1(x_n)\hat{\pi}^{(n-1)}}{p(x_n|\hat{\pi}^{(n-1)})} = \frac{f_1(x_n)\hat{\pi}^{(n-1)}}{f_1(x_n)\hat{\pi}^{(n-1)} + f_2(x_n)(1-\hat{\pi}^{(n-1)})}, \tag{6.2.6}$$

which has the same form as in the case of known π, but with π replaced by its posterior mean conditional on $x_1,\ldots,x_{n-1}$. Assuming a zero–one loss function, we would decide that $x_n \in H_1$, if $w_n \geq \frac{1}{2}$, and that $x_n \in H_2$, if $w_n < \frac{1}{2}$.

The above development is deceptively straightforward and conceals the fact that the computation of the successive $p(\pi|x^n)$, and hence of $\hat{\pi}_n$, increases in complexity as n increases, as a result of the mixture form $p(x_n|\pi)$. The computation utilizes weighted average forms involving an exponentially increasing number of terms, each requiring updating at each stage. We therefore need to seek approximations which avoid this ever-increasing computational complexity.

We begin by considering the formal Bayesian solution in the particular situation where the prior density $p_0(\pi)$ for the unknown parameter π is taken to have the form of a beta density

$$p_0(\pi) = \frac{\Gamma(\alpha_0 + \beta_0)}{\Gamma(\alpha_0)\Gamma(\beta_0)} \pi^{\alpha_0 - 1}(1 - \pi)^{\beta_0 - 1}, \tag{6.2.7}$$

which we denote by $B(\pi; \alpha_0, \beta_0)$, with $\alpha_0 > 0$, $\beta_0 > 0$. It follows from (6.2.2) that

$$p(\pi|x_1) = w_1 B(\pi; \alpha_0 + 1, \beta_0) + (1 - w_1) B(\pi; \alpha_0, \beta_0 + 1), \tag{6.2.8}$$

where

$$w_1 = \left[\frac{f_1(x_1)\alpha_0}{\alpha_0 + \beta_0}\right] \Bigg/ \left[\frac{f_1(x_1)\alpha_0}{\alpha_0 + \beta_0} + \frac{f_2(x_1)\beta_0}{\alpha_0 + \beta_0}\right], \tag{6.2.9}$$

and it is easily seen that, in general, the $p(\pi|x^n)$ build up as weighted averages of beta densities.

To avoid the expanding form for $p(\pi|x^n)$, it is natural to consider approximating (6.2.8) by a suitable beta density. At the next step, the resulting form for $p(\pi|x_1, x_2)$ would then be merely a linear combination of two beta terms and could itself be approximated by a beta density. Proceeding in this way, the necessary computation could be kept within reasonable limits.

Having just observed x_1, consider the first step. If we were informed of the true source of x_1, the distribution of π would be independent of x_1 and would be given by

$$B(\pi; \alpha_0 + \Delta_{11}, \beta_0 + \Delta_{12}), \tag{6.2.10}$$

where Δ_{1j} is 1 if $x_1 \in H_j$ and 0 otherwise, $j = 1, 2$. Since we are not informed of the true source (a situation which is referred to in the engineering literature as 'learning without a teacher'; see, for example, Agrawala, 1973), we need, in effect, to 'estimate' the unknown Δ_{ij}. Viewed from this perspective, several of the *ad hoc* solutions proposed in the literature can be examined more systematically.

(a) Decision-directed learning (DD)

This is the general name given to procedures which assign Δ_{ij} a value (0 or 1) on the basis of some decision rule. Thus, for example, Davisson and Schwartz (1970) set Δ_{11} equal to one if $w_1 \geq \frac{1}{2}$ and to zero otherwise, with $\Delta_{12} = 1 - \Delta_{11}$. This approach, in effect, assumes that the most likely H_i is, in fact, the true one. If we write

$$\hat{\Delta}_{n1} = \begin{cases} 1, & \text{if } w_n \geqslant \frac{1}{2} \\ 0, & \text{otherwise,} \end{cases}$$

the (DD) estimator for π is

$$\hat{\pi}^{(n+1)} = \frac{1}{n+1} \sum_{i=1}^{n+1} \hat{\Delta}_{i1} = \hat{\pi}^{(n)} - \frac{1}{n+1} [\hat{\pi}^{(n)} - \Delta_{(n+1)1}]. \tag{6.2.11}$$

This seemingly intuitive *ad hoc* procedure, which has been widely used, was investigated in detail by Davisson and Schwartz (1970), who demonstrated that the approach does not necessarily guarantee asymptotic unbiasedness and can also lead to problems of 'runaways', in that there are problems where the true π satisfies $0 < \pi < 1$, but $\hat{\pi}^{(n)}$ can converge to 0 to 1.

Katopis and Schwartz (1972) proposed a modified decision-directed learning procedure (MDD) which avoids the problems associated with DD, but at the cost of requiring numerical integration at each step in the recursive sequence of estimates $\hat{\pi}^{(n)}$. Kazakos and Davisson (1980) have found a fully efficient modification of the MDD approach, but their procedure is even more computationally demanding.

(b) Learning with a probabilistic teacher (PT)

This is the name given to the approach which makes a randomized choice for Δ_{11}, setting it equal to one with probability w_1 and setting it to zero otherwise, with $\Delta_{12} = 1 - \Delta_{11}$. The details of the approach are given in Agrawala (1970) and its asymptotic properties are investigated in Silverman (1979), who shows that the method has little to commend it from the efficiency point of view.

(c) Quasi-Bayes learning (QB)

Introduced in this context by Makov and Smith (1977a, b), this replaces Δ_{11} and Δ_{12} by their expectations, w_1 and $(1 - w_1)$, respectively, and so approximates $p(\pi | x_1)$ by

$$\hat{p}(\pi | x_1) = B(\pi, \alpha_0 + w_1, \beta_0 + 1 - w_1). \tag{6.2.12}$$

Subsequent updating proceeds in an identical manner: with $\hat{p}(\pi | x^{n-1}) = B(\pi; \alpha_{n-1}, \beta_{n-1})$, we update at the nth stage $(n \geqslant 1)$ to $\hat{p}(\pi | x^n) = B(\pi; \alpha_n, \beta_n)$, where

$$\alpha_n = \alpha_{n-1} + w_n,$$
$$\beta_n = \beta_{n-1} + 1 - w_n,$$

and where w_n is given by (6.2.6) with

$$\hat{\pi}^{(n-1)} = \alpha_{n-1} / (\alpha_{n-1} + \beta_{n-1}).$$

This method is thus seen to be one of 'fractional updating' (Titterington, 1976) of the parameters within the reproducing family of beta distributions. As we have noted, the required computation is minimal. Decisions regarding the sources of observations are again based upon the w_n, which are, in turn, defined by the $\hat{\pi}^{(n-1)}$. It follows that the success of the procedure depends on the convergence to π of the sequence $\hat{\pi}^{(n)}, n = 1, 2, \ldots$. We shall consider the asymptotic properties of this and other procedures in Section 6.2.3. For the present, we simply note that the $\hat{\pi}^{(n)}$ satisfy the recursive relationship

$$\hat{\pi}^{(n+1)} = \hat{\pi}^{(n)} - (\alpha_{n+1} + \beta_{n+1})^{-1}(\hat{\pi}^{(n)} - w_{n+1}). \tag{6.2.13}$$

(d) Learning with a probabilistic editor (PE)

This is the name given to the approach where Δ_{11}, Δ_{12} are chosen so that the first two moments of (6.2.10), the beta distribution approximation, are identical to those of the mixture distribution (6.2.8). Variants of this approach in more complex situations are discussed in Owen (1975), where it is referred to as 'restricted Bayes', Harrison and Stevens (1976), and Athans, Whiting and Gruber (1977). We shall consider the procedure in more detail in Section 6.5, when we discuss mixture distributions arising in the context of multiprocess Kalman filtering.

(e) The method of moments (MM)

This is defined by the simple recursion

$$\hat{\pi}^{(n+1)} = \hat{\pi}^{(n)} - \frac{1}{n+1}\left(\hat{\pi}^{(n)} - \frac{x_{n+1} - m_2}{m_1 - m_2}\right), \tag{6.2.14}$$

where m_i is the mean of the distribution corresponding to $f_i(\cdot)$, $i = 1, 2$. The method is based on the fact that $n^{-1}(x_1 + \cdots + x_n)$ converges, with probability one, to $\pi m_1 + (1 - \pi)m_2$ (see Patrick, Carayannopoulos, and Costello 1966, Johnson, 1973, and Odell and Basu, 1976, for details and applications).

We note in passing, however, that the form of recursion given in (6.2.14) could be viewed as a special case of (6.2.11), where we set

$$\hat{\Delta}_{n1} = \frac{x_n - m_2}{m_1 - m_2}.$$

6.2.2 The two-class problem: a maximum likelihood related procedure

In the case of an identifiable mixture, Kazakos (1977) has shown that the likelihood function defined by (6.1.2) is log-concave, so that, for fixed n, the maximum likelihood estimate is unique. However, its implementation for fixed n requires the numerical solution of a set of non-linear equations, and the form of solution is non-recursive in nature. To overcome this difficulty, Kazakos develops a recursive algorithm based on a Newton–Raphson-type gradient algorithm for finding the minimum of the Kullback–Leibler directed divergence measure.

Let $\hat{\pi}^{(1)}, \hat{\pi}^{(2)}, \ldots$ denote a sequence of estimates of π, where $\hat{\pi}^{(n)}$ is based on x^n, and write

$$I(\hat{\pi}^{(n)}, \pi) = \int p(x|\pi) \log\left[\frac{p(x|\pi)}{p(x|\hat{\pi}^{(n)})}\right] \mathrm{d}x$$
$$= E[\log p(x|\pi)|\pi] - E[\log p(x|\hat{\pi}^{(n)})|\pi], \tag{6.2.15}$$

where $p(x|\pi)$ is given by (6.2.1). It is easily shown that $I(\hat{\pi}^{(n)}, \pi) \geqslant 0$ for $0 \leqslant \hat{\pi}^{(n)} \leqslant 1$, with $I(\hat{\pi}^{(n)}, \pi) = 0$ only if $\hat{\pi}^{(n)} = \pi$, so that a minimal requirement for a 'good' sequence of estimates is that it should tend to the minimum of $I(\hat{\pi}^{(n)}, \pi)$. If we approach the construction of such a sequence by means of a Newton–Raphson gradient algorithm, we are led to consider a recursive sequence of the form

$$\hat{\pi}^{(n+1)} = \hat{\pi}^{(n)} - (n+1)^{-1} a_n \left[\frac{\partial}{\partial \hat{\pi}} I(\hat{\pi}, \pi)\right]_{\hat{\pi} = \hat{\pi}^{(n)}}, \tag{6.2.16}$$

where $a_1, a_2, \ldots$ is a suitable 'gain' sequence. If, in place of

$$\frac{\partial}{\partial \hat{\pi}} I(\hat{\pi}, \pi) = -E\left[\frac{\partial}{\partial \hat{\pi}} \log p(x|\hat{\pi}) \middle| \pi\right],$$

we substitute the obvious estimate

$$-[f_1(x_{n+1}) - f_2(x_{n+1})]/[\hat{\pi}^{(n)} f_1(x_{n+1}) + (1 - \hat{\pi}^{(n)}) f_2(x_{n+1})],$$

(6.2.16) suggests the general recursive form

$$\hat{\pi}^{(n+1)} = \hat{\pi}^{(n)} + (n+1)^{-1} L(\hat{\pi}^{(n)})[f_1(x_{n+1}) - f_2(x_{n+1})]$$
$$\cdot [\hat{\pi}^{(n)} f_1(x_{n+1}) + (1 - \hat{\pi}^{(n)}) f_2(x_{n+1})]^{-1}, \tag{6.2.17}$$

where $\hat{\pi}^{(n)}$ is the estimate of π after observing $x_1, \ldots, x_n$, $\hat{\pi}^{(0)}$ is an initial estimate of π, and $L(\hat{\pi}^{(n)})$ is an adjustable gain function assumed to be real-valued, positive, and bounded.

6.2.3 Asymptotic and finite-sample comparisons of the quasi-Bayes and Kazakos procedures

A convenient way to approach the study of the asymptotic properties of the various proposed procedures is through the theory of stochastic approximation, which exploits the martingale structure implicit in recursions such as (6.2.13) and (6.2.17). General accounts of available results and their application to recursive estimation are given by Hall and Heyde (1980), and Nevel'son and Has'minskii (1973): see, also, Fabian (1978). For our purposes here, it will suffice to draw attention to two specific results, which give the flavour of the kinds of theorem available.

Consider the following recursively defined sequence of random variables $X_1, X_2, \ldots$ (X_1 arbitrary):

$$X_n = X_{n-1} - a_n[M(X_{n-1}) - \alpha] - a_n Z(X_{n-1}), \tag{6.2.18}$$

where $\quad Z(X_{n-1}) = Y(X_{n-1}) - M(X_{n-1}),$

and $Y(X_{n-1})$ is a random variable such that

$$E[Y(X_{n-1})|X_1, \ldots, X_{n-1}] = M(X_{n-1}).$$

The following theorems typify the kinds of conditions under which the convergence and asymptotic normality of (6.2.18) can be established.

Theorem 6.2.1 (*Convergence*)

In the recursion (6.2.18), if θ is such that $M(\theta) = \alpha$, and

(a) $\Sigma a_n = \infty, \quad \Sigma a_n^2 < \infty,$

(b) $\inf\limits_{\varepsilon < |x-\theta| < \varepsilon^{-1}} (x - \theta)[M(x) - \alpha] > 0, \quad \forall \varepsilon > 0,$

(c) $\exists d$ such that $\forall x, E[Y^2(x)] \leqslant d(1 + x^2),$

then the sequence $X_1, X_2, \ldots$ tends to θ almost surely.

Proof: See Gladyshev (1965).

Theorem 6.2.2 (*Asymptotic normality*)

In the recursion (6.2.18), if

(a) $\Sigma a_n = \infty, \quad \Sigma a_n^2 < \infty,$

(b) $M(\theta) = \alpha, \quad (x - \theta)[M(x) - \alpha] > 0, \quad \forall x \neq \theta,$

(c) $\exists K > 0$ such that $|M(x) - \alpha| \leqslant K|x - \theta|, \quad \forall x,$

(d) $\forall x, \quad M(x) = \alpha + \alpha_1(x - \theta) + o(|x - \theta|), \quad$ with $\alpha_1 > 0,$

(e) $\sup_x E[Z^2(x)] < \infty,$

(f) $\lim_{x \to \theta} E[Z^2(x)] = S(\theta) < \infty,$

(g) $Z(X_n)$ are independent, $\forall n,$

then, if $a_n = An^{-1}$, where $A\alpha_1 > \frac{1}{2}$,

$$n^{1/2}(X_n - \theta) \to N(0, A^2 S(\theta)(2A\alpha_1 - 1)^{-1})$$

in distribution.

Proof: See Sacks (1958).

Theorem 6.2.1 is used by Makov and Smith (1977a) to establish the convergence of the quasi-Bayes procedure, and Theorem 6.2.2 is used by Kazakos (1977), who shows that, for the recursive procedure defined by (6.2.17),

$$n^{1/2}(\hat{\pi}^{(n)} - \pi) \to N(0, L^2(\pi)I(\pi)[2L(\pi)I(\pi) - 1]^{-1}), \tag{6.2.19}$$

where $$I(\pi) = \int [f_1(x) - f_2(x)]^2 [\pi f_1(x) + (1-\pi) f_2(x)]^{-1} \mathrm{d}x,$$

is the Fisher information for a single observation from $p(x|\pi) = \pi f_1(x) + (1-\pi) f_2(x)$.

Minimization of the variance term in (6.2.19) leads to the optimal choice of $L(\pi)$. This is found to be

$$L_K(\pi) = I^{-1}(\pi),$$

and forms the basis of the procedure advocated by Kazakos, which we shall denote by K. With this choice, the asymptotic variance is given by $V_K = I^{-1}(\pi)$, the Cramér–Rao lower bound, and so the K procedure is fully efficient. Thus a simple, recursive procedure is shown to achieve the same asymptotic performance as the full, intractable, maximum likelihood approach.

In order to establish the rate of convergence of the QB procedure defined by (6.2.13), we first note that this recursion is a special case of (6.2.17) with a gain function

$$\begin{aligned} L_{QB}(\pi) &= \frac{\pi(1-\pi)}{(\alpha+\beta)/n + 1} \qquad (6.2.20) \\ &\simeq \pi(1-\pi), \qquad \text{for large } n. \end{aligned}$$

Thus, using (6.2.19) and (6.2.20) we see that the asymptotic variance of the QB procedure is given by

$$V_{QB} = \frac{\pi^2(1-\pi)^2 I(\pi)}{2\pi(1-\pi)I(\pi) - 1}, \tag{6.2.21}$$

provided the denominator is greater than zero. The efficiency of the QB procedure is therefore equal to $[V_{QB} I(\pi)]^{-1}$. Clearly, the K procedure is asymptotically superior to the QB procedure. Note, however, that for f_i's whose domains of support are disjoint, $I^{-1}(\pi) = \pi(1-\pi)$ and thus $L_K(\pi)/L_{QB}(\pi) \to 1$, in which case the QB procedure is fully efficient. Clearly, the two procedures differ only in the choice of their gain functions. $L_K(\pi)$ is influenced by the value of π, the mixing parameter, and by the degree of overlap between f_1 and f_2; $L_{QB}(\pi)$ is asymptotically influenced by π only. However, its failure to take into account the overlap has only a limited effect. This obviously makes the QB procedure very attractive from the computational point of view, as it does not require integration for each n (a criticism that can be levelled, for example, against the K procedure).

For the purpose of illustration, we consider here the case of a known bipolar signal with f_1, f_2 both normal, with means m and $-m$, respectively, and equal variances σ^2. Simulation studies reported in Makov and Smith (1977a) have

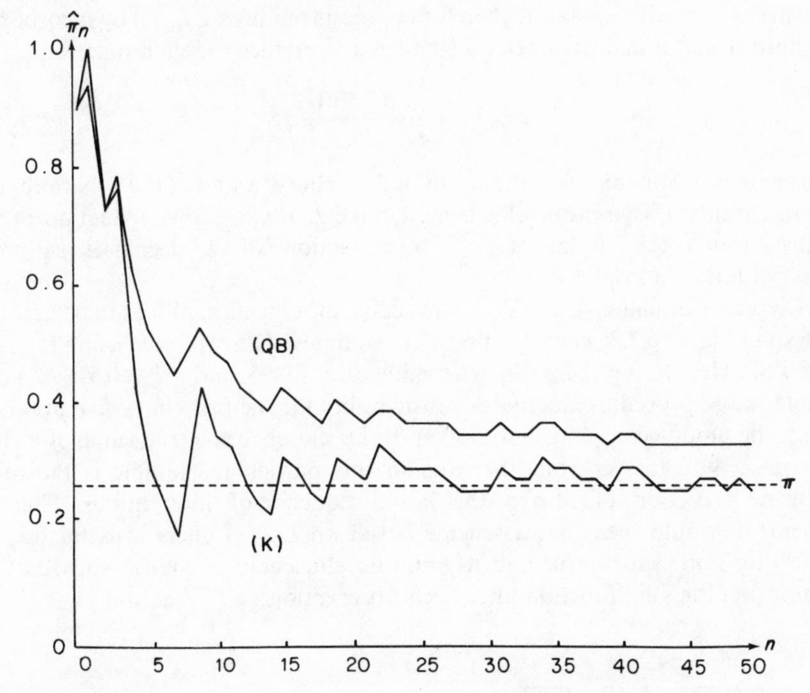

Figure 6.2.1 $\hat{\pi}$ versus the number of observations. Reproduced with permission from Makov (1980). Copyright © 1980 IEEE

shown that while for medium to high S/N (Signal to noise) ratios (m/σ), corresponding to a low degree of overlap of the components, the procedures are very similar, for small S/N the K procedure is generally faster. Figure 6.2.1 shows the first fifty estimates of π (true value 0.25), using simulated data from $m = 1$, $\sigma^2 = 1$, and prior parameters $\alpha = 0.9$, $\beta = 0.1$, which give the starting value $\hat{\pi}^{(0)} = 0.9$. The marginal superiority of the K procedure demonstrated here is typical of results observed in a range of simulation studies. For very small S/N ratios, the resulting increase in $L_K(\pi)$ can be counterproductive as it makes the scheme initially too responsive to the data. Although it certainly converges, it can fluctuate considerably for the first few observations, offering an inferior short-run performance to that of similar schemes whose gain is somewhat reduced.

Makov (1980b) proposed a modified version of $L_K(\pi)$ which checks fluctuations of this sort. To motivate the proposed modification, let us clarify the role of α and β in $L_{QB}(\pi)$ (see equation 6.2.20). For small values of n, $L_{QB}(\pi)$ is obviously influenced by the choice of α and β, whose values can drastically alter the sensitivity of the recursion to the data. In fact, one can choose an infinite number of pairs (α, β) which all correspond to the same starting point $\hat{\pi}^{(0)}$ and, from a Bayesian viewpoint, any such choice is determined by a particular prior distribution on π. Large (α, β) result in a more concentrated beta density function, which implies stronger belief in the initial estimate of π and thus in less initial sensitivity to the data. Makov therefore suggests modifying $L_K(\pi)$ by incorporating into it additional parameters, α and β, to produce the gain function

$$L_{MK}(\pi) = \frac{I^{-1}(\pi)}{(\alpha + \beta)/n + 1}.$$

This gain function takes account both of the value of π and of the S/N ratio, but also guarantees asymptotic efficiency. Unlike $L_K(\pi)$, it allows reduction of the gain through the choice of (α, β), a reduction which decreases as more observations are available.

A typical (simulated) example where such a modification proves to be useful is shown in Figure 6.2.2, where the first fifty estimates of π are plotted using L_{QB}, L_K, and L_{MK}. Here $m/\sigma = 0.5/1$, the true value of π is 0.75, and $\hat{\pi}^{(0)} = 0.5(5/(5+5))$. While the K procedure fluctuates considerably for the first thirty-five observations, the modified scheme is smoother and is closer to the true value of π. The QB procedure is even smoother and in this particular example is the most accurate procedure of all over this initial sequence of observations. Thus, in general, it would seen that a scheme based on $L_{MK}(\pi)$ offers a useful way of achieving both short-term and asymptotic efficiency, at the obvious cost of computing the gain function after each observation.

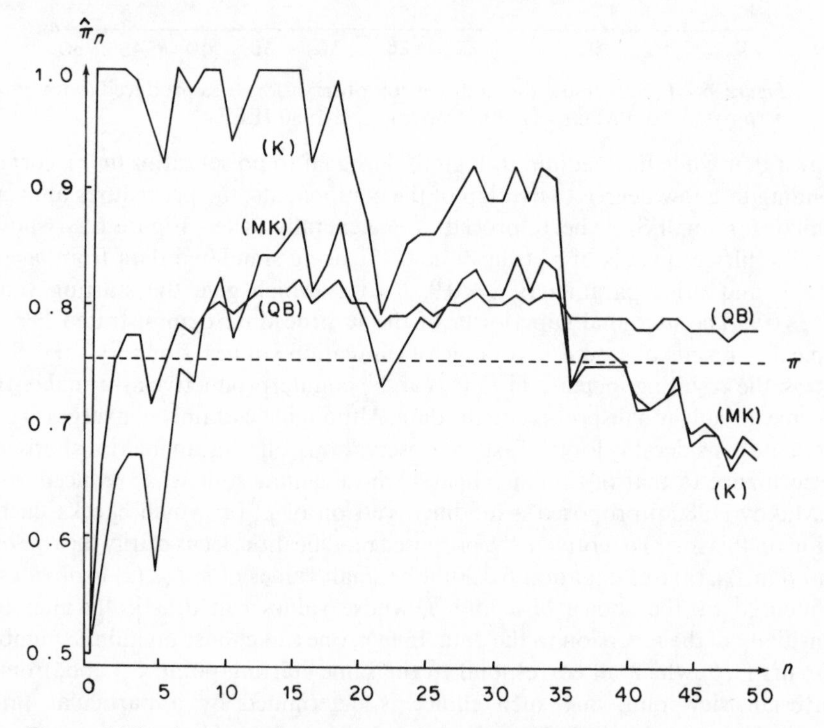

Figure 6.2.2 $\hat{\pi}$ versus the number of observations. Reproduced with permission from Makov (1980). Copyright © 1980 IEEE

6.2.4 The k-class problem: a quasi-Bayes procedure

We consider the situation in which, conditional on $\boldsymbol{\pi} = (\pi_1, \pi_2, \ldots, \pi_k)$ and density functions $f_1, f_2, \ldots, f_k$, we may assume that the random variables X_n are independent, with probability densities

$$p(x_n|\pi) = \pi_1 f_1(x_n) + \pi_2 f_2(x_n) + \cdots + \pi_k f_k(x_n), \tag{6.2.22}$$

where the π_i's are non-negative and sum to unity. The density f_i specifies the probability distribution of the observation, given that it belongs to population H_i, and π_i denotes the probability of this latter event. We further assume that the f_i are known and the π_i unknown.

As in the two-class case, the formal Bayes solution to the problem of learning about $\boldsymbol{\pi}$ and classifying the observations is deceptively straightforward.

We suppose that $p(\boldsymbol{\pi})$ denotes a prior density for $\boldsymbol{\pi}$, $p(\boldsymbol{\pi}|x^r) = p(\boldsymbol{\pi}|x_1, x_2, \ldots, x_r)$ denotes the resulting posterior density for $\boldsymbol{\pi}$ given $x_1, x_2, \ldots, x_r$, and $p_i(\boldsymbol{\pi}|x^r)$ denotes the posterior density for $\boldsymbol{\pi}$ if, in addition to $x_1, x_2, \ldots, x_r$, it were also known that the rth observation came from H_i.

By Bayes theorem, we have, for $n \geqslant 1$,

$$p(\boldsymbol{\pi}|x^n) \propto p(x_n|\boldsymbol{\pi})p(\boldsymbol{\pi}|x^{n-1}). \tag{6.2.23}$$

If we now define random variables $Y_1\ Y_2, \ldots, Y_n, \ldots$, such that $Y_n = i$ if and only if X_n belongs to H_i, $i = 1, 2, \ldots, k$, then, from (6.2.22) and (6.2.23),

$$p(\boldsymbol{\pi}|x^n) = \sum_{i=1}^{k} p(Y_n = i|x^n)p_i(\boldsymbol{\pi}|x^n), \tag{6.2.24}$$

where, for $i = 1, 2, \ldots, k$,

$$p(Y_n = i|x^n) \propto f_i(x_n)\hat{\pi}_i^{n-1}(x^{n-1}) \tag{6.2.25}$$

and

$$\hat{\pi}_i^{n-1}(x^{n-1}) = \int \cdots \int \pi_i p(\boldsymbol{\pi}|x^{n-1})\mathrm{d}\boldsymbol{\pi}. \tag{6.2.26}$$

Sequential learning about $\boldsymbol{\pi}$ takes place through $p(\boldsymbol{\pi}|x^n)$ and classification of the successive observations on the basis of $p(Y_n = i|x^n)$, $i = 1, 2, \ldots, k$. We note, incidentally, that the forms of $p(Y_n = i|x^n)$ are the same as would be obtained for known π_i, except that the latter are estimated by their expected values, given $x_1, x_2, \ldots, x_{n-1}$.

As we pointed out in previous sections, the implementation of this learning procedure raises some serious computational problems because of the mixture form which appears in (6.2.22). Successive computation of (6.2.23), or (6.2.24), introduces an ever-expanding linear combination of component posterior densities, each of which corresponds to an updating based upon a particular choice of previous classifications. At the nth stage, there are implicitly k^n such possible classifications, and calculation quickly becomes prohibitive.

As in the two-class case, there are many forms of *ad hoc* recursive procedure that one might adopt. Kazakos and Davisson (1980) consider a decision-directed approach; Kazakos (1977) extends the Newton–Raphson type algorithm to the k-class situation; and Smith and Makov (1978) extend the quasi-Bayes solution. We shall not review all these generalizations in detail. Instead, we shall outline the form of the quasi-Bayes solution and then make some general remarks about asymptotic properties.

We begin by considering the formal Bayesian solution to the problem, assuming that $p(\boldsymbol{\pi})$, the prior density for $\boldsymbol{\pi}$, has the form of a Dirichlet density,

$$p(\boldsymbol{\pi}) = \frac{\Gamma(\alpha_1^{(0)} + \alpha_2^{(0)} + \cdots + \alpha_k^{(0)})}{\Gamma(\alpha_1^{(0)})\Gamma(\alpha_2^{(0)})\cdots\Gamma(\alpha_k^{(0)})} \prod_{i=1}^{k} \pi_i^{\alpha_i^{(0)} - 1}, \tag{6.2.27}$$

which we denote by $D(\boldsymbol{\pi}; \alpha_1^{(0)}, \alpha_2^{(0)}, \ldots, \alpha_k^{(0)})$, where $\alpha_i^{(0)} \geqslant 0$, $i = 1, 2, \ldots, k$. Such a form might arise, for example, from a multinomially distributed training sample, whose correct classifications were known.

It follows from (6.2.24) that, after observing x_1,

$$p(\boldsymbol{\pi} \mid x_1) = \sum_{i=1}^{k} p(Y = i \mid x_1) D(\pi; \alpha_1^{(0)} + \delta_{i1}, \alpha_2^{(0)} + \delta_{i2}, \ldots, \alpha_k^{(0)} + \delta_{ik}), \tag{6.2.28}$$

where $p(Y_1 = i \mid x_1) \propto f_i(x_1)\alpha_i^{(0)}$ and $\delta_{ij} = 0$ if $i \neq j$, $\delta_{ij} = 1$ if $i = j$. It is natural to consider approximating (6.2.28) by a suitable Dirichlet density. The form of $p(\boldsymbol{\pi} \mid x_1, x_2)$ would then be a linear combination of k terms, as in (6.2.28), and could itself be then approximated by a Dirichlet density; and so on. Proceeding in this way, we could keep the necessary computation within reasonable bounds.

The simplest such procedure is the following. We first note that *if we were informed* which population x_1 belonged to, we should then obtain

$$p(\boldsymbol{\pi} \mid x_1) = D(\boldsymbol{\pi}; \alpha_1^{(0)} + \Delta_{11}, \alpha_2^{(0)} + \Delta_{12}, \ldots, \alpha_k^{(0)} + \Delta_{1k}), \tag{6.2.29}$$

where $\Delta_{1i} = 1$ if x_1 belongs to H_i, $\Delta_{1i} = 0$ otherwise. However, *we are not informed* of the true population; instead, we have $p(Y_i = i \mid x_1)$, the expectation (given x_1) of Δ_{1i}, when the latter is regarded as unknown. This provides the starting point for the proposed approximation: we replace (6.2.28) by

$$\begin{aligned} p(\boldsymbol{\pi} \mid x_1) = D[\boldsymbol{\pi}; \alpha_1^{(0)} + p(Y_1 = 1 \mid x_1),\quad & \alpha_2^{(0)} + p(Y_1 = 2 \mid x_1), \ldots, \\ & \alpha_k^{(0)} + p(Y_1 = k \mid x_1)]. \end{aligned} \tag{6.2.30}$$

Subsequent updating now takes place entirely within the Dirichlet family of distributions: $p(\boldsymbol{\pi} \mid x^n)$ is Dirichlet with parameters $\alpha_i^{(n)} = \alpha_i^{(n-1)} + p(Y_n = i \mid x^n)$, $i = 1, 2, \ldots, k$, where the $\alpha_i^{(n-1)}$ are the parameters of $p(\boldsymbol{\pi} \mid x^{n-1})$ and the calculation of $p(Y_n = i \mid x^n)$ proceeds through (6.2.25) and (6.2.26).

It is easily verified from standard properties of the Dirichlet distribution that the (quasi-) posterior mean for π_i, after observing $x_1, x_2, \ldots, x_n$, is given by

$$\hat{\pi}_i^{(n)}(x_n) = \frac{\alpha_i^{(n-1)} + p(Y_n = i \mid x^n)}{\alpha_0 + n} = \frac{\alpha_i^{(n)}}{\alpha_0 + n}, \tag{6.2.31}$$

where $\alpha_0 = \alpha_1^{(0)} + \alpha_2^{(0)} + \cdots + \alpha_k^{(0)}$ and $p(Y_n = i|x^n)$ is given by $f_i(x_n)\alpha_i^{(n-1)}/(\sum_i f_i(x_n)\alpha_i^{(n-1)})$. From (6.2.31), we can obtain the recurrence relations

$$\hat{\pi}_i^{(n+1)}(x^{n+1}) = \hat{\pi}_i^{(n)}(x^n) - a_{n+1}[\hat{\pi}_i^{(n)}(x^n) - p(Y_{n+1} = i|x^{n+1})], \qquad (6.2.32)$$

for $i = 1, 2, \ldots, k$, where $a_{n+1} = (\alpha_0 + n + 1)^{-1}$. Dropping the x's for notational convenience and writing $\pi_i^{(n)} = p(Y_n = i|x^n)$, we can rewrite (6.2.32) in the obvious vector form

$$\hat{\boldsymbol{\pi}}^{(n+1)} = \hat{\boldsymbol{\pi}}^{(n)} - a_{n+1}(\hat{\boldsymbol{\pi}}^{(n)} - \boldsymbol{\pi}^{(n+1)}).$$

Convergence of the sequence $\hat{\boldsymbol{\pi}}^{(n)}$ defined by (6.2.33) is established in Smith and Makov (1978) and rates of convergence are studied in Makov (1983). As in Kazakos (1977), however, it is found that these simple recursive forms are not fully efficient in the k-class case. Nevertheless, Makov (1983) reports that in a number of simulation studies the quasi-Bayes procedure has a performance similar to that of the full Bayesian procedure in small to moderate samples.

One simulation study that was undertaken (see Smith and Makov, 1978) considered three circular bivariate normal populations with unit variances, centred at $(-\mu, 0)$, $(0, \mu)$, and $(\mu, 0)$, where μ takes values 1.0, 1.5, 2.0, etc. Various

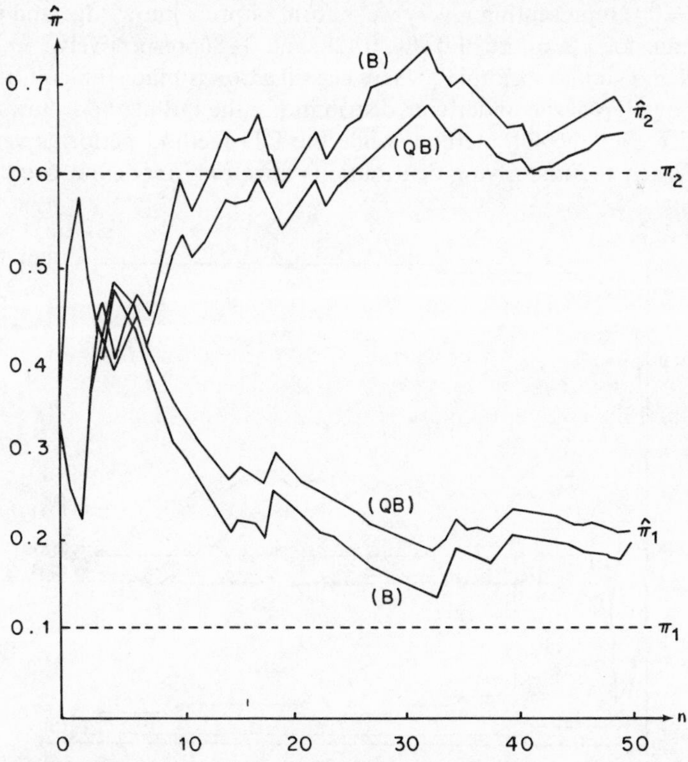

Figure 6.2.3 $\hat{\pi}_1$, $\hat{\pi}_2$ versus number of observations. (Smith and Makov, 1978). Reproduced by permission of the Editors, *Journal of the Royal Statistical Society, Series B*

combinations of the mixing probabilities π_1, π_2, and π_3 were investigated, together with various combinations of prior parameters $\alpha_1^{(0)}, \alpha_2^{(0)}$, and $\alpha_3^{(0)}$. Figure 6.2.3 summarizes the first fifty successive estimates of π_1 and π_2 for the case where $\mu = 2.0, \pi_1 = 0.1, \pi_2 = 0.6, \pi_3 = 0.3, \alpha_1^{(0)} = \alpha_2^{(0)} = \alpha_3^{(0)} = 1.0$. The plots are typical of the results obtained.

In all the cases studied, Smith and Makov (1978) report that the quasi-Bayes estimates succeed in remaining close to the Bayes estimates and perform quite well even in difficult situations; e.g. with $\mu = 1$ and $\alpha_1^{(0)} = \alpha_2^{(0)} = \alpha_3^{(0)} = 5.0$, a relatively strong, 'misleading' prior parameter (although, of course, convergence of both the Bayes and quasi-Bayes estimates becomes slower as the situation becomes more difficult).

Makov (1983) also compared the QB scheme with the DD and PT procedures, each defined in the k-class case by an obvious extension of the two-class approach.

In Figure 6.2.4 we show the paths of successive estimates of π_1, π_2 for a three-class simulated example ($k = 3$), where f_1, f_2, f_3 are circular bivariate normal distributions with all variances equal to one and means given by $(-0.5, 0), (0, 0.5)$, and $(0.5, 0)$, respectively. The prior parameters used were $\alpha_1^{(0)} = 0.1$, $\alpha_2^{(0)} = 0.15$, and $\alpha_3^{(0)} = 0.1$, representing a very weak form of prior knowledge and implying prior means for π_1, π_2, π_3 of 0.286, 0.428, and 0.286, respectively.

In this and similar examples, where classification is made difficult because of the high overlap of the underlying distributions, the QB method shows marked superiority over the PT method, while the DD method performs very badly indeed. When the underlying distributions have only moderate overlap, there

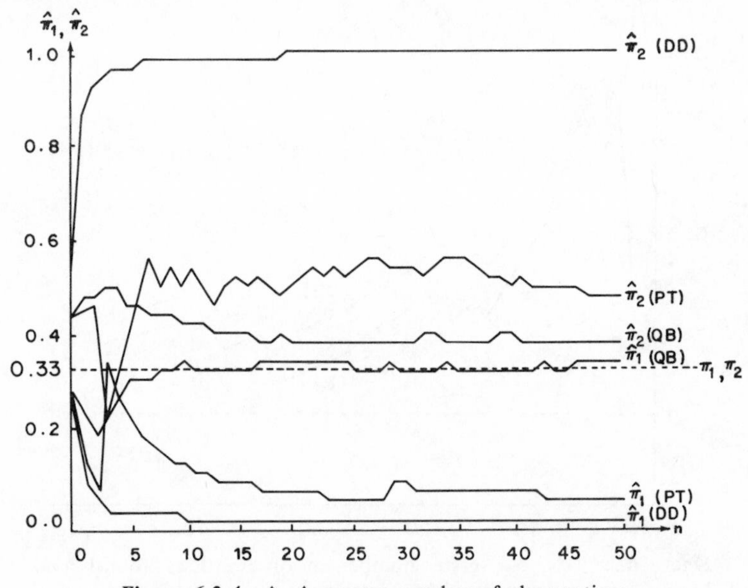

Figure 6.2.4 $\hat{\pi}_1, \hat{\pi}_2$ versus number of observations

appears little to choose between QB and PT, whereas both are markedly superior to DD.

A modified decision-directed scheme which is both asymptotically convergent and fully efficient is proposed in Kazakos and Davisson (1980), but the proposed scheme requires numerical integrations which are computationally unattractive.

6.3 APPROXIMATE SOLUTIONS: UNKNOWN COMPONENT DISTRIBUTION PARAMETERS

6.3.1 A general recursive procedure for a one-parameter mixture

Many of the sequential case B applications of greatest practical interest—which we shall discuss in detail in subsequent subsections—correspond to the situation where $p(x|\theta)$ is a mixture density with a single unknown parameter $\theta \in \Theta$ and known mixing parameters. Assuming that certain regularity conditions are satisfied, we shall investigate the properties of a general recursion of the form

$$\hat{\theta}_{n+1} = \hat{\theta}_n - n^{-1} G(\hat{\theta}_n)\gamma(\hat{\theta}_n, x_{n+1}), \tag{6.3.1}$$

where, given θ, the observations X_n are independently distributed with common density $p(x|\theta)$, $\hat{\theta}_{n+1}$ is the estimate of θ based on $x_1, \ldots, x_{n+1}$, $G(\hat{\theta}_n)$ is an adjustable gain function whose explicit form will be discussed later and $\gamma(\hat{\theta}, x)$ is defined by

$$\gamma(\hat{\theta}, x) = -\frac{\partial}{\partial \hat{\theta}} \log p(x|\hat{\theta}). \tag{6.3.2}$$

The stochastic approximation recursion (6.3.1) is clearly aimed at finding the root of $\gamma(\cdot)$, or, equivalently, through (6.3.2), the extremum of $\log p(x|\theta)$, which would coincide with the maximum likelihood estimator of θ. Patrick (1972) gives a general discussion of such recursions and we have already come across similar forms in our discussion of case A.

Defining

$$g(z, \theta) = E_\theta \gamma(z, X), \tag{6.3.3}$$

where the expectation is with respect to $p(x|\theta)$, we note the following.

Lemma 6.3.1

$g(z, \theta)$ has the following properties:

(a) $g(\theta, \theta) = 0$;
(b) there exists $\Theta' \subseteq \Theta$, a neighbourhood of θ, such that

$$\inf_z (z - \theta) g(z, \theta) > 0 \qquad \text{for } z \in \Theta', z \neq \theta. \tag{6.3.4}$$

Proof: We note that

$$g(z, \theta) = \frac{\partial}{\partial z} J(z, \theta),$$

where

$$J(z,\theta)=\int \log\left[\frac{p(x\mid\theta)}{p(x\mid z)}\right]p(x\mid\theta)\,\mathrm{d}x,$$

is the Kullback–Leibler directed divergence between $p(x\mid\theta)$ and $p(x\mid z)$: (a) and (b) follow immediately since it is well known that $J(z,\theta)\geqslant 0$, with equality if and only if $z=\theta$.

Suppose we now assume that

A1: $\theta\in(-M,M)$ for some known $M>0$, such that $(-M,M)\cap\Theta'\neq\emptyset$;
A2: $z\in(-M,M)\cap\Theta'$;

A3: $\sup_z E[\gamma^2(z,X)\mid\theta]<\infty,\qquad \theta\in\Theta'$;

A4: $G(z)$ is positive and bounded, with a bounded first derivative, and define

$$U(z,\theta)=-G(z)g(z,\theta) \tag{6.3.5}$$

and

$$R(z,x)=G(z)[\gamma(z,x)-g(z,\theta)]. \tag{6.3.6}$$

We can then establish the following properties.

Lemma 6.3.2

If assumptions A1 to A4 are satisfied, the quantities defined by (6.3.5) and (6.3.6) satisfy:

(a) $U(\theta,\theta)=0$ and $(z-\theta)U(z,\theta)<0$ for all $z\neq\theta$;
(b) $\lvert U(z,\theta)\rvert\leqslant k\lvert z-\theta\rvert$, for some $k>0$, and $\inf_z\lvert U(z,\theta)\rvert>0$, for $\eta<\lvert z-\theta\rvert<\eta^{-1}$ and for all $\eta>0$ such that assumption A2 is satisfied.
(c) $U(z,\theta)=\alpha(z-\theta)+o(\lvert z-\theta\rvert)$, for some $\alpha<0$;
(d) (i) $\sup_z E[R^2(z,X)\mid z]<\infty$,
(ii) $\lim_{z\to\theta}E[R^2(z,X)\mid z]=S(\theta)<\infty$;
(e) given z, $R(z,X_i)$, $i=1,2,\ldots$, are identically distributed.

Proof: Property (a) follows immediately from (6.3.5) and Lemma 6.3.1(b).

To establish (b), we use the Taylor expansion

$$U(z,\theta)=U(\theta,\theta)+(z-\theta)U'(z^*,\theta)=(z-\theta)U'(z^*,\theta),$$

for some z^* lying between z and θ, where $U'(z^*,\theta)$ is the derivative of $U(z,\theta)$ with respect to z, evaluated at z^*, and note that our assumptions ensure that $\lvert U'(z^*,\theta)\rvert$ is uniformly bounded. The second part of (b) follows from the remarks made in the proof of Lemma 6.3.1.

For (c), we use the expansion

$$\begin{aligned}U(z,\theta)&=-[(z-\theta)g'(\theta,\theta)+o(\lvert z-\theta\rvert)][G(\theta)+O(\lvert z-\theta\rvert)]\\&=[-G(\theta)g'(\theta,\theta)](z-\theta)+o(\lvert z-\theta\rvert),\end{aligned}$$

and note that

$$-g'(\theta,\theta) = -\int\left[-\frac{\partial^2}{\partial\theta^2}\log p(x|\theta)\right]p(x|\theta)\,\mathrm{d}x = -I(\theta),$$

where $[I(\theta)]^{-1} > 0$ is the Cramér–Rao lower bound for a single observation from $p(x|\theta)$. The choice $\alpha = -G(\theta)I(\theta) < 0$ satisfies (c).

To establish (d), we note that $E[\gamma(z,X)|z] = 0$, and so

$$E[R^2(z,X)|z] = G^2(z)\{E[\gamma^2(z,X)|z] + g^2(z,\theta)\},$$

which is bounded by virtue of assumptions A1 to A4. We note that

$$E[\gamma^2(z,X)|z] = E\left[\left\{\frac{\partial}{\partial z}\log p(X|z)\right\}^2 |z\right],$$

and hence, taking the limit as z tends to θ in the above, we see from the expression for $-g'(\theta,\theta)$ that

$$\lim_{z\to\theta} E[R^2(z,X)|z] = G^2(\theta)I(\theta).$$

The choice $S(\theta) = G^2(\theta)I(\theta) < \infty$ satisfies (d).

The final property, (e), follows straightforwardly from the assumed independence, given z, of the $X_i, i = 1, 2, \ldots$.

The following lemma will be used to establish the asymptotic properties of (6.3.1).

Lemma 6.3.3

Suppose that the conditions of Lemma 6.3.2 are satisfied and that $|\alpha| > \frac{1}{2}$. Then, for the recursion defined by (6.3.1), $n^{1/2}(\hat{\theta}_n - \theta)$ is asymptotically normally distributed with zero mean and variance

$$V = S(\theta)(2|\alpha| - 1)^{-1}.$$

Proof: This follows from an application of Theorem 6.2.2.

The main result is now the following.

Theorem 6.3.1

If in the recursion defined by (6.3.1) $G(\hat{\theta}_n) > [2I(\hat{\theta}_n)]^{-1}$, and assumptions A1 to A4 are satisfied, then $n^{1/2}(\hat{\theta}_n - \theta)$ is asymptotically normally distributed, with zero mean and variance

$$V = \frac{G^2(\theta)I(\theta)}{[2G(\theta)I(\theta) - 1]} \tag{6.3.7}$$

Proof: Truncation does not, of course, affect convergence properties (see, for example, Davisson, 1970) and the result follows immediately from Lemmas 6.3.2 and 6.3.3.

Corollary 6.3.1

A fully asymptotically efficient procedure corresponds to the choice $G(z) = [I(z)]^{-1}$.

Proof: With this choice, (6.3.7) reduces to $V_{\text{opt}} = [I(\theta)]^{-1}$.

Corollary 6.3.2

Given A1 to A4, the relative asymptotic efficiency of $\hat{\theta}_n$ with constant gain $G(z) = c > [2I(z)]^{-1}$ is given by

$$\rho = \frac{V_{\text{opt}}}{V} = \frac{2}{cI(\theta)} - \frac{1}{c^2 I^2(\theta)}. \tag{6.3.8}$$

We shall now use these properties of the general recursion (6.3.1) to examine a number of specific important practical case B problems and some proposed sequential estimation procedures.

6.3.2 Unsupervised learning for signal versus noise

We shall consider the case of a sequence of observations, $x_1, x_2, \ldots, x_n$, each of which is either a *signal*, assumed to have a normal distribution with unit variance and unknown mean θ, or *noise*, assumed to have a normal distribution with unit variance and zero mean. The corresponding normal densities will be denoted by $f_1(x|\theta)$ and $f_2(x|\theta) = f_2(x)$, respectively. The *a priori* probabilities of signal and noise will be assumed constant and known, and are denoted by π_1 and π_2 $(= 1 - \pi_1)$. Given θ, π_1, π_2, observations will be assumed independent, with common mixture density

$$p(x|\theta) = \pi_1 f_1(x|\theta) + \pi_2 f_2(x). \tag{6.3.9}$$

This problem has been treated by a number of authors using variants of the various approximation procedures which we reviewed in case A. Decision-directed schemes have been studied by Glaser (1961), Scudder (1965), Young and Farjo (1972), Young and Calvert (1974), and Farjo and Young (1976); probabilistic teacher schemes have been studied by Agrawala (1970) and Cooper (1975); and probabilistic editor schemes have been studied by Athans, Whiting, and Gruber (1977).

We shall begin our detailed discussion by considering the formal Bayesian solution when θ is assigned a normal prior density. Specifically, if $p(\theta)$, the *a priori* density for θ, is taken to be normal with mean $\hat{\theta}_0$ and variance τ^2, which we shall denote here by $N(\theta; \tau^{-2}\hat{\theta}_0, \tau^{-2})$, then it is straightforward to verify, using Bayes theorem, that the *a posteriori* density for θ, given x_1, can be written in the form

$$p(\theta|x_1) = \sum_{i=1}^{2} w_i(x_1) N(\theta; \tau^{-2}\hat{\theta}_0 + \delta_{i1}x_1, \tau^{-2} + \delta_{i1}), \tag{6.3.10}$$

where
$$w_i(x_1) = \frac{\pi_i \int f_i(x_1|\theta) p(\theta)\, \mathrm{d}\theta}{\int p(x_1|\theta) p(\theta)\, \mathrm{d}\theta} \tag{6.3.11}$$

is the probability, having observed x_1, that the first observation is a signal ($i=1$) or noise ($i=2$), respectively, and δ_{i1} is the usual Kroenecker delta. Equation (6.3.10) therefore has the form of a weighted average, with appropriate weights, of the two forms of *a posteriori* density (updating and not updating $p(\theta)$, respectively) that would be used if the true origin of x_1 (signal or noise) were known. Repeating the Bayes calculations for $p(\theta|x_1, x_2)$, we would obtain, in place of (6.3.10), a weighted average of four densities (corresponding to the four possible sequences of signal and noise), and the number of component densities increases to 2^n when we condition on $x_1, \ldots, x_n$.

In order to avoid this computational and storage problem, while trying to keep close to the spirit of the Bayesian solution, Smith and Makov (1981) proposed the following quasi-Bayes approximation. They first note that $w_i(x_1)$ involves the integration of $f_i(x_1|\theta)$ and $p(x_1|\theta)$ with respect to $p(\theta)$. Instead of using the resulting integrated forms, they propose using $f_i(x_1|\hat{\theta}_0)$ and $p(x_1|\hat{\theta}_0)$, simply conditioning on the (prior) mean of $p(\theta)$. They therefore replace $w_i(x_1)$ by

$$\hat{w}_i(x_1, \hat{\theta}_0) = \frac{\pi_i f_i(x_1|\hat{\theta}_0)}{p(x_1|\hat{\theta}_0)}. \tag{6.3.12}$$

The second part of the approximation consists in replacing (6.3.10) by a single normal distribution. Noting that the two component densities, written in terms of δ_{i1}, correspond to forms that would arise from *supervised* learning, and noting, also, that treating δ_{i1} as an indicator random variable, its expected value, given x_1, is $w_1(x_1)$, the quasi-Bayes procedure approximates (6.3.10) by

$$p(\theta|x_1) \simeq N[\theta; \tau^{-2}\hat{\theta}_0 + \hat{w}_1(x_1, \hat{\theta}_0)x_1, \tau^{-2} + \hat{w}_1(x_1, \hat{\theta}_0)]. \tag{6.3.13}$$

Subsequent updating proceeds in the same way, so that, given the reproductive property of the normal distribution, we obtain

$$p(\theta|x_1, \ldots, x_n) \simeq N\left[\theta; \tau^{-2}\hat{\theta}_0 + \sum_{j=1}^{n} \hat{w}_1(x_j, \hat{\theta}_{j-1})x_j, \tau^{-2} + \sum_{j=1}^{n} \hat{w}_1(x_j, \hat{\theta}_{j-1})\right], \tag{6.3.14}$$

where $$\hat{w}_1(x_j, \hat{\theta}_{j-1}) = \frac{\pi_1 f_1(x_j|\hat{\theta}_{j-1})}{p(x_j|\hat{\theta}_{j-1})} \tag{6.3.15}$$

and

$$\hat{\theta}_{j-1} = \left[\tau^{-2} + \sum_{i=1}^{j-1} \hat{w}_1(x_i, \hat{\theta}_{i-1})\right]^{-1} \left[\tau^{-2}\hat{\theta}_0 + \sum_{i=1}^{j-1} \hat{w}_1(x_i, \hat{\theta}_{i-1})x_i\right], \tag{6.3.16}$$

the mean of the approximating normal *a posteriori* density given $x_1, \ldots, x_{j-1}$. Equations (6.3.15) and (6.3.16) provide the basis for the proposed (QB) procedure for successive estimates of θ.

It is easily seen from (6.3.16) that the proposed estimates satisfy the general recursive relation

$$\hat{\theta}_{n+1} = \hat{\theta}_n - d_n\gamma(\hat{\theta}_n, x_{n+1}) \tag{6.3.17}$$

where $$d_n = \left[\tau^{-2} + \sum_{i=1}^{n+1} \hat{w}_1(x_i, \hat{\theta}_{i-1}) \right]^{-1} \tag{6.3.18}$$

and

$$\gamma(\hat{\theta}_n, x_{n+1}) = (\hat{\theta}_n - x_{n+1}) \hat{w}_1(x_{n+1}, \hat{\theta}_n). \tag{6.3.19}$$

Noting that, for large n,

$$d_n \approx [\tau^{-2} + (n+1)\pi_1]^{-1} \approx \frac{1}{\pi_1 n},$$

we see that (6.3.17) essentially has the general form of (6.3.1), with the particular choice of gain function

$$G_{\mathrm{QB}}(z) = \pi_1^{-1}, \tag{6.3.20}$$

whereas the fully efficient recursion would require

$$G_{\mathrm{OPT}}(z) = [I(z)]^{-1} = \left[\int \frac{\pi_1^2 (x - z)^2 f_1^2(x|z)}{p(x|z)} \mathrm{d}x \right]^{-1} \tag{6.3.21}$$

a gain function which is less attractive computationally.

Theorem 6.3.2

If $2I(\theta) > \pi_1$ and the recursion defined by (6.3.17) is truncated into the region $(-M, M) \cap \Theta'$ then

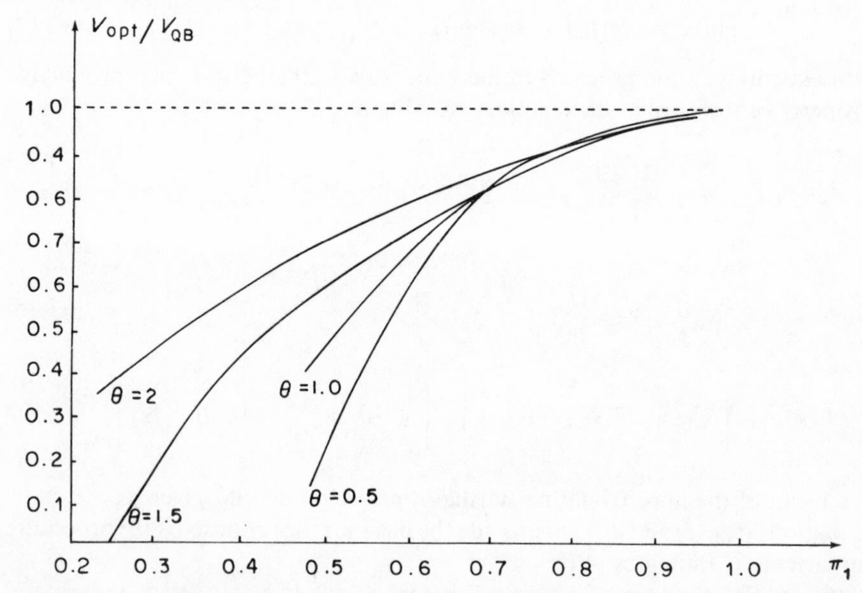

Figure 6.3.1 Efficiency of the quasi-Bayes procedure. Reproduced with permission from Smith and Makov (1981). Copyright © 1981 IEEE

(a) $n^{1/2}(\hat{\theta}_n - \theta)$ is asymptotically normally distributed, with zero mean and variance

$$V_{\text{QB}} = \pi_1^{-2} I(\theta)/[2\pi_1^{-1} I(\theta) - 1]; \tag{6.3.22}$$

(b) the relative efficiency is given by

$$\frac{V_{\text{QB}}}{V_{\text{OPT}}} = \frac{2\pi_1}{I(\theta)} - \frac{\pi_1^2}{[I(\theta)]^2}. \tag{6.3.23}$$

Proof: Assumptions A1 to A4 are satisfied for finite θ. The result follows immediately from Theorem 6.3.1 and Corollary 6.3.2, with $G_{\text{QB}}(z) = c = \pi_1^{-1}$.

Figure 6.3.1, taken from Smith and Makov (1981), shows the behaviour of (6.3.23) as a function of π_1 for some selected values of θ (recall that (6.3.7) only applies if $2I(\theta) > \pi_1$). The efficiency of the simple quasi-Bayes recursion is seen to be reasonable, even for π_1 as low as 0.65, and it is uniformly high for values of π_1 greater than 0.75.

6.3.3 A quasi-Bayes sequential procedure for the contaminated normal distribution

We shall consider the case of the contaminated normal distribution (discussed earlier in Example 2.2.1 and Section 4.4), where each of the sequence of observations to be processed may come from either a 'good' run or a 'bad' run. In particular, we shall assume that a good observation has a normal distribution with unknown mean θ and unit variance, whose density is denoted by $f_1(x|\theta)$, and that a bad observation has a normal distribution with the same mean θ, but with an inflated (known) variance $\lambda^2(>1)$, whose density is denoted by $f_2(x|\theta)$. The mixing probabilities for the good and bad components are assumed constant and known and are denoted by π_1 and $\pi_2(=1-\pi_1)$.

Given θ, π_1, π_2, observations will be assumed independent, with common mixture density

$$p(x|\theta) = \pi_1 f_1(x|\theta) + \pi_2 f_2(x|\theta). \tag{6.3.24}$$

Assuming the prior density for θ to be normal with mean $\hat{\theta}_0$ and variance τ^2, denoted by $N(\theta; \tau^{-2}\hat{\theta}_0, \tau^{-2})$, we can write the resulting posterior density for θ, given the first observation x_1, in the form

$$p(\theta|x_1) = \sum_{i=1}^{2} w_i(x_1) N[\theta; x(\delta_{i1} + \delta_{i2}\lambda^{-2}) + \hat{\theta}_0\tau^{-2}, \tau^{-2} + (\delta_{i1} + \delta_{i2}\lambda^{-2})], \tag{6.3.25}$$

where $$w_1(x_1) = \frac{\pi_1 \int f_i(x_1|\theta)p(\theta)\,d\theta}{\int p(x_1|\theta)p(\theta)\,d\theta} \tag{6.3.26}$$

is the probability, having observed x_1, that the first observation is a good one and $w_2(x_1) = 1 - w_1(x_1)$ is the probability that it is a bad one. In the case of supervised learning, δ_{i1} is the Kroenecker delta; in the unsupervised case, it is to be interpreted as an indicator random variable.

In the spirit of the quasi-Bayes procedures introduced in previous sections, we approximate (6.3.25) by

$$p(\theta|x_1) \simeq N[\theta; x_1 F(x_1, \hat{\theta}_0) + \hat{\theta}_0 \tau^{-2}, \tau^{-2} + F(x_1, \hat{\theta}_0)], \tag{6.3.27}$$

where, in general (with (6.3.27) corresponding to the case $i = 1$),

$$F(x_1, \hat{\theta}_{i-1}) = \frac{\pi_1 f_1(x_i|\hat{\theta}_{i-1})}{p(x_i|\hat{\theta}_{i-1})} + \frac{\pi_2 f_2(x_i|\hat{\theta}_{i-1})}{p(x_i|\hat{\theta}_{i-1})} \lambda^{-2}. \tag{6.3.28}$$

$F(\cdot)$ therefore approximates $(\delta_{i1} + \delta_{i2}\lambda^{-2})$ by replacing δ_{i1} and δ_{i2} by $\hat{w}_1(\cdot, \hat{\theta}_{i-1})$ and $\hat{w}_2(\cdot, \hat{\theta}_{i-1})$, respectively, where

$$\hat{w}_i(\cdot, \hat{\theta}_{i-1}) = \frac{\pi_i f_i(\cdot|\hat{\theta}_{i-1})}{p(\cdot|\hat{\theta}_{i-1})}. \tag{6.3.29}$$

Subsequent updating now proceeds in the same way and the mean of the approximating density, given $x_1, \ldots, x_{n+1}$ (the QB estimator), is

$$\hat{\theta}_{n+1} = \frac{\sum_{i=1}^{n+1} x_i F(x_i, \hat{\theta}_{i-1}) + \hat{\theta}_0 \tau^{-2}}{\sum_{i=1}^{n+1} F(x_i, \hat{\theta}_{i-1}) + \tau^{-2}} \tag{6.3.30}$$

or, written recursively,

$$\hat{\theta}_{n+1} = \hat{\theta}_n - d_n \gamma(\hat{\theta}_n, x_{n+1}) \tag{6.3.31}$$

where
$$d_n = \left[\tau^{-2} + \sum_{i=1}^{n+1} F(x_i, \hat{\theta}_{i-1}) \right]^{-1} \tag{6.3.32}$$

and

$$\gamma(\hat{\theta}, x_{n+1}) = (\hat{\theta}_n - x_{n+1}) F(x_{n+1}, \hat{\theta}_n). \tag{6.3.33}$$

The asymptotic properties of this recursion are then given by the following theorem.

Theorem 6.3.3

If $[\pi_1 + (1 - \pi_1)\lambda^{-2}]^{-1} > \frac{1}{2}[I(\theta)]^{-1}$, where $[I(\theta)]^{-1}$ is the Cramér–Rao lower bound for a single observation from (6.3.24), and if the recursion defined by (6.3.31) to (6.3.33) is truncated into the region $(-M, M)$, for some M, then

(a) $n^{1/2}(\hat{\theta}_n - \theta)$ is asymptotically normally distributed with zero mean and variance

$$\frac{[\pi_1 + (1 - \pi_1)\lambda^{-2}]^{-2} I(\theta)}{\{2[\pi_1 + (1 - \pi_1)\lambda^{-2}]\}^{-1} I(\theta) - 1}; \tag{6.3.34}$$

(b) the relative efficiency is given by

$$\frac{V_{\text{opt}}}{V_{\text{QB}}} = \frac{2[\pi_1 + (1-\pi_1)\lambda^{-2}]}{I(\theta)} - \frac{[\pi_1 + (1-\pi_1)\lambda^{-2}]^2}{[I(\theta)]^2}. \tag{6.3.35}$$

Proof: The recursion for $\hat{\theta}_n$ fits into the general form of (6.3.1) with $G(\theta) = [\pi_1 + (1-\pi_1)\lambda^{-2}]^{-1}$, since $d_n \to 1/[\pi_1 + (1-\pi_1)\lambda^{-2}]n$. The result follows from Corollary 6.3.2.

We note that for $\lambda^2 \to \infty$ the efficiency is identical to that of the signal versus noise case discussed in the previous section.

6.3.4 A quasi-Bayes sequential procedure for bipolar signal detection and related problems

A case of some interest in the field of communications engineering is that of bipolar signal detection (see, for example, Davisson and Schwartz, 1970; Rubin and Schwartz, 1970), which corresponds to a mixture, with known mixing weights, of two normal components with means $+\theta$ and $-\theta$ ($\theta > 0$) and equal, known variances (which may be taken to be unity). In fact, the structure of this problem can be generalized somewhat by considering a mixture, with known mixing weights, of k normal densities whose means are $a_1\theta, \ldots, a_k\theta$, for some known constants $\mathbf{a} = (a_1, \ldots, a_k)$, and whose variances are all unity. The bipolar signal detection problem then corresponds to $k = 2$, $a_1 = +1$, $a_2 = -1$. Moreover, the signal versus noise problem of Section 6.3.2 corresponds to $k = 2$, $a_1 = 1$, $a_2 = 0$.

A quasi-Bayes procedure for this general problem can be developed as follows. Given $\boldsymbol{\pi}$, $\boldsymbol{\theta}$ and $\mathbf{a}$, observations are assumed independent, with common mixture density

$$p(x|\theta) = \pi_1 f_1(x|a_1\theta) + \cdots + \pi_k f_k(x|a_k\theta), \tag{6.3.36}$$

where $f_i(x|a_i\theta)$ is a normal density with mean $a_i\theta$ and unit variance, $a_i \neq a_j$ for all $i \neq j$, and the a_i's are assumed bounded.

If $p(\theta)$, the prior density for θ, is taken to be normal with mean $\hat{\theta}_0$ and variance τ^2, denoted by $N(\theta; \tau^{-2}\hat{\theta}_0, \tau^{-2})$, then, using Bayes' theorem, we can verify that, in the supervised case, where the true identity of the x_i's is known, the posterior density for θ, given $x_1, \ldots, x_n$, can be written in the form

$$p(\theta|x_1, \ldots, x_n) = N\left[\theta; \tau^{-2}\hat{\theta}_0 + \sum_{j=1}^{k}\sum_{i=1}^{n} \delta_{ij}x_i a_j, \tau^{-2} + \sum_{j=1}^{k}\sum_{i=1}^{n} a_j^2\delta_{ij}\right], \tag{6.3.37}$$

where $$\delta_{ij} = \begin{cases} 1, & \text{if } E(X_i) = a_j\theta, \\ 0, & \text{otherwise.} \end{cases} \tag{6.3.38}$$

When the δ's are not known, the single posterior density for θ, (6.3.37), becomes a mixture of growing complexity, involving a combination of k^n densities. In order to avoid the resulting computational explosion as n increases, we can develop an approximation similar in spirit to the QB approximation to the signal versus noise problem, by regarding the unknown δ_{ij}'s as indicator random variables whose expectations can be estimated by

$$\hat{\delta}_{ij}(x_i, \hat{\theta}_{i-1}) = \frac{\pi_j f_j(x_i | a_j \hat{\theta}_{i-1})}{p(x_i | \hat{\theta}_{i-1})}, \tag{6.3.39}$$

where $\hat{\theta}_{i-1}$, the estimate of θ based on $x_1, \ldots, x_{i-1}$, is given below. The approximate posterior density for θ now takes the form

$$p(\theta | x_1, \ldots, x_n) \simeq N\left[\theta; \tau^{-2}\hat{\theta}_0 + \sum_{i=1}^{n} w_1(x_i, \hat{\theta}_{i-1})x_i, \tau^{-2} + \sum_{i=1}^{n} w_2(x_i, \hat{\theta}_{i-1})\right], \tag{6.3.40}$$

where $$w_1(x_i, \hat{\theta}_{i-1}) = \sum_{j=1}^{k} \hat{\delta}_{ij}(x_i, \hat{\theta}_{i-1})a_j \tag{6.3.41}$$

and

$$w_2(x_i, \hat{\theta}_{i-1}) = \sum_{j=1}^{k} \hat{\delta}_{ij}(x_i, \hat{\theta}_{i-1})a_j^2. \tag{6.3.42}$$

One can interpret $\hat{\delta}_{ij}(x_i, \hat{\theta}_{i-1})$, $w_1(x_i, \hat{\theta}_{i-1})$, and $w_2(x_i, \hat{\theta}_{i-1})$ as approximations to $E(\delta_{ij} | x_1, \ldots, x_i)$, $E(X_i/\theta | x_1, \ldots, x_{i-1})$, and $E[(X_i^2 - 1)/\theta^2 | x_1, \ldots, x_{i-1}]$, respectively.

Given $x_1, \ldots, x_n$, the proposed QB estimate of θ is the mean of the approximating normal density (6.3.40),

$$\hat{\theta}_n = \left[\tau^{-2} + \sum_{i=1}^{n} w_2(x_i, \hat{\theta}_{i-1})\right]^{-1} \left[\tau^{-2}\hat{\theta}_0 + \sum_{i=1}^{n} w_i(x_i, \hat{\theta}_{i-1})x_i\right], \tag{6.3.43}$$

which can be expressed in the recursive form

$$\hat{\theta}_{n+1} = \hat{\theta}_n - d_n \gamma(\hat{\theta}_n, x_{n+1}), \tag{6.3.44}$$

where $$d_n = \left[\tau^{-2} + \sum_{i=1}^{n+1} w_2(x_i, \hat{\theta}_{i-1})\right]^{-1} \tag{6.3.45}$$

and

$$\gamma(\hat{\theta}_n, x_{n+1}) = \hat{\theta}_n w_2(x_{n+1}, \hat{\theta}_n) - x_{n+1} w_1(x_{n+1}, \hat{\theta}_n). \tag{6.3.46}$$

Under the assumption that $\theta \leqslant M$, for some specified M, using Theorem 6.3.1, we can establish that if

$$E(A) = \sum_{i=1}^{k} \pi_i a_i \tag{6.3.47}$$

and

$$V(A) = \sum_{i=1}^{k} \pi_i a_i^2 - E^2(A), \tag{6.3.48}$$

then the relative asymptotic efficiency of $\hat{\theta}_n$ defined by (6.3.44) to (6.3.46) is given by

$$\frac{V(A)+E^2(A)}{I(\theta)}\left[2-\frac{V(A)+E^2(A)}{I(\theta)}\right], \tag{6.3.49}$$

which reduces in the bipolar signal detection case to

$$\frac{1}{I(\theta)}\left[2-\frac{1}{I(\theta)}\right],$$

where, as usual, $[I(\theta)]^{-1}$ denotes the Cramér–Rao lower bound for a single observation from the appropriate mixture density. Comparing this with the efficiency which is obtained in the signal versus noise case (6.2.23), we note that the presence of the factor π_1 renders the latter less efficient. The reason for this lies in the structure of the underlying mixture density. In the case of signal versus noise, information about θ is obtained only from a proportion π_1 of the observations; in the bipolar case *all* observations contribute information on θ. Clearly, as $\pi_1 \to 1$ the efficiencies become identical.

6.3.5 Problems with several unknown parameters

There appear to be few reports in the literature of sequential problems of case B type involving several unknown $\boldsymbol{\theta}$ parameters; see, however, Gregg and Hancock (1968), Patrick and Costello (1968), Patrick, Costello, and Monds (1970), and Farjo and Young (1976). In any case, the multiparameter extensions of the kinds of recursive algorithms we have been considering tend to apply equally well to case C. We therefore present our discussion of such algorithms in the next section.

6.4 APPROXIMATE SOLUTIONS: UNKNOWN MIXING AND COMPONENT PARAMETERS

6.4.1 A review of some pragmatic approaches

As we remarked in Section 6.3.5, there are relatively few detailed studies of case B problems involving several $\boldsymbol{\theta}$ parameters. This remark applies with even more force in case C, where the combination of unknown mixing weights, $\boldsymbol{\pi}$, and component density parameters, $\boldsymbol{\theta}$, inevitably leads to a highly parametrized form, which, in the mixture context, is inherently far more complex, mathematically, than the one- or two-parameter cases on which we have tended to concentrate for case A and case B. For this reason, there are only a limited number of published theoretical studies of sequential methods for jointly estimating $\boldsymbol{\pi}$ and the unknown parameters in $\boldsymbol{\theta}$. From a pragmatic point of view, however, many of the approximate procedures described for cases A and B can be extended in an obvious way to case C. We shall briefly mention a couple of these pragmatic procedures and then turn, in Section 6.4.2, to the theoretical study of a powerful general algorithm.

Decision-directed (DD) procedures estimate the unknown parameters by allocating an observation x to one of the underlying densities using a selected decision rule. Thus, if it is decided that x derives from the component density $f_i(x|\boldsymbol{\theta}_i)$, the $\boldsymbol{\pi}$ and $\boldsymbol{\theta}$ parameter estimates are updated on this assumption.

A commonly used decision rule in this context is that based on the maximum *a posteriori* probability, which acts as if x derives from $f_i(x|\boldsymbol{\theta}_i)$ if and only if

$$\hat{\pi}_i f_i(x|\hat{\boldsymbol{\theta}}_i) = \max_{j=1,\dots,k} \hat{\pi}_j f_j(x|\hat{\boldsymbol{\theta}}_j),$$

but other decision rules have also been suggested. In Titterington (1976), alternative DD criteria were used in the context of updating a medical databank in the presence of unconfirmed cases. However, as in most other papers dealing with DD procedures, no asymptotic analysis was provided.

We note, however, that properties of various DD procedures were thoroughly investigated in case B by Patrick and Costello (1968) and Patrick, Costello, and Monds (1970). In these papers, a mixture of two normal densities was assumed, with the unknown parameters taken to be the means of the distributions. The DD procedures were shown to be asymptotically biased, with a degree of bias related to the overlap between the two densities. In the light of these results, one should be rather circumspect about extending DD procedures to case C, where their performance is likely to be much worse than for case B, since $\boldsymbol{\pi}$ is now also assumed unknown.

Katopis and Schwartz (1972) analysed the DD solution to the problem of signal versus noise with unknown signal and noise probabilities (cf. Section 6.3.2). Although no proof of convergence is provided, the paper contains a useful discussion about the conditions under which good asymptotic properties might obtain.

The *method of moments* (MM) for estimating all the parameters of a mixture of distributions has been examined by a number of authors; see, for example, Rider (1961), Patrick, Carayannopoulos, and Costello (1966), Fu (1968), and our earlier discussion in Section 4.2. In essence, the method consists of solving the r simultaneous equations

$$E(X^i) = \sum_{j=1}^{n} \frac{x_j^i}{n}, \qquad i = 1, \dots, r,$$

subject to possible constraints imposed on the parameters to be estimated, where the number r is chosen such that a unique solution for all unknown parameters is guaranteed. For example, suppose the mixture consists of two normal densities with means θ_1 and 0, respectively, and equal variances θ_2, with mixing parameter π. If π, θ_1, and θ_2 are unknown, we solve the $r = 3$ equations:

$$E(X) = \pi\theta_1 = \bar{x}_n;$$
$$E(X^2) = \pi\theta_1^2 + \theta_2 = \sum x_i^2/n;$$
$$E(X^3) = \pi(\theta_1^3 + 3\theta_1\theta_2) = \sum x_i^3/n.$$

If we consider, for example, the implied equation for θ_1, it can be shown that for $\theta_1 \neq 0$ a unique solution exists only if $\pi = 2/3$; otherwise the problem will have multiple solutions. A unique solution is obtainable only if more information on the unknown parameters is available, or if higher moments are used. For further details see Fu (1968).

Provided the moment equations are relatively simple, the MM is computationally attractive and can be handled sequentially using stochastic approximation-type equations (Fu, 1968). If conditions for attaining a unique solution are satisfied, the method can be shown to converge to the true parameters with probability one. However, the method can be painfully inefficient.

A *maximum likelihood* approach, leading to a stochastic approximation algorithm, was proposed by Young and Coraluppi (1970).

6.4.2 A general recursion for parameter estimation using incomplete data

As we remarked in Section 3.2, one way of viewing observations from a finite mixture distribution is to regard them as incomplete data, the incompleteness referring to the absence of the indicator vectors which would identify the actual category or component membership of each observation.

In this section (based closely on Titterington, 1984), we shall begin by considering a general form of recursive estimation algorithm, together with its asymptotic properties. We shall then note the difficulties that arise in implementing this recursion in the multiparameter case with incomplete data (and, in particular, in the case of finite mixtures). Finally, exploiting links between the complete and incomplete data cases—or, in the language of earlier sections in this chapter, between 'supervised learning' and 'unsupervised learning'—we then suggest modified recursions, which are shown to have close connections with the EM algorithm. We recall that, as noted at the end of Section 6.3, although it is convenient to examine such general recursions under the heading of case C, the results apply, *a fortiori*, to appropriate case A and case B problems. Indeed, several of the recursions already discussed in previous sections will be seen to be special cases of, or closely related to, the recursions to be discussed below. Moreover, as we remarked at the beginning of Section 6.1.1, the algorithms can also be used as computational devices for non-sequential problems.

We suppose that $x_1, x_2, \ldots$ are independent observations, each with underlying probability density function $p(x \mid \psi)$, where $\psi \in \Psi \subset R^s$, for some s. Let $\mathbf{S}(x, \psi)$ denote the vector of scores,

$$S_j(x, \psi) = \frac{\partial}{\partial \psi_j} \log p(x \mid \psi), \qquad j = 1, \ldots, s.$$

Let $\mathbf{D}^2(x, \psi)$ denote the matrix of second derivatives of $\log p(x \mid \psi)$ and let $I(\psi)$ denote the Fisher information matrix corresponding to one observation. It is assumed that all derivatives and expected values exist and that

$$E_{\psi}\mathbf{S}(x, \boldsymbol{\psi}) = \int \mathbf{S}(x, \boldsymbol{\psi})p(x|\boldsymbol{\psi})\,\mathrm{d}x = \mathbf{0},$$

$$I(\boldsymbol{\psi}) = E_{\psi}[\mathbf{S}(x, \boldsymbol{\psi})\mathbf{S}^{\mathrm{T}}(x, \boldsymbol{\psi})] = -E_{\psi}\mathbf{D}^2(x, \boldsymbol{\psi}).$$

Consider the recursion

$$\boldsymbol{\psi}_{n+1}^* = \boldsymbol{\psi}_n^* + [nI(\boldsymbol{\psi}_n^*)]^{-1}\mathbf{S}(x_{n+1}, \boldsymbol{\psi}_n^*), \qquad n = 0, 1, \ldots, \tag{6.4.1}$$

which is recognizable as a stochastic approximation procedure. Under regularity conditions over and above those implicitly assumed thus far, as $n \to \infty$,

$$\sqrt{n}(\boldsymbol{\psi}_n^* - \boldsymbol{\psi}_0) \to N[\mathbf{0}, I(\boldsymbol{\psi}_0)^{-1}] \tag{6.4.2}$$

in distribution, where $\boldsymbol{\psi}_0$ denotes the true parameter value. This result appears in Sacks (1958), Fabian (1968), Nevel'son and Has'minskii (1973, Chapter 8), and Fabian (1978).

The following conditions are required for the most useful version of the result in Fabian (1978).

C1: *Continuity*

(a) $\int [\mathbf{S}(x, \boldsymbol{\delta}) - \mathbf{S}(x, \boldsymbol{\psi})]^{\mathrm{T}}[\mathbf{S}(x, \boldsymbol{\delta}) - \mathbf{S}(x, \boldsymbol{\psi})]p(x|\boldsymbol{\psi})\,\mathrm{d}x \to 0$

as $\boldsymbol{\delta} \to \boldsymbol{\psi}$ in Ψ.

(b) If, as $n \to \infty$, $\boldsymbol{\psi}_n^* \to \boldsymbol{\psi}_0$, then

$$[I(\boldsymbol{\psi}_n^*)]^{-1} \to [I(\boldsymbol{\psi}_0)]^{-1}.$$

C2: *Definiteness*

$$-(\boldsymbol{\delta} - \boldsymbol{\psi})^{\mathrm{T}}I(\boldsymbol{\delta})^{-1}E_{\psi}\mathbf{S}(x, \boldsymbol{\delta}) > 0 \qquad \text{for } \boldsymbol{\delta} \neq \boldsymbol{\psi}. \tag{6.4.3}$$

C3: *Boundedness*

$$E_{\psi}\| I(\boldsymbol{\delta})^{-1}\mathbf{S}(x, \boldsymbol{\delta})\|^2 \leqslant C(1 + \|\boldsymbol{\delta} - \boldsymbol{\psi}\|^2), \tag{6.4.4}$$

where $\|\mathbf{u}\|^2 = \mathbf{u}^{\mathrm{T}}\mathbf{u}$ and C is independent of $\boldsymbol{\delta}$.

One further comment should be made. Theoretical results are based on the assumption that $\boldsymbol{\psi}_n^* \in \Psi$, for all n. In practice, (6.4.2) may have to be modified slightly to ensure that this condition holds. For instance, if ψ reduces to a mixing weight an additional constraint should be added, such as: $\varepsilon < \psi_n^* < 1 - \varepsilon$, for all n and some small positive ε. Given these conditions and modifications, (6.4.2) is guaranteed.

If (6.4.2) holds for (6.4.1) then it will also hold for

$$\boldsymbol{\psi}_{n+1}^* = \boldsymbol{\psi}_n^* + [(n+1)I(\boldsymbol{\psi}_n^*)]^{-1}\mathbf{S}(x_{n+1}, \boldsymbol{\psi}_n^*), \qquad n = 0, 1, \ldots, \tag{6.4.5}$$

which is a particularly elegant form to use in that it provides the exact recursion obeyed by maximum likelihood estimates in exponential family models (exercise for the reader; or see Titterington, 1984).

However, in the case of multiparameter finite mixture models, complications arise in applying recursions (6.4.1) or (6.4.5) in the computation and inversion of $I(\psi_n^*)$. Numerical integration is often necessary and the fact that we are dealing with incomplete data will add to the complications. Suppose, with reference to (6.4.1), that we write

$$V_n = [nI(\psi_n^*)]^{-1}.$$

Then the following alternatives to V_n^{-1} suggest themselves as more easily calculated approximations.

(a) $nI(\psi')$, where ψ' is fixed at an initial parameter estimate, or is only updated infrequently, rather than at each iteration.

(b) $\sum_{i=1}^{n} J_i(\psi_n^*)$, where $J_i(\cdot)$ denotes the sample information matrix from the ith observation.

(c) $\sum_{i=1}^{n} I(\psi_i^*)$.

(d) $\sum_{i=1}^{n} J_i(\psi_i^*)$.

Suggestion (a) corresponds to a familiar modification to the method of scoring for obtaining maximum likelihood estimates. Suggestion (b) is similar to Newton's method for the same purpose. Suggestions (c) and (d) would be very useful in providing recursive calculation of $\{V_n^{-1}\}$. If (c) is used, for example, we obtain

$$V_n^{-1} = V_{n-1}^{-1} + I(\psi_n^*). \tag{6.4.6}$$

Recursion (6.4.1), with exactly this modification, was used by Walker and Duncan (1967) in the recursive estimation of parameters in a linear logistic model for quantal response.

Theoretical and practical investigation of all these modifications would be of interest, but we shall just concentrate on the following modification of (6.4.1), which is suggested by considering the link between complete and incomplete data problems:

$$\tilde{\psi}_{n+1} = \tilde{\psi}_n + [nI_c(\tilde{\psi}_n)]^{-1}\mathbf{S}(x_{n+1}, \tilde{\psi}_n), \qquad n = 0, 1, \ldots, \tag{6.4.7}$$

where $I_c(\psi)$ denotes the Fisher Information matrix corresponding to a *complete* observation.

The modification of (6.4.7) analogous to (6.4.5) is given by

$$\tilde{\psi}_{n+1} = \tilde{\psi}_n + [(n+1)I_c(\tilde{\psi}_n)]^{-1}\mathbf{S}(x_{n+1}, \tilde{\psi}_n), \qquad n = 0, 1, \ldots. \tag{6.4.8}$$

Although such recursions do not lead to full asymptotic efficiency, it is possible in some cases to guarantee $\sqrt{n}$-consistency and asymptotic normality. The following theorem, stated in its univariate version, is taken from Sacks (1958) and Fabian (1968).

Theorem 6.4.1

Given conditions corresponding to those above and provided $2I(\psi_0)I_c(\psi_0)^{-1} > 1$,

$$\sqrt{n}(\tilde{\psi}_n - \psi_0) \to N\{0, I_c(\psi_0)^{-2}I(\psi_0)/[2I(\psi_0)I_c(\psi_0)^{-1} - 1]\}$$

in distribution, as $n \to \infty$.

As will become clear later, it does not always happen that $2I(\psi_0) > I_c(\psi_0)$. Suppose

$$0 < \beta < 2I(\psi_0)/I_c(\psi_0) < 1$$

and we consider the recursion

$$\tilde{\psi}_{n+1} = \tilde{\psi}_n + n^{-(1+\beta)/2}I_c(\tilde{\psi}_n)^{-1}S(x_{n+1}, \tilde{\psi}_n), \qquad n = 0, 1, \ldots, \tag{6.4.9}$$

Then, according to Fabian (1968),

$$n^{\beta/2}(\tilde{\psi}_n - \psi_0) \to N\{0, I_c(\psi_0)^{-2}I(\psi_0)/[2I(\psi_0)I_c(\psi_0)^{-1} - \beta]\}$$

in distribution, as $n \to \infty$.

Thus, provided there is some information in the incomplete data ($I(\psi_0) > 0$), a modified version of (6.4.7) leads to a consistent, asymptotically normal estimator.

Multidimensional versions of these results are, of course, required for many applications of interest, but details will not be given here: see Sacks (1958) and Fabian (1968).

The important practical advantage of recursions (6.4.7), (6.4.8), and (6.4.9) is that $I_c(\boldsymbol{\psi})$ will usually be much easier to evaluate and, in the case of an information matrix, much easier to invert than $I(\boldsymbol{\psi})$.

In the following, we derive versions of some of these recursions for two examples involving finite mixtures. We denote by $x_1, x_2, \ldots$ a sequence of incomplete observations and by $y_1, y_2, \ldots$ corresponding 'complete' versions. Thus, given x, y belongs to a subset $\mathcal{Y}(x)$ of the overall sample space x and, if $g(y|\boldsymbol{\psi})$ denotes the p.d.f. of y, then

$$p(x|\boldsymbol{\psi}) = \int_{\mathcal{Y}(x)} g(y|\boldsymbol{\psi})\,\mathrm{d}y$$

(see Dempster, Laird, and Rubin, 1977).

6.4.3 Illustrations of the general recursion

Example 6.4.1 Estimation of mixing weights

We consider first the case of a mixture of k known densities $f_j(\cdot)$, $j = 1, \ldots, k$:

$$p(x|\boldsymbol{\pi}) = \sum_{j=1}^{k-1} \pi_j f_j(x) + \left(1 - \sum_{j=1}^{k-1} \pi_j\right) f_k(x),$$

where the $\pi_1, \ldots, \pi_k$ are all non-zero probabilities. Then

$$S_j(x|\boldsymbol{\pi}) = [f_j(x) - f_k(x)]/p(x|\boldsymbol{\pi}), \qquad j = 1, \ldots, k-1,$$

$$\mathbf{D}^2_{jr}(x|\boldsymbol{\pi}) = -[f_j(x) - f_k(x)][f_r(x) - f_k(x)]/[p(x|\boldsymbol{\pi})]^2,$$

$$j = 1, \ldots, k-1, \; r = 1, \ldots, k-1,$$

and

$$I_{jr}(\boldsymbol{\pi}) = \int [f_j(x) - f_k(x)][f_r(x) - f_k(x)]p(x|\boldsymbol{\pi})^{-1}\,\mathrm{d}x, \quad j, r = 1, \ldots, k-1.$$

Verification of the regularity conditions is subsumed in Kazakos (1977) and Smith and Makov (1978).

For the special case of $k = 2$, with $\pi_1 = \pi$, we obtain, for (6.4.1), as in Kazakos (1977),

$$\pi^*_{n+1} = \pi^*_n + [nI(\pi^*_n)]^{-1}[f_1(x_{n+1}) - f_2(x_{n+1})]/p(x_{n+1}|\pi^*_n), \qquad n = 1, 2, \ldots,$$

with $\quad I(\pi) = \int [f_1(x) - f_2(x)]^2 p(x|\pi)^{-1}\,\mathrm{d}x.$

We concentrate our discussion on the case $k = 2$. Here the incompleteness is caused by ignorance of the source of an observed x; is it component 1 or component 2? We may write

$$y = (x, \mathbf{z}),$$

where $\quad \mathbf{z}^{\mathrm{T}} = (1, 0)$ or $(0, 1)$ according to the source. Thus

$$\log g(y|\pi) = \mathbf{z}^{\mathrm{T}}\mathbf{u}(\pi) + \mathbf{z}^{\mathrm{T}}\mathbf{v}(\pi)$$

where $\quad \mathbf{u}^{\mathrm{T}}(\pi) = [\log \pi, \log(1 - \pi)]$

and

$$\mathbf{v}^{\mathrm{T}}(\pi) = [\log f_1(x), \log f_2(x)].$$

Thus (cf. Section 3.2), $I_c(\pi) = 1/\pi(1 - \pi)$ and (6.4.7) becomes

$$\tilde{\pi}_{n+1} = \tilde{\pi}_n + n^{-1}\tilde{\pi}_n(1 - \tilde{\pi}_n)[f_1(x_{n+1}) - f_2(x_{n+1})]/p(x_{n+1}|\tilde{\pi}_n). \qquad (6.4.10)$$

Asymptotically, if $I(\pi) > \frac{1}{2}I_c(\pi)$, Theorem 6.4.1 holds. Otherwise, strong consistency can still be guaranteed (see Makov and Smith, 1977; Smith and Makov, 1978) and recursions like (6.4.9) may also be used (cf. Section 6.2).

Example 6.4.2 Mixture of two univariate normals

Let

$$p(x|\boldsymbol{\pi}, \boldsymbol{\mu}, \mathbf{v}) = \pi_1\phi(x|\mu_1, \sqrt{v_1}) + \pi_2\phi(x|\mu_2, \sqrt{v_2})$$

$$= \pi_1 f_1(x) + \pi_2 f_2(x),$$

where $0 < \pi_1 = 1 - \pi_2 < 1$. Then the components of the score vector are

$$\partial \log p(x)/\partial \pi_1 = [f_1(x) - f_2(x)]/p(x),$$

$$\partial \log p(x)/\partial \mu_j = (x - \mu_j)w_j(x)/v_j, \qquad j = 1, 2,$$

$$\partial \log p(x)/\partial v_j = [(x - \mu_j)^2 - v_j]w_j(x)/2v_j^2, \quad j = 1, 2,$$

where $w_j(x) = \pi_j f_j(x)/p(x), j = 1, 2$. Note that, for $j = 1, 2$, $w_j(x)$ is the conditional probability that an observation comes from component j, given its realized value, x.

We shall not give the details of the verification of conditions (6.4.3) and (6.4.4). These are very complicated, as is application of the recursion (6.4.1) itself, due to the complex nature of the information matrix, even for univariate mixtures, and the requirement for numerical integration (see Behhoodian, 1972a, and Section 3.2).

In fact, the difficulty of applying (6.4.1) is the main reason for reconsidering this example and provides strong motivation for the use of recursions like (6.4.7). For this we require $I_c(\pi_1, \boldsymbol{\mu}, \boldsymbol{\nu})$.

Again $y = (x, \mathbf{z})$ and now

$$\log g(y|\boldsymbol{\pi}, \boldsymbol{\mu}, \boldsymbol{\nu}) = \mathbf{z}^T\mathbf{u}(\boldsymbol{\pi}) + \mathbf{z}^T\mathbf{v}(\boldsymbol{\mu}, \boldsymbol{\nu}),$$

where, for instance,

$$v_1(\boldsymbol{\mu}, \boldsymbol{\nu}) = \log \phi(x|\mu_1, \sqrt{\nu_1}).$$

If the parameters are ordered as $\pi_1, \mu_1, \mu_2, \nu_1, \nu_2$, then

$$I_c(\pi_1, \boldsymbol{\mu}, \boldsymbol{\nu}) = \operatorname{diag}[\pi_1^{-1}(1-\pi_1)^{-1}, \pi_1/\nu_1, (1-\pi_1)/\nu_2, \pi_1/2\nu_1^2, (1-\pi_1)/2\nu_2^2]$$

and recursion (6.4.7) assumes the following simple form:

$$\tilde{\pi}_1^{(n+1)} = \tilde{\pi}_1^{(n)} + n^{-1}[w_1^{(n)}(x_{n+1}) - \tilde{\pi}_1^{(n)}]$$
$$\tilde{\mu}_j^{(n+1)} = \tilde{\mu}_j^{(n)} + (n\pi_j^{(n)})^{-1} w_j^{(n)}(x_{n+1})(x_{n+1} - \mu_j^{(n)})$$
$$\tilde{\nu}_j^{(n+1)} = \tilde{\nu}_j^{(n)} + (n\pi_j^{(n)})^{-1} w_j^{(n)}(x_{n+1})[(x_{n+1} - \mu_j^{(n)})^2 - \nu_j^{(n)}]$$

where

$$w_j^{(n)}(x) = \pi_j^{(n)}\phi(x|\mu_j^{(n)}, \sqrt{\nu_j^{(n)}})/p(x|\boldsymbol{\pi}^{(n)}, \boldsymbol{\mu}^{(n)}, \boldsymbol{\nu}^{(n)}), \qquad j = 1, 2.$$

6.4.4 Connections with the EM algorithm

As was pointed out by Fabian (1978, Section 5.8), there is a strong relationship between recursion (6.4.1) and the method of scoring. Recursion (6.4.7), on the other hand, is similarly linked to the EM algorithm.

Suppose $y_1, \ldots, y_n$ represent n independent complete observations, corresponding to $x_1, \ldots, x_n$. Define

$$Q(\boldsymbol{\psi}, \boldsymbol{\psi}') = E_{\boldsymbol{\psi}'}\left[\sum_{i=1}^{n} \log g(y_i|\boldsymbol{\psi})|x_1, \ldots, x_n\right].$$

As we saw in Section 4.3, the EM algorithm generates a sequence $\{\boldsymbol{\psi}_n\}$ of parameter estimates by repeating the following double step:

E step: Evaluate $Q(\boldsymbol{\psi}, \boldsymbol{\psi}_n)$.
M step: Choose $\boldsymbol{\psi} = \boldsymbol{\psi}_{n+1}$ to maximize $Q(\boldsymbol{\psi}, \boldsymbol{\psi}_n)$.

Consider the following recursive version. At stage $n + 1$, with current estimate $\tilde{\boldsymbol{\psi}}_n$, define

$$L_{n+1}(\psi) = E_{\tilde{\psi}_n}[\log g(y_{n+1}|\psi)|x_{n+1}] + L_n(\psi). \tag{6.4.11}$$

Choose $\psi = \tilde{\psi}_{n+1}$ to maximize $L_{n+1}(\psi)$. Finally, estimate ψ_0 by $\tilde{\psi}_n$.

In addition to providing a procedure for maximum likelihood estimation, both the EM algorithm and its recursive version may be used in Bayesian analysis for the computation of posterior modes. In the latter case, we can initialize the recursion (6.4.11) using

$$L_0(\psi) = \log p(\psi),$$

where $p(\cdot)$ is the prior density for ψ, with mode $\tilde{\psi}_0$.

Theorem 6.4.2

Given appropriate regularity conditions, recursion (6.4.11) can be approximated by

$$\hat{\psi}_{n+1} = \hat{\psi}_n + [(n+1)I_c(\hat{\psi}_n)]^{-1}S(x_{n+1}, \hat{\psi}_n),$$

which is the recursion we called (6.4.8) in Section 6.4.1.

Proof: See Titterington (1984).

Whenever it is practicable, recursion (6.4.1) is the ideal choice, but it is likely to be complicated to apply in multiparameter contexts, where the modified recursions (6.4.7) and (6.4.8) promise to be much easier to implement. Although only a few examples have been described here, in the context of mixture problems, Titterington and Jiang (1983) show that such recursions have much wider applicability to missing data problems. They also provide numerical details about the relative performance of some of these procedures and discuss the order effect introduced when the sequential approximation is applied to a non-sequential problem; see also Anderson (1979). Makov (1980a) points out, in the context of Example 6.4.1 with $k = 2$, that recursion (6.4.1) may be unsatisfactorily unstable, relative to (6.4.7) or (6.4.8), particularly in the early stages.

We conclude this section with a final comment about the EM algorithm. Recursion (6.4.1) is related to the method of scoring, which generates a sequence of estimates $\{\hat{\psi}_r\}$ according to the recursion

$$\hat{\psi}_{r+1} = \hat{\psi}_r + [nI(\hat{\psi}_r)]^{-1} \sum_{i=1}^{n} S(x_i, \hat{\psi}_r), \quad r = 0, 1, \ldots,$$

where $x_1, \ldots, x_n$ denote n independent observations.

It is easy to show, using the methods of Theorem 6.4.2, that the EM algorithm is given, approximately, by

$$\hat{\psi}_{r+1} = \hat{\psi}_r + [nI_c(\hat{\psi}_r)]^{-1} \sum_{i=1}^{n} S(x_i, \hat{\psi}_r), \quad r = 0, 1, \ldots,$$

which we have encountered previously as Equation (4.3.10).

6.5 APPROXIMATE SOLUTIONS: DYNAMIC LINEAR MODELS

6.5.1 Dynamic linear models and finite mixture Kalman filters

Unsupervised learning (i.e. incomplete data) problems also arise in a number of important areas of application where the underlying situation requires modelling in the form of a *time series*. In this section we shall outline briefly one possible approach to such problems based on dynamic linear model (DLM) time-series representations (see, for example, Harrison and Stevens, 1976).

We assume a discrete stochastic process representation of the underlying system evolution in the form

$$\boldsymbol{\theta}_t = A_t \boldsymbol{\theta}_{t-1} + \mathbf{w}_t, \tag{6.5.1}$$

where, with subscripts denoting time, $\boldsymbol{\theta}(p \times 1)$ is the unknown state of the process, $A(p \times p)$ is a known transition matrix, and $\mathbf{w}(p \times 1)$ is the process noise, with a normal distribution

$$\mathbf{w}_t \sim N(\mathbf{0}, Q_t), \tag{6.5.2}$$

with mean 0 and known covariance matrix Q_t. Equation (6.5.1) is referred to as the *system equation*.

However, we cannot observe the underlying system directly; we can merely observe some aspects of the system and our observations will be further subjected to measurement error. In general, therefore, we have, in addition to (6.5.1), a *measurement equation*, which we assume to be of the form

$$\mathbf{x}_t = H_t \boldsymbol{\theta}_t + \mathbf{v}_t, \tag{6.5.3}$$

where, with subscripts again denoting time, $\mathbf{x}(r \times 1)$, $r \leqslant p$, is the observation, $H(r \times p)$ is a known matrix, and $\mathbf{v}(r \times 1)$ is the measurement noise, with a normal distribution

$$\mathbf{v}_t \sim N(\mathbf{0}, R_t), \tag{6.5.4}$$

with mean 0 and known covariance matrix R_t.

From (6.5.1) and (6.5.3) it is easily shown, using Bayes' theorem and induction, that the posterior distribution for $\boldsymbol{\theta}_t$, conditional on $\mathbf{x}_1, \ldots, \mathbf{x}_t$, is $N(\mathbf{m}_t, \mathbf{C}_t)$, where

$$\mathbf{m}_t = A_t \mathbf{m}_{t-1} + C_t H_t^{\mathrm{T}} R_t^{-1} (\mathbf{x}_t - H_t A_t \mathbf{m}_{t-1}) \tag{6.5.5}$$

$$C_t^{-1} = H_t^{\mathrm{T}} R_t^{-1} H_t + (A_t C_{t-1} A_t^{\mathrm{T}} + Q_t)^{-1}, \tag{6.5.6}$$

and it is assumed that, initially, $\boldsymbol{\theta}_0 \sim N(\mathbf{m}_0, \mathbf{C}_0)$. The recursive equations (6.5.5) and (6.5.6) are known as the Kalman filter updating equations (see Jazwinski, 1970).

We have already discussed, in Example 2.2.3, the use of Gaussian sums to model non-normal noise distribution in DLMs. The following examples illustrate other kinds of practical problems which, using variants of (6.5.1) and (6.5.3), lead to finite mixture forms.

Example 6.5.1 Signal versus noise

With appropriate interpretations of $\boldsymbol{\theta}_t$ and $\mathbf{x}_t$ as spatial coordinates and velocities, together with choices of A_t and H_t which reflect target and observer characteristics, (6.5.1) and (6.5.3) serve to model many target tracking situations. See, for example, Singer, Sea, and Housewright (1974), Bar-Shalom and Tse (1973), Bar-Shalom (1978), and Gauvrit (1984).

In such situations, where the observations $(\mathbf{x}_t)$ represent radar or similar signals, there is always the possibility that a particular $\mathbf{x}_t$ is simply 'noise', which may, for example, be a signal from another unrelated target whose track is of no interest. If 'noise' is assumed to generate an $\mathbf{x}_t$ having an $N(\mathbf{0}, S_t)$ distribution, with S_t known, then the original measurement equation defined by (6.5.3) and (6.5.4) should be replaced by

$$\mathbf{x}_t \sim \pi N(H_t\boldsymbol{\theta}_t, R_t) + (1-\pi)N(\mathbf{0}, S_t),$$

where the mixing weight π denotes the proportion of observations which are true signals.

Once $\mathbf{x}_1$ has been observed, the posterior distribution for $\boldsymbol{\theta}_1$ is easily seen to be a mixture of two normal distributions. The first corresponds to updating on the basis of an assumed signal and has mean and covariance matrix as given by (6.5.5) and (6.5.6); the second corresponds to no updating, on the assumption that the observation was pure noise and has mean and covariance matrix $\mathbf{m}_{t-1}$ and C_{t-1}. The weight on the first posterior component is the posterior probability that $\mathbf{x}_1$ really was a signal.

By extension, given observations $\mathbf{x}_1, \ldots, \mathbf{x}_n$, the corresponding posterior distribution is a mixture of 2^n components, each corresponding to a particular assumed history of the process (i.e. string of signal or noise identifications).

Example 6.5.2 Tracking a manoeuvring or wandering target

A useful model for the system states of a target which can 'manoeuvre' (i.e. can introduce sudden changes—for example, in velocity or position—in its system evolution) or 'wander' (i.e. is subject to externally generated perturbations to its system evolution) is obtained by extending (6.5.1) to the more general form

$$\boldsymbol{\theta}_t = A_t\boldsymbol{\theta}_{t-1} + B_t\mathbf{u}_t + \mathbf{w}_t, \tag{6.5.7}$$

where B_t is a known matrix and $\mathbf{u}_t$ represents the possible 'jumps' of the system.

In some cases, the limited manoeuvrability of the target implies that $\mathbf{u}_t$ is one from a finite set of options (including a 'null' option which corresponds to 'no manoeuvre'). In other cases, 'wandering' may be caused by sudden external impulses, so that $\mathbf{u}_t$ may be modelled as a normally distributed 'shock' to the system.

For any given version of (6.5.7), recursive learning is simply accomplished using the Kalman filter updating equations. However, any uncertainty about manoeuvres or impulses implies, in effect, that (6.5.7) should be replaced by a finite mixture model, with weights reflecting the relative plausibility of the various manoeuvres, etc., that might obtain.

Of course, yet more complications can be introduced! Firstly, we could combine Examples 6.5.1 and 6.5.2, so that in addition to the uncertain nature of the various possible signals that might be received, we acknowledge that we might be receiving pure noise. Secondly, knowledge of the target's manoeuvring tactics might lead to an additional Markov chain structure on the mixing weights.

Example 6.5.3 Monitoring renal transplants

Smith *et al.* (1983) and Smith and West (1983) have shown that, in the context of renal transplant monitoring, a suitably transformed biochemical series can be well represented as a particular case of a DLM with possible jumps.

The additional feature of the analysis given in Smith and West (1983) is that the covariance structures appearing in (6.5.1) and (6.5.3) also contain an unknown parameter. This means that, in addition to the growing finite mixture form of posterior distribution for the system parameters conditional on the covariance parameter, there is also a growing mixture form in the posterior distribution for the unknown covariance parameter.

In the following section, we shall briefly outline some of the approximation strategies that have been suggested. Almost all of these are, in effect, extensions to the DLM context of the various suggestions we reviewed earlier in cases A and B. However, the cumbersome algebraic descriptions which result in the DLM case do not lend themselves easily to brief descriptions. We shall therefore content ourselves, in the main, with providing representative references to the published literature.

6.5.2 An outline of suggested approximation procedures

For concreteness, we shall try to give the flavour of the forms of approximation that have been suggested for finite mixture Kalman filtering by considering the special case of signal versus noise as outlined in Example 6.5.1.

For this example, if, after the single observation $\mathbf{x}_1$, we were informed of the true source of the observation (signal or noise), Equations (6.5.5) and (6.5.6) would become

$$\mathbf{m}_1 = \Delta A_1 \mathbf{m}_0 + C_1 H_1^{\mathrm{T}} R_1^{-1}(\mathbf{x}_1 - H_1 A_1 \mathbf{m}_0) + (1-\Delta)\mathbf{m}_0 \tag{6.5.8}$$

$$C_1^{-1} = \Delta[H_1^{\mathrm{T}} R_1^{-1} H_1 + (A_1 C_0 A_1^{\mathrm{T}} + Q_1)^{-1}] + (1-\Delta)C_0^{-1}, \tag{6.5.9}$$

where $\Delta = 1$ if the observation is a signal, $\Delta = 0$ if it is noise.

In fact, however, we do not know the value of Δ. Instead, we have available the quantity $\Pr(\Delta = 1 | \mathbf{x}_1)$, given by

$$\Pr(\Delta = 1 | \mathbf{x}_1) = \frac{p(\mathbf{x}_1 | \Delta = 1)\Pr(\Delta = 1)}{\sum_{j=0}^{1} p(\mathbf{x}_1 | \Delta = j)\Pr(\Delta = j)}, \tag{6.5.10}$$

where $\Pr(\Delta = 1)$ denotes the prior probability that the first observation is actually a signal and $p(\mathbf{x}_1 | \Delta = j)$ is the predictive density obtained as the convolution of (6.5.1) and (6.5.3) (for $t = 1$). The various approximation strategies that have been proposed then reduce, essentially, to suggestions for using (6.5.10) to approximate Δ in (6.5.8) and (6.5.9).

The *decision-directed* schemes involve cut-off rules based on (6.5.10), or modifications thereof, in order to assign either the value $\Delta = 1$ or $\Delta = 0$. See, for example, Eisenstein (1972), Sanyal (1974), and Willsky and Jones (1976).

The *probabilistic teacher* schemes randomly choose $\Delta = 1$ or $\Delta = 0$ with probabilities given by (6.5.10). See, for example, Rajasekaran and Srinath (1973) and Chang and Srinath (1976).

The *probabilistic editor* schemes replace the mixture posterior distribution by a single normal distribution whose mean and covariance matrix are equal to those of the mixture (taking Δ to be a random variable with distribution given by 6.5.10). See, for example, Harrison and Stevens (1976), Athans, Whiting, and Gruber (1977), and Smith and West (1983), who extend the idea in the context of learning about a variance parameter by approximating a mixture of gamma distributions by a single gamma.

The *quasi-Bayes* scheme replaces Δ in (6.5.8) and (6.5.9) by its expectation (i.e. by the value of 6.5.10). See Makov and Smith (1977b), Makov (1980c), and Smith and Makov (1980).

All these schemes then proceed for subsequent observations in an obvious recursive way—in effect, collapsing the mixture posterior distribution to a single distribution or, at least, a smaller mixture at each stage (either after each individual observation or perhaps after a batch of observations).

There are, of course, a great variety of situations that can be modelled within the unsupervised DLM framework and so it is difficult to arrive at a global view regarding the performance of these various suggested procedures. However, limited experience with simulation studies suggests that the probabilistic editor approach has much to commend it. Smith and Makov (1980) suggest that this conclusion is strengthened if it is possible to process observations in batches, so that the posterior mixture is allowed to grow somewhat before being collapsed back, using the probabilistic editor scheme, to a smaller mixture, or to a single distribution.

References

Abraham, B., and Box, G. E. P. (1978). Linear models and spurious observations. *Appl. Statist.*, **27**, 131–138.

Ageno, M., and Frontali, C. (1963). Analysis of frequency distribution curves in overlapping Gaussians. *Nature*, **198**, 1294–1295.

Agrawala, A. K. (1970). Learning with a probabilistic teacher. *IEEE Trans. Inform. Th.*, **IT-16**, 373–379.

Agrawala, A. K. (1973). Learning with various types of teachers. *Proc. First Int. Joint Conf. Pattern Recognition*, **1973**, 453–461.

Aitchison, J. (1955). On the distribution of a positive random variable having a discrete probability mass at the origin. *J. Amer. Statist. Assoc.*, **50**, 901–908.

Aitchison, J., and Dunsmore, I. R. (1975). *Statistical Prediction Analysis*. Cambridge University Press.

Aitkin, M., Anderson, D., and Hinde, J. (1981). Statistical modelling of data on teaching styles (with discussion). *J. R. Statist. Soc. A*, **144,** 419–461.

Aitkin, M., and Rubin, D. B., (1985). Estimation and hypothesis testing in finite mixture models. *J.R. Statist. Soc. B*, **47**: to appear.

Aitkin, M., and Tunnicliffe Wilson, G. (1980). Mixture models, outliers and the EM algorithm. *Technometrics*, **22**, 325–332.

Akaike, H. (1974). A new look at statistical model identification. *IEEE Trans. Automat. Contr.*, **AC-19**, 716–723.

Allen, G. C., and McMeeking, R. F. (1978). Deconvolution of spectra by least-squares fitting. *Anal. Chim. Acta*, **103**, 73–108.

Alspach, D. L. (1974). Gaussian sum approximations in nonlinear filtering and control. *Inform. Sciences*, **7**, 271–290.

Alspach, D. L. (1977). A Gaussian sum Bayesian approach to passive bearings only track. Air Force Office of Scientific Research/NM, AFOSR-TR-77-1285.

Alspach, D. L., and Sorenson, H. W. (1972). Nonlinear Bayesian estimation using Gaussian sum approximation. *IEEE Trans. Autom. Contr.*, **AC-17**, 439–448.

Anderson, J. A. (1979). Multivariate logistic compounds. *Biometrika*, **66**, 17–26.

Anderson, J. A., and Blair, V. (1982). Penalized maximum likelihood estimation in logistic regression and discrimination. *Biometrika*, **69,** 123–136.

Anderson, T. W. (1966). Some nonparametric multivariate procedures based on statistically equivalent blocks. In *Multivariate Analysis* (Ed. P. R. Krishnaiah), Vol. I, pp. 5–27. Academic Press, New York.

Andrews, D. F. (1972). Plots of high-dimensional data. *Biometrics*, **28**, 125–136.

Andrews, D. F., Bickel, P. J., Hampel, F. R., Huber, P. J., Rogers, W. H., and Tukey, J. W. (1972). *Robust Estimates of Location.* Princeton University Press.

Andrews, D. F., and Mallows, C. L. (1974). Scale mixtures of normal distributions. *J. R. Statist. Soc. B,* **36**, 99–102.

Antelman, G. R. (1972). Interrelated Bernoulli processes. *J. Amer. Statist. Assoc.*, **67**, 831–841.

Antoniak, C. E. (1974). Mixtures of Dirichlet processes with applications to Bayesian nonparametric problems. *Ann. Statist.*, **2**, 1152–1174.

Aroian, L. A. (1948). The fourth degree exponential distribution function. *Ann. Math. Statist.*, **19**, 589–592.

Ashford, J. R., and Walker, P. J. (1972). Quantal response analysis for a mixture of populations. *Biometrics*, **28**, 981–988.

Ashton, W. D. (1971). Distribution for gaps in road traffic. *J. Inst. Maths. Applics.*, **7**, 37–46.

Athans, M., Whiting, R. H., and Gruber, H. (1977). A suboptimal estimation algorithm with probabilistic editing for false measurements with applications to target tracking with wake phenomena. *IEEE Trans. Automat. Contr.*, **AC-22**, 372–384.

Bagnold, R. A. (1941). *The Physics of Blown Sand and Desert Dunes.* Methuen, London.

Baker, G. A. (1932). Distribution of the means divided by the standard deviations of samples from non-homogeneous populations. *Ann. Math. Statist.*, **3**, 1–9.

Barlow, R. E., Bartholomew, D. J., Bremner, J. M., and Brunk, H. D. (1972). *Inference under Order Restrictions.* Wiley, New York.

Barndorff-Nielsen, O. (1965). Identifiability of mixtures of exponential families. *J. Math. Anal. Appl.*, **12**, 115–121.

Barndorff-Nielsen, O. (1979). Models for non-Gaussian variation, with applications to turbulence. *Proc. Roy. Soc. A*, **353**, 401–419.

Barndorff-Nielsen, O., Kent, J., and Sorensen, M. (1982). Normal variance-mean mixtures and *z* distributions. *Int. Statist. Rev.*, **50**, 145–159.

Barnett, V. (1975). Probability plotting method and order statistics. *Appl. Statist.*, **24**, 95–108.

Barnett, V. (1976). Convenient probability plotting positions for the normal distribution. *Appl. Statist.*, **25**, 47–50.

Barnett, V., and Lewis, T. (1978). *Outliers in Statistical Data.* Wiley, New York.

Bar-Shalom, Y. (1978). Tracking methods in a multitarget environment. *IEEE Trans. Automat. Contr.*, **AC-23**, 618–626.

Bar-Shalom, Y., and Tse, E. (1973). Tracking in a cluttered environment with probabilistic data association. *Proc. fourth Symp. Nonlinear Estimation Theory and Its Applications,* **1973**, 13–22.

Bartholomew, D. J. (1969). Sufficient conditions for a mixture of exponentials to be a probability density function. *Ann. Math. Statist.*, **40**, 2183–2188.

Bartholomew, D. J. (1980). Factor analysis for categorical data (with discussion). *J. R. Statist. Soc. B*, **42**, 293–321.

Bartlett, M. S., and Macdonald, P. D. M. (1968). Least squares estimation of distribution mixtures. *Nature*, **217**, 195–196.

Barton, D. E., and Dennis, K. E. (1952). The conditions under which the Gram–Charlier and Edgeworth curves are positive definite and unimodal. *Biometrika*, **39**, 425–427.

Baum, L. E., and Eagon, J. A. (1967). An inequality with applications to statistical estimation for probabilistic functions of Markov processes and to a model for ecology. *Bull. Amer. Math. Soc.*, **73**, 360–363.

Baum, L. E., Petrie, T., Soules, G., and Weiss, N. (1970). A maximization technique occurring in the statistical analysis of probabilisitic functions of Markov chains. *Ann. Math. Statist.*, **41**, 164–171.

Beale, E. M. L., and Mallows, C. L. (1959). Scale mixing of symmetric distributions with zero means. *Ann. Math. Statist.*, **30**, 1145–1151.

Beall, G. (1940). The fit and significance of contagious distributions when applied to observations on larval insects. *Ecology*, **21**, 460–474.

Bebbington, A. C. (1978). A method of bivariate trimming for robust estimation of the correlation coefficient. *Appl. Statist.*, **27**, 221–226.

Beckman, R. J., and Cook, R. D. (1983). Outliers (with discussion). *Technometrics*, **25**, 119–163.

Behboodian, J. (1970a). On a mixture of normal distributions. *Biometrika*, **57**, 215–217.

Behboodian, J. (1970b). On the modes of a mixture of two normal distributions. *Technometrics*, **12**, 131–139.

Behboodian, J. (1972a). Information matrix for a mixture of two normal distributions. *J. Statist. Comp. Simul.*, **1**, 295–314.

Behboodian, J. (1972b). On the distribution of a symmetric statistic from a mixed population. *Technometrics*, **14**, 919–923.

Behboodian, J. (1975). Structural properties and statistics of finite mixtures. In *Statistical Distributions in Scientific Work* (Eds G. P. Patil *et al.*), Vol. 1, pp. 103–112. Reidel, Dordrecht.

Bellman, R. E. (1960). On the separation of exponentials. *Boll. Un. Mat. Ital.*, **15**, 38–39.

Beran, R. (1977). Minimum Hellinger distance estimates for parametric models. *Ann. Statist.*, **5**, 445–463.

Berry, E. R., and Chanutin, A. (1955). Detailed electrophoretic analyses of sera of healthy young men. *J. Clin. Invest.*, **34**, 115–135.

Bezdek, J. C., and Dunn, J. C. (1975). Optimal fuzzy partitions: a heuristic for estimating the parameters in a mixture of normal distributions. *IEEE Trans. Computers*, **C-24**, 835–838.

Bhattacharya, C. G. (1967). A simple method of resolution of a distribution into Gaussian components. *Biometrics*, **23**, 115–135.

Bignami, A., and de Matteis, A. (1971). A note on sampling from combinations of distributions. *J. Inst. Maths. Applics.*, **8**, 80–81.

Binder, D. A. (1978a). Bayesian cluster analysis. *Biometrika*, **65**, 31–38.

Binder, D. A. (1978b). Comments on a paper by Quandt and Ramsey. *J. Amer. Statist. Assoc.*, **73**, 746–747.

Blåfield, E. (1980). Clustering of observations from finite mixtures with structural information. *Univ. Jyväskylä Studies in Comp. Sci. Econ. and Statist.*, No. 2, Finland.

Blakley, G. R. (1967). Darwinian natural selection acting within populations. *J. Theoret. Biol.*, **17**, 252–281.

Blischke, W. R. (1962). Moment estimation for the parameters of a mixture of two binomial distributions. *Ann. Math. Statist.*, **33**, 444–454.

Blischke, W. R. (1964). Estimating the parameters of mixtures of binomial distributions. *J. Amer. Statist. Assoc.*, **59**, 510–528.

Blischke, W. R. (1965). Mixtures of discrete distributions. In *Classical and Contagious Discrete Distributions* (Ed. G. P. Patil), pp. 351–372. Pergamon, New York.

Blischke, W. R. (1978). Mixtures of distributions. In *International Encyclopaedia of Statistics* (Eds. W. H. Kruskal and J. M. Tanur), pp. 174–180. The Free Press, New York.

Blum, J. R., and Susarla, V. (1977). Estimation of a mixing distribution function. *Ann. Statist.*, **5**, 200–209.

Blumenthal, S., and Govindarajulu, Z. (1977). Robustness of Stein's 2-stage procedure for mixtures of normal populations. *J. Amer. Statist. Assoc.*, **72**, 192–196.

Boes, D. C. (1966). On the estimation of mixing distributions. *Ann. Math. Statist.*, **37**, 177–188.

Boes, D. C. (1967). Minimax unbiased esimator of mixing distribution for finite mixtures. *Sankhya A*, **29**, 417–420.

Bõhning, D. (1982). Convergence of Simar's algorithm for finding the maximum likelihood estimate of a compound Poisson process. *Ann. Statist.*, **10**, 1006–1008

Boneva, L. I., Kendall, D. G., and Stefanov, I. (1971). Spline transformations. *J. R. Statist. Soc. B*, **33**, 1–70.

Bowman, K. O., and Shenton, L. R. (1973). Space of solutions for a normal mixture. *Biometrika*, **60**, 629–636.

Box, G. E. P., and Tiao, G. C. (1968). A Bayesian approach to some outlier problems. *Biometrika*, **55**, 119–129.

Boyles, R. A. (1983). On the convergence of the EM algorithm. *J. R. Statist. Soc. B*, **45**, 47–50.

Brazier, S., Sparks, R. S. J., Carey, S. N., Sigurdsson, H., and Westgate, J. A. (1983). Biomodal grain size distribution and secondary thickening in air-fall ash layers. *Nature*, **301**, 115–119.

Broadbent, D. E. (1966). A difficulty in assessing bimodality in certain distributions. *Brit. J. Math. Stat. Psychol.*, **19**, 125–126.

Brown, D. J. P. (1978). A study of Harding's method for the decomposition of a mixture into normal components. B.Sc. Dissertation, Univ. St. Andrews, Scotland.

Brown, G. H. (1976). Combining estimates of category and subcategory proportions in a mixed population. *Biometrics*, **32**, 453–457.

Brownell, G. L., and Callaghan, A. B. (1963). Transform methods for tracer data analysis. *Ann. New York Acad. Sci.*, **108**, 172–181.

Brownie, C., Habicht, J-P., and Robson, D. S. (1983). An estimation procedure for the contaminated normal distributions arising in clinical chemistry, *J. Amer. Statist. Assoc.*, **78**, 228–237.

Bryant, J. L., and Paulson, A. S. (1979). Some comments on characteristic function-based estimations. *Sankhya A*, **41**, 109–116.

Bryant, J. L., and Paulson, A. S. (1983). Estimation of mixing proportions via distance between characteristic functions. *Commun. Statist. A*, **12**, 1009–1029.

Bryant, P., and Williamson, J. A. (1978). Asymptotic behaviour of classification ML estimates. *Biometrika*, **65**, 273–281.

Buchanan-Wollaston, H. J., and Hodgson, W. C. (1928). A new method of treating frequency curves in fishery statistics, with some results. *J. Cons. Perm. Intern. Expl. Mer.*, **4**, 207–225.

Burridge, J. (1982). Some unimodality properties of likelihoods derived from grouped data. *Biometrika*, **69**, 145–151.

Campbell, N. A. (1984). Mixture models and atypical values. *Math. Geol.* **16**, 465–477.

Cassie, R. M. (1954). Some uses of probability paper in the analysis of size frequency distributions. *Austral. J. Mar. Fish. Freshw. Res.*, **5**, 513–522.

Cassie, R. M. (1962). Frequency distribution models in the ecology of plankton and other organisms. *J. Anim. Ecol.*, **31**, 65–92.

Chandra, S. (1977). On the mixtures of probability distributions. *Scand, J. Statist.*, **4**, 105–112.

Chang, F. S., and Srinath, M. D. (1976). Target tracking over fading channels. *IEEE Trans. Commun.*, **COM-24**, 432–437.

Chang, W. C. (1976). The effects of adding a variable in dissecting a mixture of two normal populations with a common covariance matrix. *Biometrika*, **63**, 676–678.

Chang, W. C. (1979). Confidence interval estimation and transformation of data in a mixture of two multivariate normal distributions with any given large dimension. *Technometrics*, **21**, 351–355.

Charlier, C. V. L. (1906). Researches into the theory of probability. *Lunds Univ. Ars. Ny foljd*, Afd 2.1, No. 5.

Charlier, C. V. L., and Wicksell, S. D. (1924). On the dissection of frequency functions. *Arkiv f. Matematik Astron. och Fysik.*, Bd. 18, No. 6.

Choi, K. (1969a). Estimators for the parameters of a finite mixture of distributions. *Ann. Inst. Statist. Math.*, **21**, 107–116.

Choi, K. (1969b). Empirical Bayes procedure for (pattern) classification with stochastic learning. *Ann. Inst. Statist. Math.*, **21**, 117–125.

Choi, K., and Bulgren, W. B. (1968). An estimation procedure for mixtures of distributions. *J. R. Statist. Soc. B*, **30**, 444–460.

Choi, S. C. (1979). Two sample tests for compound distributions for homogeneity of mixing proportions. *Technometrics*, **21**, 361–365.

Clark, I. (1977). ROKE, a computer program for non-linear least-squares decomposition of mixtures of distributions. *Comput & Geosc.*, **3**, 245–256.

Clark, M. W. (1976). Some methods for statistical analysis of multimodal distributions and their application to grain-size data. *J. Math. Geol.*, **8**, 267–282.

Clark, M. W. (1977). GETHEN: a computer program for the decomposition of mixtures of two normal distributions by the method of moments. *Comput. Geosci.*, **3**, 257–267.

Clark, V. A., Chapman, J. M., Coulson, A. H., and Hasselblad, V. (1968). Dividing the blood pressures from the Los Angeles heart study into two normal distributions. *Johns Hopkins. Med. J.*, **122**, 77–83.

Clarke, B. R., and Heathcote, C. R. (1978). Comment on a paper by Quandt and Ramsey. *J. Amer. Statist. Assoc.*, **73**, 749.

Cobb, L., Koppstein, P., and Chen, N. H. (1983). Estimation and moment recursion relations for multimodal distributions of the exponential family. *J. Amer. Statist. Assoc.*, **78**, 124–130.

Cohen, A. C. (1960). Extension of a truncated Poisson distribution. *Biometrics*, **16**, 446–450.

Cohen, A. C. (1965). Estimation in mixtures of discrete distributions. In *Classical and Contagious Discrete Distributions* (ed. G. P. Patil), pp. 373–378. Pergamon, New York.

Cohen, A. C. (1966a). A note on certain discrete mixed distributions. *Biometrics*, **22**, 566–571.

Cohen, A. C. (1966b). Discussion of a paper by V. Hasselblad. *Technometrics*, **8**, 445–446.

Cohen, A. C. (1967). Estimation in mixtures of two normal distributions. *Technometrics*, **9**, 15–28.

Cooper, D. B., and Cooper, P. W. (1964). Nonsupervised adaptive signal detection and pattern recognition. *Inform. Control*, **7**, 416–444.

Cooper, D. B. (1975). On some convergence properties of learning with a probabilistic teacher algorithms. *IEEE Trans. Inform. Th.*, **IT-21**, 699–703.

Cornell, R. G. (1962). A method for fitting linear combinations of exponentials. *Biometrics*, **18**, 104–113.

Covey-Crump, P. A. K. (1970). Statistical analysis of granule size in the granular cells of the magnum of the hen oviduct. *Q. J. Exp. Physiol.*, **55**, 233–237.

Cox, D. R. (1966). Notes on the analysis of mixed frequency distributions. *Br. J. Math. Stat. Psychol.*, **19**, 39–47.

Cox, D. R., and Hinkley, D. V. (1974). *Theoretical Statistics*. Chapman and Hall, London.

Dalal, S. R. (1978). A note on the adequacy of mixtures of Dirichlet processes. *Sankhya A*, **40**, 185–191.

Dalal, S. R., and Hall, W. J. (1983). Approximating priors by mixtures of natural conjugate priors. *J. R. Statist. Soc. B*, **45**, 278–286.

Dallaville, J. M., Orr, C., and Blocker, H. G. (1951). Fitting bimodal particle size distribution curves. *Ind. Eng. Chem.*, **43**, 1377–1380.

Daniels, H. E. (1961). Mixtures of geometric distributions. *J. R. Statist. Soc. B*, **23**, 409–413.

Davies, R. B. (1977). Hypothesis testing when a nuisance parameter is present only under the alternative. *Biometrika*, **64**, 247–254.

Davis, D. J. (1952). An analysis of some failure data. *J. Amer. Statist. Assoc.*, **47**, 113–150.
Davisson, L. D. (1970). Convergence probability bounds for stochastic approximation. *IEEE Trans. Inform. Th.*, **IT-16**, 680–685.
Davisson, L. D., and Schwartz, S. C. (1970). Analysis of decision-directed receiver with unknown priors. *IEEE Trans. Inform. Th.*, **IT-16**, 270–276.
Dawid, A. P. (1976). Properties of diagnostic data distributions. *Biometrics*, **32**, 647–658.
Dawid, A. P., and Skene, A. M. (1979). Maximum likelihood estimation of observer error-rates using the EM algorithm. *Appl. Statist.*, **28**, 20–28.
Day, N. E. (1969). Estimating the components of a mixture of normal distributions. *Biometrika*, **56**, 463–474.
Dean, P. N., and Jett, J. H. (1974). Mathematical analysis of DNA distributions derived from flow microfluorometry. *J. Cell. Biol.*, **60**, 523–527.
Deely, J. J., and Kruse, K. L. (1968). Construction of sequences estimating the mixing distribution. *Ann. Math. Statist.*, **39**, 286–288.
Deely, J. J., and Lindley, D. V. (1981). Bayes empirical Bayes. *J. Amer. Statist. Assoc.*, **76**, 833–841.
Defares, J. G., Sneddon, I. N., and Wise, M. E. (1973). *An Introduction to the Mathematics of Medicine and Biology.* North-Holland, London.
DeGroot, M. H. (1970). *Optimal Statistical Decisions*, McGraw-Hill, New York.
Dempster, A. P., Laird, N. M., and Rubin, D. B. (1977). Maximum likelihood estimation from incomplete data via the EM algorithm (with discussion). *J. R. Statist. Soc. B,* **39**, 1–38.
Diaconis, P., and Ylvisaker, D. (1985). Quantifying prior opinion. In *Bayesian Statistics 2* (Eds. J. M. Bernardo *et al.*), pp. 133–156. North Holland, Amsterdam.
Dick, N. P., and Bowden, D. C. (1973). Maximum likelihood estimation for mixtures of two normal distributions. *Biometrics*, **29**, 781–790.
Di Gesu, V., and Maccarone, M. C. (1984). The Bayesian direct convolution method: properties and applications. *Signal Processing*, **6**, 201–211.
Dixon, W. J. (1950). Analysis of extreme values. *Ann. Math. Statist.*, **21**, 488–506.
Do, K., and McLachlan, G. J. (1984). Estimation of mixing proportions: a case study. *Appl. Statist.*, **33**, 134–140.
Doetsch, G. (1928). Die elimination der dopplereffekts bei spektroskopischen feinstrukturen und exakte bestimmung der komponenten. *Zeitschrift. f. Phys.*, **49**, 705–730.
Doetsch, G. (1936). Zerlegung einer Funktion in Gausche Fehlerkurven und zeitliche Zuruckverfolgung eines Temperatur-zustandes. *Math. Zeitschrift.*, **41**, 283–318.
Draper, N. R., and Tierney, D. E. (1972). Regions of positive and unimodal series expansion of the Edgeworth and Gram-Charlier approximations. *Biometrika*, **59**, 463–465.
Duda, R. O., and Hart, P. E. (1973). *Pattern Recognition and Scene Analysis.* Wiley, New York.
Durairajan, T. M., and Kale, B. K. (1979). Locally most powerful test for the mixing proportion. *Sankhya*, **41**, 91–100.
Efron, B. (1975). The efficiency of logistic regression compared to normal discriminant analysis. *J. Amer. Statist. Assoc.*, **70**, 892–898.
Efron, B., and Olshen, R. (1978). How broad is the class of normal scale mixtures? *Ann. Statist.*, **6**, 1159–1164.
Eisenberger, I. (1964). Genesis of bimodal distributions. *Technometrics*, **6**, 357–363.
Eisenstein, B. A. (1972). Decision directed estimation/detection of repetitive signals. *Proc. IEEE Conf. Decision and Control*, **1972**.
Elashoff, J. D. (1972). A model for quadratic outliers in linear regression. *J. Amer. Statist. Assoc.*, **67**, 478–489.
Engleman, L., and Hartigan, J. A. (1969). Percentage points for a test for clusters. *J. Amer. Statist. Assoc.*, **64**, 1647–1648.

Everett, G. V. (1973). The rainbow trout Salmo gairdneri (Rich.) fishery of Lake Titicaca *J. Fish. Biol.*, **5**, 429–440.

Everitt, B. S. (1978). *Graphical Techniques for Multivariate Data.* Heinemann, London.

Everitt, B. S. (1980). *Cluster Analysis*, 2nd ed. Heinemann, London.

Everitt, B. S. (1981a). A Monte Carlo investigation of the likelihood ratio test for the number of components in a mixture of Normal distributions. *Multivar. Behav. Res.*, **16**, 171–189.

Everitt, B. S. (1981b). Bimodality and the nature of depression. *Brit. J. Psychiat.*, **138**, 336–339.

Everitt, B. S. (1984). Maximum likelihood estimation of the parameters in a mixture of two univariate normal distributions; a comparison of different algorithms. *The Statistician*, **33**, 205–215.

Everitt, B. S., and Hand, D. J. (1981). *Finite Mixture Distributions.* Chapman and Hall, London.

Fabi, F., and Rossi, C. (1983). Decomposition of mixtures via a generalized E. M. method. *Metron*, **41**, 133–146.

Fabian, V. (1968). On asymptotic normality in stochastic approximation. *Ann. Math. Statist.*, **39**, 1327–1332.

Fabian, V. (1978). On asymptotically efficient recursive estimation. *Ann. Statist.*, **6**, 854–866.

Falls, L. W. (1970). Estimation of parameters in compound Weibull distributions. *Technometrics*, **12**, 399–407.

Falmagne, J. C. (1968). Note on a simple fixed-point property of binary mixtures. *Brit. J. Math. Stat. Psychol.*, 131–132.

Farjo, A. A., and Young, T. Y. (1976). Analysis and design of decision-directed learning schemes using stochastic approximation. *Inform. Sciences*, **10**, 199–215.

Feder, P. I. (1975). The log likelihood ratio in segmented regression *Ann Statist.*, **3**, 84–97.

Feller, W. (1943). On a class of 'contagious' distributions. *Ann. Math. Statist.*, **14**, 389–400.

Ferguson, T. S. (1983). Bayesian density estimation via mixtures of normal distributions. In *Recent Advances in Statistics*, pp. 287–302. Academic Press, New York.

Fielding, A. (1977). Latent structure analysis. In *Exploring Data Structures* (Eds. C. A. O'Muircheartaigh and C. Payne), pp. 125–157. Wiley, New York.

Fisher, L., and Yakowitz, S. J. (1970). Estimating mixing distributions in metric spaces. *Sankhya A*, **32**, 411–418.

Fisher, R. A. (1921). On the mathematical foundation of theoretical statistics. *Phil. Trans. Roy. Soc. A*, **222**, 309–368.

Fisher, R. A. (1936). The use of multiple measurements in taxonomic problems. *Ann. Eugenics*, **7**, 179–188.

Fletcher, R. (1971). A general quadratic programming algorithm. *J. Inst. Maths. Applics.*, **7**, 76–91.

Fletcher, R., and Reeves, C. M. (1964). Function minimization by conjugate gradients. *Computer J.*, **7**, 149–154.

Folk, R. L. (1971). Longitudinal dunes of the northwestern edge of the Simpson Desert, Northern Territory, Australia. I. Geomorphology and quasi-size relationships. *Sedimentology*, **16**, 5–54.

Fowlkes, E. B. (1979). Some methods for studying the mixture of two normal (lognormal) distributions. *J. Amer. Statist. Assoc.*, **74**, 561–575.

Fraser, M. D., Hsu, Y-S., and Walker, J. J. (1981). Identifiability of finite mixtures of von Mises distributions. *Ann. Statist.*, **9**, 1130–1131.

Fraser, R. D. B., and Suzuki, B. (1966). Resolution of overlapping absorption bands by least squares procedures. *Anal. Chem.*, **38**, 1770–1773.

Freeman, P. R. (1981). On the number of outliers in data from a linear model. In *Bayesian Statistics* (Eds. J. M. Bernardo *et al.*), pp. 349–365. Univ. Press, Valencia.
French, C. S., Towner, G. H., Bellis, D. R., Cook, R. M., Fair, W. R., and Holt, W. W. (1954). A curve analysis and general purpose graphical computer. *Rev. Sci. Inst.*, **25**, 765–775.
Fryer, J. G., and Holt, D. (1970). On the robustness of the standard estimates of the exponential mean to contamination. *Biometrika*, **57**, 641–648.
Fryer, J. G., and Robertson, C. A. (1972). A comparison of some methods for estimating mixed normal distributions. *Biometrika*, **59**, 639–648.
Fu, K. S. (1968). *Sequential Methods in Pattern Recognition and Machine Learning.* Academic Press, New York.
Fukunaga, K., and Flick, T. E. (1983). Estimation of the parameters of a Gaussian mixture using the method of moments. *IEEE Trans. Patt. Anal. Match. Intell.*, **PAMI-5**, 410–416.
Ganesalingam, S., and McLachlan, G. J. (1978). The efficiency of a linear discriminant function based on unclassified initial samples. *Biometrika*, **65**, 658–662.
Ganesalingam, S., and McLachlan, G. J. (1979a). Small sample results for a linear discriminant function estimated from a mixture of normal populations. *J. Statist. Comput. Simul.*, **9**, 151–158.
Ganesalingam, S., and McLachlan, G. J. (1979b). A case study of two clustering methods based on maximum likelihood. *Stat. Neer.*, **33**, 81–90.
Ganesalingam, S., and McLachlan, G. J. (1981). Some efficiency results for the estimation of the mixing proportion in a mixture of two normal distributions. *Biometrics*, **37**, 23–33.
Gardner, D. G. (1963). Resolution of multi-component exponential decay curves using Fourier transforms. *Ann. New York Acad. Sci.*, **108**, 195–203.
Gauvrit, M. (1984). Bayesian adaptive filter for tracking with measurements of uncertain origin. *Automatica*, **20**, 217–224.
Gelfand, A. E., and Solomon, N. (1975). Analysing the decision making process of the American jury. *J. Amer. Statist. Assoc.*, **70**, 305–310.
Ghose, B. K. (1970). Statistical analysis of mixed fossil populations. *J. Math. Geol.*, **2**, 265–276.
Ghosh, P. (1978). A characterization of a bimodal distribution. *Commun. Statist. A*, **7**, 475–477.
Gladyshev, E. G. (1965). On stochastic approximation. *Theory of Prob. and Its Appl.*, **10**, 275–278.
Glaser, E. N. (1961). Signal detection by adaptive filters. *IEEE Trans. Information Th.*, **IT-7**, 87–98.
Gnandesikan, R. (1977). *Methods for Statistical Data Analysis of Multivariate Observations.* Wiley, New York.
Godambe, A. V. (1977). On representation of Poisson mixtures as Poisson sums and a characterization of the gamma distribution. *Math. Proc. Camb. Phil. Soc.*, **82**, 297–300.
Goldfeld, S. M., and Quandt, R. E. (1973). A Markov model for switching regressions. *J. Econometrics*, **1**, 3–16.
Goldfeld, S. M., and Quandt, R. E. (1976). *Studies in Nonlinear Estimation.* Ballinger, Cambridge, MA.
Goodman, L. A. (1974). Exploratory latent structure models using both identifiable and unidentifiable models. *Biometrika*, **61**, 215–331.
Gordon, A. D. (1981). *Classification.* Chapman and Hall, London.
Gordon, A. D., and Prentice, I. C. (1977). Numerical methods in quaternary palaeoecology. IV. Separating mixtures of morphologically similar pollen taxa. *Rev. Palaeobot. Palynol.*, **23**, 359–372.
Gottschalk, V. H. (1948). Symmetric bimodal frequency curves. *J. Franklin Inst.*, **245**, 245–252.

Grannis, G. F., and Lott, J. A. (1978). A technique for determining the probability of abnormality. *Clin. Chem.*, **24**, 640–651.

Greenwood, J. A., and Hartley, H. O. (1962). *Guide to Tables in Mathematical Statistics.* Oxford Univ. Press.

Gregg, W. D., and Hancock, J. C. (1968). An optimum decision directed scheme for Gaussian mixtures. *IEEE Trans. Inform. Th.*, **IT-14**, 451–461.

Gregor, J. (1969). An algorithm for the decomposition of a distribution into Gaussian components. *Biometrics*, **25**, 79–93.

Gridgeman, N. T. (1970). A comparison of two methods of analysis of mixtures of normal distributions. *Technometrics*, **12**, 823–833.

Gumbel, E. J. (1940). La dissection d'une repartition. *Ann. Univ. Lyon, 3A*, **2**, 39–51.

Gupta, A. K., and Miyawaki, T. (1978). On a uniform mixture model. *Biom. J.*, **20**, 631–637.

Gurland, J. (1957). Some interrelations among compound and generalized distributions. *Biometrika*, **44**, 265–268.

Guseman, L. F., and Walton, J. R. (1977). An application of linear feature selection to estimation of proportions. *Commun. Statist., A*, **6**, 611–617.

Guseman, L. F., and Walton, J. R. (1978). Methods for estimating proportions of convex combinations of normals using linear feature selection. *Commun. Statist., A*, **7**, 1439–1450.

Guttman, I., Dutter, R., and Freeman, P. R. (1978). Care and handling of univariate outliers in the general linear model to detect spuriosity—a Bayesian approach. *Technometrics*, **20**, 187–193.

Hald, A. (1952). *Statistical Theory with Engineering Applications.* Wiley, New York.

Hald, A. (1960). The compound hypergeometric distribution and a system of single sampling inspection plans based on prior distributions and costs. *Technometrics*, **2**, 275–340.

Haldane, J. B. S. (1952). Simple tests for bimodality and bitangentiality. *Ann. Eugenics*, **16**, 359–364.

Hall, P. (1981). On the non-parametric estimation of mixture proportions. *J. R. Statist. Soc. B*, **43**, 147–156.

Hall, P., and Heyde C. C. (1980). *Martingale Limit Theory and Its Applications.* Academic Press, New York.

Hall, P., and Titterington, D. M. (1984). Efficient nonparametric estimation of mixture proportions. *J. R. Statist. Soc. B*, **46**, 465–473.

Hall, P., and Titterington, D. M. (1985). The use of uncategorized data to improve the performance of a nonparametric estimator of a mixture density. *J. R. Statist. Soc. B*, **47**, 155–163.

Hall, P., and Welsh, A. H. (1983). A test for normality based on the characteristic function. *Biometrika*, **70**, 485–489.

Han, C.-P. (1978). Estimating means when a group of observations is classified by a linear discriminant function. *J. Amer. Statist. Assoc.*, **73**, 661–665.

Hand, D. J. (1981). *Discrimination and Classification.* Wiley, New York.

Harding, J. P. (1949). The use of probability paper for the graphical analysis of polymodal frequency distributions. *J. Marine Biol. Assoc.*, **28**, 141–153.

Hardy, G. H., Littlewood, J. E., and Polya, G. (1952). *Inequalities.* Cambridge University Press.

Harris, C. M. (1983). On finite mixtures of geometric and negative binomial distributions. *Commun. Statist. A*, **12**, 987–1007.

Harris, D. (1968). A method of separating two superimposed normal distributions using arithmetic probability paper. *J. Anim. Ecol.*, **37**, 315–319.

Harris, H., and Smith, C. A. B. (1949). The sib-sib age of onset correlation among individuals suffering from the same hereditary syndrome produced by more than one gene. *Ann. Eugenics*, **14**, 309–318.

Harrison, P. J., and Stevens, C. F. (1976). Bayesian forecasting (with discussion). *J. R. Statist. Soc. B*, **38**, 205–247.

Hartigan, J. A. (1977). Distribution problems in clustering. In *Classification and Clustering* (Ed. J. van Ryzin), pp. 45–71. Academic Press, New York.

Hartigan, J. A., and Hartigan, P. M. (1985). The dip test of unimodality. *Ann. Statist.*, **13**, 70–84.

Hartley, M. J. (1978). Comments on a paper by Quandt and Ramsey. *J. Amer. Statist. Assoc.*, **73**, 738–741.

Hasselblad, V. (1966). Estimation of parameters for a mixture of normal distributions. *Technometrics*, **8**, 431–444.

Hasselblad, V. (1969). Estimation of finite mixtures of distributions from the exponential family. *J. Amer. Statist. Assoc.*, **64**, 1459–1471.

Hathaway, R. J. (1983). Constrained maximum likelihood estimation for a mixture of multivariate normal densities. Tech. Rep. **92**, Dept. Math. Stat., Univ. S. Carolina, Columbia, S. C.

Hathaway, R. J. (1985). A constrained formulation of maximum-likelihood estimation for normal mixture distributions. *Ann. Statist.*, **13**, 795–800.

Hawkins, R. H. (1972). A note on multiple solutions to the mixed distribution problem. *Technometrics*, **14**, 973–976.

Heathcote, C. R. (1977). Integrated mean square error estimation of parameters. *Biometrika*, **64**, 255–264.

Helguero, F. de (1904). Sui massimi delle curve dimofriche. *Biometrika*, **31**, 84–98.

Hermans, J., and Habbema, J. D. F. (1975). Comparison of five methods to estimate posterior probabilities. *EDV in Medizin und Biologie*, **1/2**, 14–19.

Heyde, C. C., and Johnstone, I. M. (1979). On asymptotic posterior normality for stochastic processes. *J. R. Statist. Soc. B*, **41**, 184–189.

Hill, B. M. (1963). Information for estimating the proportions in mixtures of exponential and normal distributions. *J. Amer. Statist. Assoc.*, **58**, 918–932.

Hill, D. L., Saunders, R., and Land, P. W. (1980). Maximum likelihood estimation for mixtures. *Canad. J. Statist.*, **8**, 87–93.

Ho, Y. C., and Agrawala, A. K. (1968). On pattern classification algorithms; introduction and survey. *Proc. IEEE*, **56**, 2102–2114.

Hoadley, B. (1969). The compound multinomial distribution and Bayesian analysis of categorical data from finite populations. *J. Amer. Statist. Assoc.*, **64**, 216–229.

Holgate, P. (1970). The modality of some compound Poisson distributions. *Biometrika*, **57**, 666–667.

Holgersson, M., and Jorner, U. (1979). Decomposition of a mixture into normal components: a review. *Int. J. Biomed. Comput.*, **9**, 367–392.

Hosmer, D. W. (1973a). A comparison of iterative maximum likelihood estimates of the parameters of a mixture of two normal distributions under three different types of sample. *Biometrics*, **29**, 761–770.

Hosmer, D. W. (1973b). On MLE of the parameters in a mixture of two normal distributions when the sample size is small. *Commun. Statist.*, **1**, 217–227.

Hosmer, D. W. (1974). Maximum likelihood estimates of the parameters of a mixture of two regression lines. *Commun. Statist.*, **3**, 995–1006.

Hosmer, D. W. (1978a). Comments on a paper by Quandt and Ramsey. *J. Amer. Statist. Assoc.*, **73**, 741–744.

Hosmer, D. W. (1978b). A use of mixtures of two normal distributions in a classification problem. *J. Statist. Comput. Simul.*, **6**, 281–294.

Hosmer, D. W., and Dick, N. P. (1977). Information and mixtures of two normal distributions. *J. Statist. Comput. Simul.*, **6**, 137–148.

Hoxter, G., Wajchenberg, B. L., and Mungioli, R. (1957). Analysis of electrophoretic patterns. *Nature*, **179**, 423–424.

Huber, P. J. (1964). Robust estimation of a location parameter. *Ann. Math. Statist.*, **35**, 73–101.

Huber, P. J. (1981). *Robust Statistics.* Wiley, New York.

Hyrenius, H. (1950). Distribution of 'Student'–Fisher *t* in samples from compound normal function. *Biometrika*, **37**, 429–442.

Hyrenius, H. (1952). Sampling from bivariate non-normal universes by means of compound normal distributions. *Biometrika*, **39**, 238–246.

Ibragimov, I. A. (1956). On the composition of unimodal distributions. *Theor. Prob. Applics*, **1**, 255–260.

Ifram, A. F. (1970). On mixtures of distributions with applications to estimation. *J. Amer. Statist. Assoc.*, **65**, 749–754.

Isii, K. (1958). Note on a characterization of unimodal distributions. *Ann. Inst. Statist. Math.*, **9**, 173–184.

James, I. R. (1978). Estimation of the mixing proportion in a mixture of two normal distributions with simple rapid measurements. *Biometrics*, **34**, 265–275.

Jazwinski, A. H. (1970). *Stochastic Processes and Filtering Theory.* Academic Press, New York.

Jewell, N. P. (1982). Mixtures of exponential distributions. *Ann. Statist.*, **10**, 479–484.

Joffe, A. D. (1964). Mixed exponential estimation by the method of half moments. *Appl. Statist.*, **13**, 91–98.

John, S. (1970a). On analyzing mixed samples. *J. Amer. Statist. Assoc.*, **65**, 755–760.

John, S. (1970b). On identifying the population of origin of each observation in a mixture of observations from two normal populations. *Technometrics*, **12**, 553–563.

John, S. (1970c). On identifying the population of origin of each observation in a mixture of observations from two gamma populations. *Technometrics*, **12**, 565–568.

Johnson, D. E., McGuire, S. A., and Milliken, G. A. (1978). Estimating σ^2 in the presence of outliers. *Technometrics*, **20**, 441–455.

Johnson, N. L. (1973). Some simple tests of mixtures with symmetrical components. *Commun. Statist.*, **1**, 17–25.

Johnson, N. L. (1978). Comments on a paper by Quandt and Ramsey. *J. Amer. Statist. Assoc.*, **73**, 750.

Johnson, N. L., and Kotz, S. (1969). *Discrete Distributions.* Wiley, New York.

Johnson, N. L., and Kotz, S. (1970a). *Continuous Univariate Distributions*, Vol. 1. Wiley, New York.

Johnson, N. L., and Kotz, S. (1970b). *Continuous Univariate Distributions*, Vol. 2. Wiley, New York.

Johnson, N. L., and Kotz, S. (1972). *Distributions in Statistics: Continuous Multivariate Distributions.* Wiley, New York.

Jones, T. A. (1969). Determination of '*n*' in weight frequency data. *J. Sed. Pet.*, **39**, 1473–1476.

Jones, T. A., and James, W. R. (1972). Analysis of bimodal orientation data. *J. Math. Geol.*, **1**, 129–135.

Kabir, A. B. M. L. (1968). Estimation of parameters of a finite mixture of distributions. *J. R. Statist. Soc. B*, **30**, 472–482.

Kanno, R. (1975). Estimation of parameters for a mixture of two normal distributions. *Rep. Stat. Appl. Res. JUSE*, **22**, 1–15.

Kao, J. H. K (1959). A graphical estimation of mixed Weibull parameters in life-testing electron tubes. *Technometrics*, **1**, 389–407.

Katopis, A., and Schwartz, S. (1972). Decision-directed learning using stochastic approximation. *Proc. of the Modelling and Simulation Conf.*, **1972**, 473–481.

Katti, S. K., and Gurland, J. (1961). The Poisson Pascal distribution. *Biometrics*, **17**, 527–538.

Kaufman, G. M., and King, B. (1973). A Bayesian analysis of nonresponse in dichotomous processes. *J. Amer. Statist. Assoc.*, **68**, 670–678.

Kazakos, D. (1977). Recursive estimation of prior probabilities using a mixture. *IEEE Trans. Inform. Th.*, **IT-23**, 203–211.

Kazakos, D., and Davisson, L. D. (1980). An improved decision-directed detector. *IEEE Trans. Inform. Th.*, **IT-26**, 113–116.

Keilson, J., and Gerber, H. (1971). Some results for discrete unimodality. *J. Amer. Statist. Assoc.*, **66**, 386–389.

Keilson, J., and Steutel, F. W. (1974). Mixtures of distributions, moment inequalities and measures of exponentiality and normality. *J. Appl. Prob.*, **2**, 112–130.

Kent, J. T. (1983). Identifiability of finite mixtures for directional data. *Ann. Statist.*, **11**, 984–988.

Khinchin, A. Y. (1938). On unimodal distributions (in Russian). *Izv. Nauchno-Issled. Inst. Mat. Mekh. Tomsk. Gos. Univ.*, **2**, 1–7.

Kiefer, J., and Wolfowitz, J. (1956). Consistency of the maximum likelihood estimator in the presence of infinitely many nuisance parameters. *Ann. Math. Statist.*, **27**, 887–906.

Kiefer, N. M. (1978a). Discrete parameter variation: efficient estimation of a switching regression model. *Econometrica*, **46**, 427–434.

Kiefer, N. M. (1978b). Comment on a paper by Quandt and Ramsey. *J. Amer. Statist. Assoc.*, **73**, 744–745.

Kingman, J. F. C. (1966). The algebra of queues. *J. Appl. Prob.*, **3**, 285–326.

Kocherlakota, S., and Kocherlakota, S. (1981). On the distribution of r in samples from the mixtures of bivariate normal populations. *Commun. Statist. A*, **10**, 1943–1946.

Konstantellos, A. C. (1980). Unimodality conditions for Gaussian sums. *IEEE Trans. Automat. Contr.*, **AC-25**, 838–839.

Korn, G. A., and Korn, T. M. (1968). *Mathematical Handbook for Scientists and Engineers*. McGraw-Hill, New York.

Kornbrot, D. E. (1983). Binary mixture collections: cumulant methods for estimation of parameters and distribution type. Personal Communication.

Krikelis, N. J. (1977). Bayes decision rules for a finite mixed population. *J. Statist. Comput. Simul.*, **6**, 11–18.

Krolikowska, K. (1975). Estimation of the parameters of the mixture of an arbitrary finite number of Poisson's distributions. *Z. Nauk. Politech. Lodz.*, **7**, 17–23.

Kullback, S., and Leibler, R. A. (1951). On information and sufficiency. *Ann. Math. Statist.*, **22**, 79–86.

Kumar, K. D., Nicklin, E. H., and Paulson, A. S. (1979). Comment on a paper by Quandt and Ramsey. *J. Amer. Statist. Assoc.*, **74**, 52–55.

Lachenbruch, P. A. (1975). *Discriminant Analysis*. Collier-Macmillan, London.

Lachenbruch, P. A., and Broffitt, B. (1980). On classifying observations when one population is a mixture of normals. *Biom. J.*, **22**, 295–301.

Laird, N. M. (1978a). Nonparametric maximum likelihood estimation of a mixing distribution. *J. Amer. Statist. Assoc.*, **73**, 805–811.

Laird, N. M. (1978b). Empirical Bayes methods for two-way contingency tables. *Biometrika*, **65**, 581–590.

Lambert, D., and Tierney, L. (1984). Asymptotic properties of maximum likelihood estimates in the mixed Poisson model. *Ann. Statist.*, **12**, 1388–1399.

Larkin, K. P. (1979). An algorithm for assessing bimodality vs. unimodality in a univariate distribution. *Behav. Res. Meth. Instr.*, **11**, 467–468.

Lee, A. F. S., and D'Agostino, R. B. (1976). Levels of significance of some two-sample tests when observations are from compound normal distributions. *Commun. Statist. A*, **5**, 325–342.

Lee, A. F. S., and Gurland, J. (1977). One-sample t-test when sampling from a mixture of normal distributions. *Ann. Statist.*, **5**, 803–807.

Lehmann, E. L. (1983). *Theory of Point Estimation.* Wiley, New York.

Lepeltier, C. (1969). A simplified statistical treatment of geochemical data by graphical representation. *Econ. Geol.*, **64**, 538–550.

Leytham, K. M. (1984). Maximum likelihood estimates for the parameters of mixture distributions. *Water Resour. Res.*, **20**, 896–902.

Li, L. A., and Sedransk, N. (1985). Mixtures of distributions: a topological approach. Typescript.

Lindgren, G. (1978). Markov regime models for mixed distributions and switching regressions. *Scand. J. Statist.*, **5**, 81–91.

Lindley, D. V., and Smith, A. F. M. (1972). Bayes estimates for the linear model (with discussion). *J. R. Statist. Soc. B*, **34**, 1–41.

Lindsay, B. G. (1981). Properties of the maximum likelihood estimator of a mixing distribution. In *Statistical Distributions in Scientific Work* (Eds. C. Taillie *et al.*), Vol. 5, pp. 95–110.

Lindsay, B. G. (1983a). The geometry of mixing likelihoods: a general theory. *Ann. Statist.*, **11**, 86–94.

Lindsay, B. G. (1983b). The geometry of mixing likelihoods, part II: the exponential family. *Ann. Statist.*, **11**, 783–792.

Lindsay, B. G. (1983c). Efficiency of the conditional score in a mixture setting. *Ann. Statist.*, **11**, 486–497.

Lingappaiah, G. (1975). On the mixture of exponential distributions. *Metron.*, **33**, 403–411.

Lipscomb, W. N., Rubin, T. R., and Sturdivant, J. H. (1947). An investigation of a method for the analysis of smokes according to particle size. *J. Appl. Physics*, **18**, 72–79.

Little, R. J. A. (1978). Consistent regression methods for discriminant analysis with incomplete data. *J. Amer. Statist. Assoc.*, **73**, 319–322.

Little, R. J. A., and Rubin, D. B. (1983). On jointly estimating parameters and missing values by maximizing the complete data likelihood. *Amer. Statist.*, **37**, 218–220.

Longsworth, L. G. (1942). Recent advances in the study of protein by electrophoresis. *Chem. Rev.*, **30**, 323–340.

Louis, T. (1982). Finding the observed information matrix when using the EM algorithm. *J. R. Statist. Soc. B*, **44**, 226–233.

Macdonald, P. D. M. (1969). FORTRAN programs for statistical estimation of distribution mixtures: some techniques for statistical analysis of length–frequency data. *Fish Res. Bd. Can. Tech. Rep.* 129.

Macdonald, P. D. M. (1971). Comment on a paper by Choi and Bulgren. *J. R. Statist. Soc. B*, **33**, 326–329.

Macdonald, P. D. M. (1975). Estimation of finite distribution mixtures. In *Applied Statistics* (Ed. R. P. Gupta), pp. 231–245. North Holland, Amsterdam.

Macdonald, P. D. M., and Pitcher, T. J. (1979). Age groups from size-frequency data: a versatile and efficient method of analyzing distribution mixtures. *J. Fish. Res. Bd. Can.*, **36**, 987–1001.

Maceda, E. C. (1948). On the compound and generalized Poisson distributions. *Ann. Math. Statist.*, **19**, 414–416.

McGuire, J. U., Brindley, T. A., and Bancroft, T. A. (1957). The distribution of European Corn borer larvae *pyrausta nubilalis* (H. N.) in field corn. *Biometrics*, **13**, 65–78.

McLachlan, G. J. (1975). Iterative reclassification procedure for constructing an asymptotically optimal rule of allocation in discriminant analysis. *J. Amer. Statist. Assoc.*, **70**, 365–369.

McLachlan, G. J. (1977). Estimating the linear discriminant function from initial samples containing a small number of unclassified observations. *J. Amer. Statist. Assoc.*, **22**, 403–406.

McLachlan, G. J., Lawoko, C. R. O., and Ganesalingam, S. (1982). On the likelihood ratio test for compound distributions for homogeneity of mixing proportions. *Technometrics*, **24**, 331–334.

Makov, U. E. (1980a). The statistical problems of unconfirmed cases in medical statistics. In *Teoria delle Decisioni in Medicina* (Ed. E. Girelli-Bruni), pp. 149–162. Bertani, Verona.

Makov, U. E. (1980b). On the choice of gain functions in recursive estimation of prior probabilities. *IEEE Trans. Inform. Th.*, **IT-26**, 497–498.

Makov, U. E. (1980c). Approximations to unsupervised filters. *IEEE Trans. Automat. Contr.*, **AC-25**, 842–847.

Makov, U. E. (1983). Approximate Bayesian solutions to some unsupervised learning problems. Unpublished Ph.D. Dissertation, London.

Makov, U. E., and Smith, A. F. M. (1977a). A quasi-Bayes unsupervised learning procedure for priors. *IEEE Trans. Inform. Th.*, **IT-23**, 761–764.

Makov, U. E., and Smith, A. F. M. (1977b). A quasi-Bayes approximation to unsupervised filters. *Proc. Conf. Measurement and Control*, (**MECO 77**).

Mancini, P., and Pilo, A. (1970). A computer program for multiexponential fitting by the peeling method. *Comput. Biomed. Res.*, **3**, 1–14.

Mandelbaum, J., and Harris, C. M. (1982). Parameter estimation under progressive censoring conditions for a finite mixture of Weibull distributions. *TIMS/Studies in Management Sciences*, **19**, 239–260.

Mardia, K. V. (1972). *Statistics of Directional Data*. Academic Press, London.

Mardia, K. V. (1975). Assessment of multinormality and the robustness of Hotelling's T^2 test. *Appl. Statist.*, **24**, 163–171.

Mardia, K. V., and Sutton, T. W. (1975). On the modes of the mixture of two von Mises distributions. *Biometrika*, **62**, 699–701.

Marks, R. G., and Rao, P. V. (1979). An estimation procedure for data containing outliers with a one-directional shift in the mean. *J. Amer. Statist. Assoc.*, **74**, 614–620.

Marriott, F. H. C. (1975). Separating mixtures of normal distributions. *Biometrics*, **31**, 767–769.

Marsaglia, G. (1961). Expressing a random variable in terms of uniform random variables. *Ann. Math. Statist.*, **32**, 894–898.

Martin, E. S. (1936). A study of an Egyptian series of mandibles. With special reference to mathematical methods of sexing. *Biometrika*, **28**, 149–178.

Martin, R. D., and Schwartz, S. C. (1972). On mixture, quasi-mixture and nearly normal random processes. *Ann. Math. Statist.*, **43**, 948–967.

Masuyama, M. (1977). A mixture of two gamma distributions applied to rheumatoid arthritis. *Rep. Stat. Appl. Res. JUSE*, **24**, 28–31.

Matz, A. W. (1978). Maximum likelihood parameter estimation for the quartic exponential distribution. *Technometrics*, **20**, 475–484.

Medgyessy, P. (1977). *Decomposition of Superpositions of Density Functions and Discrete Distributions*. Adam Hilger, Bristol.

Meeden, G. (1972). Bayes estimation of the mixing distribution, the discrete case. *Ann. Math. Statist.*, **43**, 1993–1999.

Mendenhall, W., and Hader, R. J. (1958). Estimation of parameters of mixed exponentially distributed failure time distributions from censored life test data. *Biometrika*, **45**, 504–520.

Mill, G. M. (1983). A comparison of density estimation techniques based on incomplete data. M.Sc. Thesis, University of Glasgow.

Minder, C. E. (1980). On inspecting multiparameter likelihood surfaces. *Commun. Statist. A*, **9**, 1931–1939.

Molenaar, W. (1965). Survey of estimation methods for a mixture of two normal distributions. *Stat. Neer*, **19**, 249–263.

Molenaar, W., and van Zwet, W. R. (1966). On mixtures of distributions. *Ann. Math. Statist*, **37**, 281–283.

Mosimann, J. E. (1962). On the compound multinomial distribution, the multivariate beta distribution, and correlations among proportions. *Biometrika*, **49**, 64–82.

Mudholkar, G. S., and Trivedi, C. T. (1981). A Gaussian approximation to the distribution of the sample variance for nonnormal populations. *J. Amer. Statist. Assoc.*, **76**, 479–485.

Muench, H. (1936). The probability distribution of protection test results. *J. Amer. Statist. Assoc.*, **31**, 677–689.

Muench, H. (1938). Discrete frequency distributions arising from mixtures of several single probability values. *J. Amer. Statist. Assoc.*, **33**, 390–398.

Mundry, E. (1972). On the resolution of mixed frequency distributions into normal components. *J. Math. Geol.*, **4**, 55–60.

Murphy, E. A. (1964). One cause? Many causes? The argument from the bimodal distribution. *J. Chron. Dis.*, **17**, 301–324.

Murray, G. D., and Titterington, D. M. (1978). Estimation problems with data from a mixture. *Appl. Statist.*, **27**, 325–334.

NAG (1978). *Fortran Library, Mark* 7. Numerical Algorithms Group, Oxford.

Nagy, G. (1972). Digital image processing in remote sensing for earth resources. *Proc. IEEE*, **60**, 1177–1200.

Namera, T., and Stubberud, A. R. (1983). Gaussian sum approximation for non-linear fixed-point prediction. *Int. J. Control*, **38**, 1047–1053.

Naylor, J. C., and Smith, A. F. M. (1982). Applications of a method for the efficient computation of posterior distributions. *Appl. Statist.*, **31**, 214–225.

Naylor, J. C., and Smith, A. F. M. (1983). A contamination model in clinical chemistry: an illustration of a method for the efficient computation of posterior distributions. *Statistician*, **32**, 82–87.

Nelder, J. A. and Mead, R. (1965). A simplex method for function minimization. *Computer J.*, **7**, 308–313.

Nevel'son, M. B., and Has'minskii, R. Z. (1973). Stochastic approximation and recursive estimation. *Translations of Mathematical Monographs*, No. 47, Amer. Math. Soc., Rhode Island.

Newcomb, S. (1886). A generalized theory of the combination of observations so as to obtain the best result. *Amer. J. Math.*, **8**, 343–366.

Neyman, J. (1939). On a new class of 'contagious distributions' applicable in entomology and bacteriology. *Ann. Math. Statist.*, **10**, 35–57.

Neyman, J. (1959). Optimal asymptotic tests of composite statistical hypotheses. In *Probability and Statistics: The Harold Cramér Volume* (Ed. U. Grenander), pp. 213–234. Wiley, New York.

Noble, F. W., Hayes, J. E., and Eden, M. (1959). Repetitive analog computer for analysis of sums of distribution functions. *Proc. IRE*, **47**, 1952–1956.

Odell, P. L., and Basu, J. P. (1976). Concerning several methods for estimating crop acreages using remote sensing data. *Commun. Statist., A*, **5**, 1091–1114.

Oja, H. (1981). Two location and scale-free goodness-of-fit tests. *Biometrika*, **68**, 637–640.

Oja, H. (1983). New tests for normality. *Biometrika*, **70**, 297–299.

Oka, M. (1954). Ecological studies on the kidai by the statistical method. II. On the growth of kidai (*Taitus turnifrons*). *Bull. Fac. Fish. Nagasaki*, **2**, 8–25.

Olshen, R. A., and Savage, L. J. (1970). A generalized unimodality. *J. Appl. Prob.*, **7**, 21–34.

O'Neill, T. J. (1978). Normal discrimination with unclassified observations. *J. Amer. Statist. Assoc.*, **73**, 821–826.

Orchard, T., and Woodbury, M. A. (1972). A missing information principle: theory and applications. *Proc. Sixth Berkeley Symp. Math. Stat. & Prob.*, **1**, 697–715.

Ord, J. K. (1972). *Families of Frequency Distributions*. Griffin: London.

O'Toole, A. L. (1933a). On the system of curves for which the method of moments is the best method of fitting. *Ann. Math. Statist.*, **4**, 1–29.

O'Toole, A. L. (1933b). A method for determining the constants in the bimodal fourth degree exponential function. *Ann. Math. Statist.*, **4**, 79–93.

Owen, J. R. (1975). A Bayesian sequential procedure for quantal response in the context of adaptive mental testing. *J. Amer. Statist. Assoc.*, **70**, 351–356.

Parr, W. C. (1981). Minimum distance estimation: a bibliography. *Commun. Statist. A*, **10**, 1205–1224.

Parsons, D. H. (1968). Biological problems involving sums of exponential functions of time: a mathematical analysis that reduces experimental time. *Math. Biosci.*, **2**, 123–128.

Parsons, D. H. (1970). Biological problems involving sums of exponential functions of time: an improved method of calculation. *Math. Biosci.*, **9**, 37–47.

Parzen, E. (1962). On the estimation of a probability density function and the mode. *Ann. Math. Statist.*, **33**, 1065–1076.

Patil, G. P., and Bildikar, S. (1966). Identifiability of countable mixtures of discrete probability distributions using methods of infinite matrices. *Proc. Camb. Phil. Soc.*, **62**, 485–494.

Patrick, E. A. (1972). *Fundamentals of Pattern Recognition.* Prentice Hall, New Jersey.

Patrick, E. A., Carayannopoulos, G. L., and Costello, J. P. (1966). Five results on unsupervised learning systems. Purdue Univ. Report, TR-EE 66-21.

Patrick, E. A., and Costello, J. P. (1968). Asymptotic probability of error using two decision-directed estimators for two unknown mean vectors. *IEEE Trans. Inform. Th.*, **IT-14**, 140–162.

Patrick, E. A., Costello, J. P., and Monds, F. C. (1970). Decision-directed estimation of a two-class decision boundary. *IEEE Trans. Comp.*, **C-19**, 197–205.

Patrick, E. A., and Hancock, J. C. (1966). Nonsupervised sequential classification and recognition of patterns. *IEEE Trans. Inform. Th.*, **IT-12**, 362–372.

Paull, A. E. (1978). A generalized compound Poisson model for consumer purchase panel data analysis. *J. Amer. Statist. Assoc.*, **73**, 706–713.

Paulson, A. S., Holcomb, E. W., and Leitch, R. A. (1975). The estimation of the parameters of the stable laws. *Biometrika*, **62**, 163–170.

Paulson, A. S., and Nicklin E. H. (1983). Integrated distance estimators for linear models applied to some published data sets. *Appl. Statist.*, **32**, 32–50.

Pearson, K. (1894). Contribution to the mathematical theory of evolution. *Phil. Trans. Roy. Soc. A*, **185**, 71–110.

Pearson, K. (1914). A study of Trypanosome strains. *Biometrika*, **10**, 85–143.

Pearson, K. (1915). On certain types of compound frequency distributions in which the individual components can be individually described by binomial series. *Biometrika*, **11**, 139–144.

Perlman, M. D. (1970). On the strong consistency of approximate maximum likelihood estimators. *Proc. Sixth Berkeley Symp. Math. Statist. & Prob.*, **1**, 263–282.

Peters, B. C., and Coberly, W. A. (1976). The numerical evaluation of the maximum-likelihood estimate of mixture proportions. *Commun. Statist. A*, **5**, 1127–1135.

Peters, B. C., and Walker, H. F. (1978a). An iterative procedure for obtaining maximum-likelihood estimates of the parameters for a mixture of normal distributions. *SIAM J. Appl. Math.*, **35**, 362–378.

Peters, B. C., and Walker, H. F. (1978b). The numerical evaluation of the maximum likelihood estimate of a subset of mixture proportions. *SIAM J. Appl. Math.*, **35**, 447–452.

Peterson, A. V., and Kronmal, R. A. (1982). On mixture methods for the computer generation of random variables. *Amer. Statist.*, **36**, 184–191.

Pettit, L. I., and Smith, A. F. M. (1984). Bayesian model comparisons in the presence of outliers. *Proc. 44th ISI Session*, Madrid.

Pettit, L. I., and Smith, A. F. M. (1985). Outliers and influential observations in linear models. In *Bayesian Statistics 2* (Eds. J. M. Bernardo *et al.*), pp. 473–494. North-Holland, Amsterdam.

Policello, G. (1981). Conditional maximum likelihood estimation in Gaussian mixtures. In *Statistical Distributions in Scientific Work* (Eds. C. Taillie *et al.*), Vol. 5 pp. 111–125.

Pollard, H. S. (1934). On the relative stability of the median and the arithmetic mean, with particular reference to certain frequency distributions which can be dissected into two normal distributions. *Ann. Math. Statist.*, **5**, 227–262.

Postaire, J-G., and Vasseur, C. P. (1981). An approximate solution to normal mixture identification with application to unsupervised pattern classification. *IEEE Trans. PAMI*, **PAMI-3**, 163–179.

Poulik, M. D., and Pinteric, L. (1955). An electronic computer for the evaluation of results of filterpaper electrophoresis. *Nature*, **176**, 1226–1227.

Press, S. (1968). A modified compound Poisson process with normal compounding. *J. Amer. Statist. Assoc.*, **63**, 607–613.

Preston, E. J. (1953). A graphical method for the analysis of statistical distributions into two normal populations. *Biometrika*, **40**, 460–464.

Quandt, R. E. (1972). A new approach to estimating switching regressions. *J. Amer. Statist. Assoc.*, **67**, 306–310.

Quandt, R. E., and Ramsey, J. B. (1978). Estimating mixtures of normal distributions and switching regressions. *J. Amer. Statist. Assoc.*, **73**, 730–738.

Rajasekaran, P. K., and Srinath, M. D. (1973). Unsupervised learning structure and parameter adaptive pattern recognition with discrete data. *Inform. Sciences*, **5**, 247–264.

Rao, C. R. (1948). The utilization of multiple measurements in problems of biological classification. *J. R. Statist. Soc. B*, **10**, 159–203.

Rao, C. R. (1965). *Linear Statistical Inference and Its Applications.* Wiley, New York.

Raper, L. P., Balkau, B., Taylor, R., Milner, B., Collins, V., and Zimmet, P. (1982). Plasma glucose distributions in two Pacific populations: the bimodality phenomenon. Typescript.

Rayment, P. R. (1972). The identification problem for a mixture of observations from two normal populations. *Technometrics*, **14**, 911–918.

Redner, R. A. (1981). Note on the consistency of the maximum-likelihood estimate for nonidentifiable distributions. *Ann. Statist.*, **9**, 225–228.

Redner, R. A., and Walker, H. F. (1984). Mixture densities, maximum likelihood and the EM algorithm. *SIAM Rev.*, **26**, 195–239.

Rennie, R. R. (1972). On the interdependence of the identifiability of finite multivariale mixtures and the identifiability of the marginal mixtures. *Sankhya, A*, **34**, 449–452.

Rennie, R. R. (1974). An identification algorithm for finite mixtures of distributions with common central moments. *Sankhya A*, **36**, 315–320.

Rider, P. R. (1961). The method of moments applied to a mixture of two exponential distributions. *Ann. Math. Statist.*, **32**, 143–147.

Rider, P. R. (1962). Estimating the parameters of mixed Poisson, binomial and Weibull distributions by the method of moments. *Bull. ISI*, **39**, Part 2, 225–232.

Robbins, H. (1964). The empirical Bayes approach to statistical decision problems. *Ann. Math. Statist.*, **35**, 1–20.

Robbins, H., and Pitman, E. J. G. (1949). Application of the method of mixtures to quadratic forms in normal variates. *Ann. Math. Statist.*, **20**, 552–560.

Robertson, C. A., and Fryer, J. G. (1969). Some descriptive properties of normal mixtures. *Skand. Aktur. Tidskr.*, **52**, 137–146.

Rolph, J. E. (1968). Bayesian estimation of mixing distributions. *Ann. Math. Statist.*, **39**, 1289–1302.

Rosenblatt, M. (1956). Remarks on some nonparametric estimates of a density function. *Ann. Math. Statist.*, **27**, 832–837.

Rubin, I., and Schwartz, S. (1970). Runaway bounds for decision directed receivers. *Proc. Conf. Decision and Control.*

Ruhe, A. (1980). Fitting empirical data by positive sums of exponentials. *SIAM J. Sci. Stat. Comput.*, **1**, 481–498.

Rushforth, N. B., Bennett, P. H., Steinberg, A. G., Burch, T. A., and Miller, M. (1971). Diabetes in the Pima Indians. Evidence of bimodality in glucose tolerance distributions. *Diabetes*, **20**, 756–765.

Sacks, J. (1958). Asymptotic distribution of stochastic approximation procedures. *Ann. Math. Statist.*, **29**, 373–405.

Saleh, A. K. M. E. (1981). Decomposition of finite mixture of distributions by minimum chi-square method. *Aligarh. J. Statist.*, **1**, 86–97.

Samaniego, F. J. (1976). A characterization of convoluted Poisson distributions with applications to estimation. *J. Amer. Statist. Assoc.*, **71**, 475–479.

Sammon, J. (1968). An adaptive technique for multiple signal detection and identification. In *Pattern Recognition* (Ed. L. N. Kanal), pp. 409–439. Thompson, Washington, D. C.

Sandor, T., Sridhar, B., and Hollenberg, N. K. (1978). Multiexponential fit of data by using the maximum likelihood method on a minicomputer. *Comput. Biomed. Res.*, **11**, 35–40.

Sanyal, P. (1974). Bayes' detection rule for rapid detection and adaptive estimation scheme with space applications. *IEEE Trans. Automat. Contr.*, **AC-19**, 228–231.

Schilling, W. (1947). A frequency distribution represented as the sum of two Poisson distributions. *J. Amer. Statist. Assoc.*, **42**, 407–424.

Schmidt, P. (1982). An improved version of the Quandt–Ramsey MGF estimator for mixtures of normal distributions and switching regressions. *Econometrica*, **50**, 501–516.

Schwartz, G. (1978). Estimating the dimension of a model. *Ann. Statist.*, **6**, 461–464.

Schweder, T. (1981). On the dispersion of mixtures. Personal Communication.

Sclove, S. C. (1977). Population mixture models and clustering algorithms. *Commun. Statist. A*, **6**, 417–434.

Sclove, S. C. (1983). Application of the conditional population mixture model to image segmentation. *IEEE Trans. Patt. Anal. Mach. Intell.*, **PAMI-5**, 428–433

Scott, A. J., and Symons, M. J. (1971). Clustering methods based on likelihood ratio criteria. *Biometrics*, **27**, 238–397.

Scudder, H. J. (1965). Probability of error of some adaptive pattern recognition machines. *IEEE Trans. Inform. Th.*, **IT-11**, 363–371.

Sen, N. (1922). Uber den Einfluss des Dopplereffekts auf spektroskopische Feinstrukturen und seine Elimination. *Phys. Zeitschr*, **23**, 397–399.

Shaked, M. (1980). On mixtures from exponential families. *J. R. Statist. Soc. B*, **42**, 192–198.

Shapiro, C. P. (1974). Bayesian classification: asymptotic results. *Ann. Statist.*, **2**, 763–774.

Shenton, L. R., and Bowman, K. O. (1967). Remarks on large sample estimators for some discrete distributions. *Technometrics*, **9**, 587–598.

Sichel, H. S. (1975). On a distribution law for word frequencies. *J. Amer. Statist. Assoc.*, **70**, 542–547.

Silverman, B. W. (1978). Density ratios, empirical likelihood, and cot death. *Appl. Statist.*, **27**, 26–33.

Silverman, B. W. (1979). Some asymptotic properties of the probabilistic teacher. *IEEE Trans. Inform. Th.*, **IT-26**, 246–249.

Silverman, B. W. (1981). Using kernel density estimates to investigate multimodality. *J. R. Statist. Soc. B*, **43**, 97–99.

Silverman, B. W. (1983). Some properties of a test for multimodality based on kernel density estimates. In *Probability, Analysis and Statistics* (Eds. J. F. C. Kingman and G. E. H. Reuter), pp. 248–259. LMS Lecture Notes No. 79. Cambridge University Press.

Silvey, S. D. (1975). *Statistical Inference*. Chapman and Hall, London.

Silvey, S. D. (1980). *Optimal Design*. Chapman and Hall, London.

Silvey, S. D., and Titterington, D. M. (1973). A geometric approach to optimal design theory. *Biometrika*, **60**, 21–32.

Simar, L. (1976). Maximum likelihood estimation of a compound Poisson process. *Ann. Statist.*, **4**, 1200–1209.

Singer, R. A., Sea, R. G., and Housewright, K. B. (1974). Derivation and evaluation of improved tracking filters for use in dense multi-target environments. *IEEE Trans. Inform. Th.*, **IT-20**, 423–432.

Skellam, J. G. (1948). A probability distribution derived from the binomial distribution by regarding the probability of success as variable between sets of trials. *J. R. Statist. Soc. B*, **10**, 257–261.

Skene, A. M. (1978). Discrimination using latent structure models. In *COMPSTAT 1978* (Eds. L. Corsten and J. Hermans), pp. 199–204. Physica-Verlag, Vienna.

Skene, A. M. (1980). Discussion of a paper by D. J. Bartholomew. *J. R. Statist. Soc. B*, **42**, 314–315.

Smith, A. F. M. (1984). Present position and potential developments: some personal views, Bayesian statistics. *J. R. Statist. Soc. A.*, **147**, 245–259.

Smith, A. F. M., and Makov, U. E. (1978). A quasi-Bayes sequential procedure for mixtures. *J. R. Statist. Soc. B*, **40**, 106–111.

Smith, A. F. M., and Makov, U. E. (1980). Bayesian detection and estimation of jumps in linear systems. In *Analysis and Optimization of Stochastic Systems* (Ed. O. L. R. Jacobs), pp. 333–345. Academic Press, London.

Smith, A. F. M., and Makov, U. E. (1981). Unsupervised learning for signal versus noise. *IEEE Trans. Inform. Th.*, **IT-27**, 498–500.

Smith, A. F. M., Skene, A. M., Shaw, J. E. H., Naylor, J., and Dransfield, M. (1985). The implementation of the Bayesian paradigm. *Commun. Statist.*: to appear.

Smith, A. F. M., and West, M. (1983). Monitoring renal transplants: an application of the multi-process Kalman filter. *Biometrics*, **39**, 867–878.

Smith, A. F. M., West, M., Gordon, K., Knapp, M. S., and Trimble, I. M. G. (1983). Monitoring kidney transplant patients. *Statistician*, **32**, 46–54.

Smith, D. M., and Bartlet, J. C. (1961). Calculation of the areas of isolated or overlapping normal probability curves. *Nature*, **191**, 688–689.

Smith, M. R., Cohn-Sfetcu, S., and Buckmaster, H. A. (1976). Decomposition of multicomponent exponential decays by spectral analysis techniques. *Technometrics*, **18**, 467–482.

Sorenson, H. W., and Alspach, D. L. (1971). Recursive Bayesian estimation using Gaussian sums. *Automatica*, **7**, 465–479.

Spiegelhalter, D. J. (1983). Diagnostic tests of distributional shape. *Biometrika*, **70**, 401–409.

Stanat, D. S. (1968). Unsupervised learning of mixtures of probability functions. In *Pattern Recognition* (Ed. L. N. Kanal), pp. 357–389. Thompson, Washington, D. C.

Steutel, F. W. (1967). Note on the infinite divisibility of exponential mixtures. *Ann. Math. Statist.*, **38**, 1303–1305.

Subrahmaniam, K. (1972). On quadratic forms from mixtures of two normal populations. *S. Afr. Statist. J.*, **6**, 103–120.

Subrahmaniam, K., Subrahmaniam, K., and Messeri, J. Y. (1975). On the robustness of some tests of significance in sampling from a compound normal population. *J. Amer. Statist. Assoc.*, **70**, 435–438.

Suchindran, C. M., and Lachenbruch, P. A. (1974). Estimates of parameters in a probability model for first livebirth interval. *J. Amer. Statist. Assoc.*, **69**, 507–513.

Sundberg, R. (1972). Maximum likelihood theory and applications for distributions generated when observing a function of an exponential family variable. Dissertation, Inst. Math. Statist., Univ. Stockholm.

Sundberg, R. (1974). Maximum likelihood theory for incomplete data from an exponential family. *Scand. J. Statist.*, **1**, 49–58.

Sundberg, R. (1976). An iterative method for solution of the likelihood equations for incomplete data from exponential families. *Commun. Statist. Simul. Comput.*, *B*, **5**, 55–64.

Svedberg, T., and Pedersen, K. O. (1940). *The Ultracentrifuge.* Clarendon, Oxford.

Symons, M. J. (1981). Clustering criteria and multivariate normal mixtures. *Biometrics*, **37**, 35–43.

Tallis, G. M. (1969). The identifiability of mixtures of distributions. *J. Appl. Prob.*, **6**, 389–398.

Tallis, G. M., and Chesson, P. (1982). Identifiability of mixtures. *J. Austral. Math. Soc. A*, **32**, 339–348.

Tallis, G. M., and Light, R. (1968). The use of fractional moments for estimating the parameters of a mixed exponential distribution. *Technometrics*, **10**, 161–175.

Tan, W. Y. (1978). On the distribution of the sample covariance matrix from a mixture of normal densities. *S. Afr. Statist. J.*, **12**, 47–55.

Tan, W. Y. (1980). On probability distributions from mixtures of multivariate densities. *S. Afr. Statist. J.*, **14**, 47–59.

Tan, W. Y., and Chang, W. C. (1972a). Some comparisons of the method of moments and the method of maximum likelihood in estimating parameters of a mixture of two normal densities. *J. Amer. Statist. Assoc.*, **67**, 702–708.

Tan, W. Y., and Chang, W. C. (1972b). Convolution approach to genetic analysis of quantitative characters of self-fertilized population. *Biometrics*, **28**, 1073–1090.

Tanaka, S. (1962). A method of analysing a polymodal frequency distribution and its application to the length distribution of the porgy, *Taius tumifrons* (T & S). *J. Fish. Res. Bd. Can.*, **19**, 1143–1159.

Tanner, W. F. (1959). Sample components obtained by the method of differences. *J. Sed. Pet.*, **29**, 408–411.

Tanner, W. F. (1962). Components of the hypsometric curve of the earth. *J. Geophys. Res.*, **67**, 2841–2843.

Tapia, R. A., and Thompson, J. R. (1978). *Nonparametric Probability Density Estimation.* Johns Hopkins Univ. Press, Baltimore, Md.

Tarter, M., and Silvers, A. (1975). Implementation and application of bivariate Gaussian mixture decomposition. *J. Amer. Statist. Assoc.*, **70**, 47–55.

Taylor, B. J. R. (1965). The analysis of polymodal frequency distributions. *J. Anim. Ecol.*, **34**, 445–452.

Teicher, H. (1960). On the mixture of distributions. *Ann. Math. Statist.*, **31**, 55–73.

Teicher, H. (1961). Identifiability of mixtures. *Ann. Math. Statist.*, **32**, 244–248.

Teicher, H. (1963). Identifiability of finite mixtures. *Ann. Math. Statist.*, **34**, 1265–1269.

Teicher, H. (1967). Identifiability of mixtures of product measures. *Ann. Math. Statist.*, **38**, 1300–1302.

Teichroew, D. (1957). The mixtures of normal distributions with different variances. *Ann. Math. Statist.*, **28**, 510–512.

Thomas, E. A. C. (1969). Distribution free tests for mixed probability distributions. *Biometrika*, **56**, 475–484.

Thornton, J. C., and Paulson, A. S. (1977). Asymptotic distribution of characteristic function-based estimators for the stable laws. *Sankhya A*, **39**, 341–354.

Tiago de Oliveira, J. (1965). Some elementary tests of mixtures of discrete distributions. In *Classical and Contagious Discrete Distributions* (Ed. G. P. Patil), pp. 379–384. Pergamon, New York.

Tiao, G. C., and Ali, M. M. (1971). Analysis of correlated random effects linear model with two random components. *Biometrika*, **58**, 37–51.

Tierney, L., and Lambert, D. (1984). The asymptotic efficiency of estimators of functionals of mixed distributions. *Ann. Statist.*, **12**, 1380–1387.
Tiselius, A., and Kabat, E. A. (1939). Electrophoretic study of immune sera and purified antibody preparations. *J. Exp. Med.*, **69**, 119–131.
Titterington, D. M. (1976). Updating a diagnostic system using unconfirmed cases. *Appl. Statist.*, **25**, 238–247.
Titterington, D. M. (1978). Estimation of correlation coefficients by ellipsoidal trimming. *Appl. Statist.*, **27**, 227–234.
Titterington, D. M. (1983). Minimum distance non-parametric estimation of mixture proportions. *J. R. Statist. Soc. B*, **45**, 37–46.
Titterington, D. M. (1984). Recursive parameter estimation using incomplete data. *J. R. Statist. Soc. B*, **46**, 257–267.
Titterington, D. M., and Jiang, J. M. (1983). Recursive estimation procedures for missing-data problems. *Biometrika*, **70**, 613–624.
Tubbs, J. D., and Coberly, W. A. (1976). An empirical sensitivity study of mixture proportion estimators. *Commun. Statist. A*, **5**, 1115–1125.
Tucker, H. G. (1963). An estimate of the compounding distribution of a compound Poisson distribution. *Theor. Prob. Applics.*, **8**, 195–200.
Umbach, D. (1981). On inference for a mixture of a Poisson and a degenerate distribution. *Commun. Statist.*, *A*, **10**, 299–306.
Usinger, H. (1975). Pollenanalytische und stratigraphische Untersuchungen an zwei Spatglazial-Vorkammen in Schleswig-Holstein. *Mitt. Arbeitsgem. Geobot. Schleswig-Holstein, Hamburg*, **25**.
Van Andel, T. H. (1973). Texture and disposal of sediments in the Panama Basin. *J. Geol.*, **81**, 434–457.
Van Ryzin, J. (1973). A histogram method of density estimation. *Commun. Statist.*, **2**, 493–506.
Wadhams, P. D. (1981). Sea-ice topography of the Arctic Ocean. *Phil. Trans. Roy. Soc. A*, **302**, 45–85.
Wahba, G., and Wold, S. (1975). A completely automatic French curve: fitting spline functions by cross-validation. *Commun. Statist.*, **4**, 125–141.
Walker, H. F. (1980). Estimating the proportions of two populations in a mixture using linear maps. *Commun. Statist.*, *A*, **9**, 837–849.
Walker, S. H., and Duncan, D. B. (1967). Estimation of the probability of an event as a function of several independent variables. *Biometrika*, **54**, 167–179.
Walsh, G. R. (1975). *Methods of Optimization.* Wiley, New York.
Weiner, S. (1962). Samples from mixed-exponential populations. Mimeo, ARINC Res. Corp., Washington, D. C.
Wessels, J. (1964). Multimodality in a family of probability densities with application to a linear mixture of two normal densities. *Stat. Neer.*, **18**, 267–282.
Whittaker, J. (1973). The Bhattacharyya matrix for the mixture of two distributions. *Biometrika*, **60**, 201–202.
Wild, A. E. (1965). Protein concentration of the rabbit foetal fluids. *Proc. Roy. Soc. B*, **163**, 95–115.
Wilkins, C. A. (1961). A problem concerned with weighting of distributions. *J. Amer. Statist. Assoc.*, **56**, 281–292.
Willsky, A. S., and Jones, H. L. (1976). A generalized likelihood ratio approach to the detection and estimation of jumps in linear systems. *IEEE Trans. Automat. Contr.*, **AC-21**, 108–112.
Wilson, D. L., and Sargent, R. G. (1979). Some results of Monte Carlo experiments in estimating the parameters of the finite mixed exponential distribution. *Proc. Twelfth Ann. Symp. Interface* (Ed. J. F. Gentleman), pp. 461–465. Univ. Waterloo, Canada.

Wolfe, J. H. (1965). A computer program for the maximum likelihood analysis of types. Tech. Bull. 65-15, US Nav. Pers. Res. Act., San Diego.

Wolfe, J. H. (1970). Pattern clustering by multivariate mixture analysis. *Multivar. Behav. Res.*, **5**, 329–350.

Wolfe, J. H. (1971). A Monte Carlo study of the sampling distribution of the likelihood ratio for mixtures of multinomial distributions. Tech. Bull. STB 72-2, Nav. Pers. & Tran. Res. Lab., San. Diego.

Wolfowitz, J. (1957). The minimum distance method. *Ann. Math. Statist.*, **28**, 75–88.

Woodward, W. A., Parr, W. C., Schucany, W. R., and Lindsey, H. (1984). A comparison of minimum distance and maximum likelihood estimation of a mixture proportion. *J. Amer. Statist. Assoc.*, **79**, 590–598.

Wu, C. F. (1983). On the convergence properties of the EM algorithm. *Ann. Statist.*, **11**, 95–103.

Yakowitz, S. J. (1969). A consistent estimator for the identification of finite mixtures. *Ann. Math. Statist.*, **40**, 1728–1735.

Yakowitz, S. J. (1970). Unsupervised learning and the identification of finite mixtures. *IEEE Trans. Inform. Th.*, **IT-16**, 330–338.

Yakowitz, S. J., and Spragins, J. D. (1968). On the identifiability of finite mixtures. *Ann. Math. Statist.*, **39**, 209–214.

Young, T. Y., and Calvert, T. W. (1974). *Classification, Estimation and Pattern Recognition.* American Elsevier, New York.

Young, T. Y., and Coraluppi, G. (1970). Stochastic estimation of a mixture of normal density functions using an information criterion. *IEEE Trans. Inform. Theory,* **IT-16**, 258–263.

Young, T. Y., and Farjo, A. A. (1972). On decision-directed estimation and stochastic approximation. *IEEE Trans. Inform. Th.*, **IT-18**, 671–673.

Zierler, K. (1981). A critique of compartmental analysis. *Ann. Rev. Biophys. Bioeng.*, **10**, 531–562.

Zlokazov, V. B. (1978). UPEAK-spectro-oriented routine for mixture decomposition. *Comput. Phys. Commun.*, **13**, 389–398.

Index

Akaike's criterion, 159
Andrews curves, 68–71
Applications, direct, 2, 8–22
  agriculture, 18
  botany, 18
  economics, 17
  electrophoresis, 11–12, 18, 133–135
  fisheries research, 6, 9–11, 16, 148
  geology, 7, 18
  grain sizes (sedimentology), 6, 8–9, 18
  hypsometric curve, 14
  medical diagnosis, 13, 148, 159, 168, 178–179
  medicine, 17, 18
  miscellaneous, 20
  monitoring renal transplants, 214
  palaentology, 18
  pattern recognition, 7, 176
  psychology, 18
  reliability, 19
  remote sensing, 7, 14, 178
  structure of stellar cluster, 15, 22
  switching regressions, 12–13
  table of references, 16–21
  tracking a signal, 6, 176, 213–214
  zoology, 18–19
Applications, indirect, 2, 22–34
  approximation by non-mixture densities, 29–34
  cluster analysis, 4, 25–27, 149, 151
  empirical Bayes, 27–28, 103
  Gaussian sums, 24–25, 212
  kernel density estimation, 28–29, 101–102, 166–167
  latent structure, 25–27, 149, 157
  modelling prior densities, 27
  outliers, 22–23, 108–113
  random variate generation, 29
  robust estimators, 23
  robustness studies, 23–24
Asymptotic properties
  CF method, 129–130
  EM-type recursion, 207–208
  Kazakos' method, 185–186
  maximum likelihood, 92–94
  method of moments, 71
  MGF method, 128–129
  minimum distance methods, 114, 117, 122, 126
  quasi-Bayes method, 185, 190–192, 200–201
  stochastic approximation, 184–185, 193–196, 206

Bayesian methods, 5, 82, 106–114, 177, 179–181, 189, 196–197, 212–214, 215
  calculation of posterior modes, 82
  cluster analysis, 113–114
  computational explosion, 6, 179, 181, 189, 197, 201, 213, 214
  existence and number of outliers, 110–113
  mixture of known distributions, 107–108
  outlier models, 108–113
  sequential methods, 177, 179–181
  see also Quasi-Bayes method
Bhattacharya information, 48
Bhattacharya's method, 55–57, 149
Bimodality, 48, 160–161, 164
Bitangentiality, 48, 161–162
Bootstrap, 167
Buchanan–Wollaston and Hodgson's method, 57–58, 71

CF method, *see* Minimum distance methods
Change point inference, 157
Characteristic function (CF), 127

Cluster analysis, 4, 25–27, 149, 151
 Bayesian approach, 113–114
 maximum likelihood, 104–106
 optimum partitioning, 151
Component densities, 1
Compound distributions, 2, 50–51
Compound gamma, 98, 104
Compound hypergeometric, 51
Compound multinomial, 51
 method of moments, 81
Compound Poisson, 50, 98
 EM algorithm, 104
 method of moments, 81
Confusion matrix, 73
Contagious disease models, 50
Contaminated normal, 22–23, 34
 Bayesian methods, 108–113
 sequential methods, 200–201
Cramér-von Mises statistic, 122

Data
 birth rates, 31
 blood chloride levels, 33
 copepod lengths, 60
 Darwin's plant heights data, 111–113
 death notices data, 89
 hypsometric curve, 14–15
 iris data, 174
 Mount St. Helens ashfall, 9
 pike lengths, 11
 snapper lengths, 11
 three-component normal mixture, 53
Decision-directed learning, 181, 190, 196, 204, 215
Discriminant analysis, use of uncategorized cases in, 5–6, 148, 168–175
 logistic method, 173
 mixture of multinomials, 170
 mixture of two multivariate normals with common covariance matrix, 170–174
 nonparametric method, 175
 sampling and diagnostic paradigms, 168–169
Dynamic linear models, 212–215

Edgeworth expansion, 33–34, 164
Electrophoresis, 11–12, 18, 133–135
EM algorithm, 84–97
 comparison with NR and MS, 88–91
 convergence, 92, 95–97
 grouped data, 100
 kernel method, 102
 Markov chain mixture, 100–101
 maximum penalized likelihood, 102
 mixture of exponential family densities, 85–86
 mixture of known densities, 85
 mixture of multivariate normals, 87
 mixture of normals, 91
 mixture of two univariate normals, 86–87
 monotonic property, 88
 recursive version, 208–211
 relationship with method of scoring, 211
 simulation study, 97
 slow convergence, 89–91
 speeding up, 91, 93
Empirical Bayes, 27–28
 maximum likelihood, 99, 103

Finite mixture distributions, 1
Fisheries research, 5, 6, 9–11, 16, 148
Fletcher–Reeves method, 97
Forms of sampling, 3
Fowlkes' graphical method, 64

Gaussian sums, 24–25, 212
General mixtures, 2, 50–51
 identifiability, 41
 maximum likelihood, 103–104
  EM algorithm, 104
General normal mixtures, 51
Geology, 7
Grain-size distributions, 5, 8–9, 18, 148
Gram–Charlier expansion, 32–34
 modality, 164–165
Graphical methods, 52–71
 based on cumulative distribution function, 58–67
 based on density function, 52–58
 Bhattacharya's method, 55–57, 149
 Buchanan–Wollaston and Hodgson's method, 57–58, 71
 Fowlkes' method, 64
 mixture of binomials, 66
 mixture of exponentials, 67
 mixture of lognormals, 52
 mixture of multivariate normals, 67–71
 mixture of normals, 52–65
 mixture of Poissons, 67
 mixture of uniforms, 66–67
 mixture of Weibulls, 65
 probability plots, 59
 Tanner's method, 54–55, 149

Huber density, 24
Hypsometric curve, 14–15

Identifiability, 35–42, 115
  definition, 36
  finite scale mixtures, 38
  general mixtures, 41
  mixture of binomials, 35, 40–41
  mixture of Cauchys, 38
  mixture of gammas, 39
  mixture of uniforms, 36, 67
  mixture of univariate normals, 38
  mixture of von Mises, 42
  multivariate models assuming independence, 41
  necessary and sufficient conditions, 37
  sufficient conditions, 39
  translation parameter mixtures, 38
Information, 42–48
  Bhattacharya information, 48
  Fisher information matrix, 42
  mixture of two known densities, 43–45, 48
  mixture of two known exponentials, 44
  mixture of two known normals, 44–45
  mixture of two multivariate normals, 47–48
  mixture of two univariate normals, 45–47
Isotonic fitting, 166

Kalman filter, 25, 212–215
Kazakos' recursive method, 183–188, 190
  bipolar signal, 186–188
Kernel density estimation, 28–29, 101–102
  assessment of modality, 166–167

Latent structure, 25–27, 149, 157
Linear programming, 120
Locally optimal score test, 157, 158
  mixture of exponentials, 158
  mixture of known densities, 157
  two-parameter mixture, 157–158
Lognormal distribution, 29–30

Markov chain mixtures
  maximum likelihood, EM, 100–101
Markov models, 13
Maximum likelihood, 5, 82–106, 115, 183, 205
  asymptotic properties, 92–94
  cluster method, 104–106
    biases, 105–106
  comparison with CF method, 132
  comparison with method of moments, 95
  comparison with MGF method, 129
  comparison with minimum distance method, 122–123
  EM algorithm, 84–97
    see also EM algorithm
  general parametric mixtures, 103–104
    link with optimal design, 103–104
  grouped data, 100
  Markov chain mixture, 100–101
  mixture of binomials, 86
  mixture of exponential family distributions, 85–86, 95
  mixture of exponentials, 86, 91
  mixture of geometrics, 86
  mixture of multinomials with fully-categorized data, 97, 100
  mixture of multivariate normals, 86
  mixture of Poissons, 86
  mixture of two known densities, 82–83, 90–91
  mixture of two univariate normals, 83
  mixture of univariate normals, 86, 91
  nonparametric methods, 101–102
  practical efficiency, 95–97
  recursive approximation, 183–188, 190, 205–207
  relation to Kullback–Leibler measure, 115, 124–125, 184
  singularities, 83, 89, 93–94, 97
  table of references,, 98–99
Maximum penalized likelihood, 102, 175
Medgyessy's method, 138–142, 145–146, 149
  mixture of binomials, 139, 146
  mixture of exponentials, 139, 145–146
  mixture of geometrics, 139
  mixture of lognormals, 145
  mixture of Poissons, 139
  mixture of two univariate normals, 139–140
  mixture of univariate normals, 139, 141–142, 146
Medical diagnosis, 13, 17–18, 148
Method of moments, 5, 71–81, 183, 204–205
  asymptotic theory, 71
  comparison with maximum likelihood, 95

comparison with MGF method, 133
compound binomial, 81
compound multinomial, 81
compound Poisson, 81
fractional moments, 75, 79
lack of efficiency, 79
mixture of binomials, 80
mixture of exponentials, 80
mixture of gammas, 81
mixture of multivariate normals, 78
mixture of negative binomials, 81
mixture of Poissons, 79, 81, 90
mixture of two exponentials, 74–75
mixture of two known densities, 72–74, 183
mixture of two univariate normals, 75–78
mixture of uniforms, 79
mixture of Weibulls, 79, 81
optimal moment, 73, 74
Method of scoring (MS), 88, 121, 207
comparison with EM and NR, 88–89
relation to EM, 211
MGF method, *see* Minimum distance methods
Minimum chi-squared method, 5
see also Minimum distance methods
Minimum distance methods, 5, 114–133, 135
asymptotic theory, 114
Boes' method, 124–125, 132
CF method, 129–132
asymptotic theory, 129–130
comparison with ML and Boes method, 132
mixture of known distributions, 130–131
mixture of two known normals, 131–132
chi-squared distance, 115, 120, 123
comparison with ML, 122–123
distance measures, 114–117
Hall's approach, 120–121
Hellinger distance, 116
Kullback–Leibler measure, 115, 124–125, 184
Levy distance, 115
L^2-norm, 115
MGF method, 127–130
asymptotic theory, 128–129
comparison with ML and method of moments, 129
comparison with method of moments, 133
mixture of two univariate normals, 127–128
minimum chi-squared method, 5, 115, 120, 123
mixture of binomials, 123
mixture of Poissons, 123
mixture of two univariate normals, 123
mixture of known distributions, 117–120
modified chi-squared distance, 115, 120
sup-norm, 116, 120
Mixing weights, 1
Mixture of binomials
graphical method, 67
identifiability, 35, 40–41
maximum likelihood, 86, 98
Medgyessy's method, 139, 146
method of moments, 80
minimum chi-squared method, 123
Mixture of Cauchys, 38
Mixture of components from different families, 6
Mixture of Dirichlet processes, 51
Mixture of exponential family distributions
convergence of EM, 95
general mixture, 104
maximum likelihood, 85–86
method of moments, 81
Mixture of exponentials
graphical methods, 66
information about mixing weight, 44
maximum likelihood, 86, 91, 98
Medgyessy's method, 139, 145–146
method of moments, 74–75, 80
number of components, 158
numerical decomposition, 137–138
with negative weights, 50
Mixture of gammas
identifiability, 39
maximum likelihood, 98
method of moments, 81
Mixture of geometrics, 51
maximum likelihood, 86, 98
Medgyessy's method, 139
Mixture of *k* unknown densities with fully-categorized data, 120–121
Mixture of known distributions
Bayesian methods, 107–108
Bhattacharya informatin, 48
CF method, 130–131

information about mixing weight, 43–45
maximum likelihood, 82–83, 99
method of moments, 72–74, 183
MGF method, 133
minimum distance methods, 114–117
number of components, 150–154, 157
sequential methods, 178, 179–193, 208–209
Mixture of lognormals
graphical methods, 52
Medgyessy's method, 145
Mixture of multinomials, 97, 100, 170
Mixture of multivariate normals, 26, 178
graphical method, 67–71
information about mixing weight, 44
maximum likelihood, 86, 91, 99
Medgyessy's method, 139, 145–146
modality, 162–163
number of components, 154
signal versus noise, 213, 214–215
Tarter and Silvers' method, 142–145
Mixture of multivariate normals with common covariance matrix
discriminant analysis, 170–174
information matrix, 45–47
maximum likelihood, 87–88, 99
cluster method, 105
method of moments, 78
Mixture of negative binomials, 81
Mixture of normals, 7, 22
Mixture of Poissons
graphical method, 67
maximum likelihood, 86, 89–90, 98
Medgyessy's method, 139
method of moments, 79, 81, 90
minimum chi-squared method, 123
number of components, 152
Mixture of two known normals
CF method, 131–132
information about mixing weight, 44–45
maximum likelihood, 90–91
MGF method, 133
Mixture of two univariate normals
bimodality and bitangentiality, 48–49
decision-directed learning, 204
EM-type recursion, 209–210
information matrix, 45–47
maximum likelihood, 83, 86–87, 98–99
EM algorithm, 86–87
singularities, 83, 89, 93–94, 97
Medgyessy's method, 139
method of moments, 75–78
MGF method, 127–128
minimum chi-squared method, 123
modality, 160–161
number of components, 151
signal versus noise, 196–199
Mixture of uniforms
graphical method, 66–67
identifiability, 36, 67
maximum likelihood, 98
method of moments, 79
Mixture of univariate normals, 1
graphical methods, 52–65
identifiability, 38
maximum likelihood, 86
Medgyessy's method, 139, 141–142, 146
number of components, 149–150, 156–157
Taylor's method, 147
Mixture of von Mises distribution
identifiability, 42
modality, 163, 166
Mixture of Weibulls
graphical method, 65
method of moments, 79, 81
Mixtures with negative weights, 50
Modality, assessment of, 165–167
asymptotic theory, 165
kernel-based approach, 166–167
mixture of von Mises density, 166
Modality, structure of, 159–165
Edgeworth expansion, 164
Gram–Charlier expansion, 164–165
mixture of two multivariate normals, 162–163
mixture of two univariate normals, 160–161
mixture of two von Mises, 163
quartic exponential distribution, 163–164
see also Bimodality, Bitangentiality, Multi-modality, Unimodality
Model choice, 158–159
Akaike's criterion, 159
Bayes factors, 158
penalized likelihoods, 158
Modified decision-directed learning, 182, 193
Moment generating function (MGF), 127
Multimodality, 48–50, 54, 149, 159–167
see also Bimodality

Negative binomial as mixture, 50–51
Nelder–Mead method, 97
Newton–Raphson (NR) method, 88, 121, 124, 184, 207
  comparison with EM and MS, 88–91
  failure to converge, 89
  modified NR method, 122
Non-central chi-squared as mixture, 51
Number of components, testing for, 4–5, 148–167
  GLF test, 152–154
  locally optimal scores test, 157–158
  mixture of multivariate normals, 154
  mixture of Poissons, 152
  mixture of two known densities, 150–151, 152–154
  mixture of two univariate normals, 151
  mixture of univariate normals, 149–150, 156–157
  model choice, 158–159
  Monte Carlo testing, 157
  see also Modality, assessment of
Numerical decomposition, 12, 133–147
  Medgyessy's method, 138–142, 145–146
  mixture of exponentials, 137–138
  Pedersen's technique, 136
  Tarter and Silvers' method, 142–145
  Taylor's method, 147

Outliers, 22–23
  Bayesian methods, 108–113
Overlap biases, 62–63

Pascal distribution, 50
Pattern recognition, 7, 176
Probabilistic editor, 183, 196, 215
Probabilistic teacher, 182, 192–193, 196, 215
Probability generating function, 127

Quadratic programming, 118
Quantile contour plot, 70
Quartic exponential distribution, 30–32, 163–164
Quasi-Bayes method
  asymptotic theory, 185, 190–192, 200–201
  bipolar signal, 186–188, 201–203
  contaminated normal, 199–201
  Kalman filter, 215
  k-class problem, 189–193
  signal versus noise, 197–199
  two-class problem, 182–183, 186–188

Random variate generation, 29
Remote sensing, 7, 14, 178
Robust estimation, 23
Robustness studies, 23–24
Runaways, 182

Sequential problems, 6, 176–215
  recursive updating, 6, 179
  see also Quasi-Bayes methods
Signal versus revise, 13
  Kalman filter version, 213, 214–215
  sequential approach, 196–199, 204
Singularities in likelihood, 83, 89, 93–94, 97
Statistically equivalent blocks, 125
Stochastic approximation, 184–185, 193–196, 206
Switching regressions, 12–13

Tanner's graphical method, 54–55, 149
Tracking a signal, 6, 176, 213–214
Translation parameter mixture, 38, 81

Ultracentrifuge, 136
Unimodality, 49, 149
  necessary and sufficient conditions, 159–160
  test of, 166
Unsupervised learning, 176
  'learning without a teacher', 181
Use of fully categorized data
  efficiency of maximum likelihood, 95
  in discriminant analysis, 168, 171–173
  in Hall's method, 120–121
  mixture of multinomials, 97, 100

Variance dilation, 51

Zeroth order moments, 72, 76

IMAN and CONOVER • Modern Business Statistics
JAGERS • Branching Processes with Biological Applications
JESSEN • Statistical Survey Techniques
JOHNSON and KOTZ • Distributions in Statistics
Discrete Distributions
Continuous Univariate Distributions—1
Continuous Univariate Distributions—2
Continuous Multivariate Distributions
JOHNSON and KOTZ • Urn Models and Their Application: An Approach to Modern Discrete Probability Theory
JOHNSON and LEONE • Statistics and Experimental Design in Engineering and the Physical Sciences, Volumes I and II, *Second Edition*
JUDGE, HILL, GRIFFITHS, LÜTKEPOHL and LEE • Introduction to the Theory and Practice of Econometrics
JUDGE, GRIFFITHS, HILL, LÜTKEPOHL and LEE • The Theory and Practice of Econometrics, *Second Edition*
KALBFLEISCH and PRENTICE • The Statistical Analysis of Failure Time Data
KISH • Survey Sampling
KUH, NEESE, and HOLLINGER • Structural Sensitivity in Econometric Models
KEENEY and RAIFFA • Decisions with Multiple Objectives
LAWLESS • Statistical Models and Methods for Lifetime Data
LEAMER • Specification Searches: Ad Hoc Inference with Nonexperimental Data
LEBART, MORINEAU, and WARWICK • Multivariate Descriptive Statistical Analysis: Correspondence Analysis and Related Techniques for Large Matrices
McNEIL • Interactive Data Analysis
MAINDONALD • Statistical Computation
MANN, SCHAFER and SINGPURWALLA • Methods for Statistical Analysis of Reliability and Life Data
MARTZ and WALLER • Bayesian Reliability Analysis
MIKÉ and STANLEY • Statistics in Medical Research: Methods and Issues with Applications in Cancer Research
MILLER • Beyond ANOVA, Basics of Applied Statistics
MILLER • Survival Analysis
MILLER, EFRON, BROWN, and MOSES • Biostatistics Casebook
MONTGOMERY and PECK • Introduction to Linear Regression Analysis
NELSON • Applied Life Data Analysis
OSBORNE • Finite Algorithms in Optimization and Data Analysis
OTNES and ENOCHSON • Applied Time Series Analysis: Volume I, Basic Techniques
OTNES and ENOCHSON • Digital Time Series Analysis
PANKRATZ • Forecasting with Univariate Box-Jenkins Models: Concepts and Cases
PIELOU • Interpretation of Ecological Data: A Primer on Classification and Ordination
POLLOCK • The Algebra of Econometrics
PRENTER • Splines and Variational Methods
RAO and MITRA • Generalized Inverse of Matrices and Its Applications
RIPLEY • Spatial Statistics
SCHUSS • Theory and Applications of Stochastic Differential Equations
SEAL • Survival Probabilities: The Goal of Risk Theory
SEARLE • Linear Models
SEARLE • Matrix Algebra Useful for Statistics
SPRINGER • The Algebra of Random Variables
STEUER • Multiple Criteria Optimization

(*continued from front*)